D0843160

Manufacturing of
Gene Therapeutics

Manufacturing of Gene Therapeutics

Methods, Processing, Regulation, and Validation

Edited by

G. Subramanian
Littlebourne, Kent, England

Withdrawn
University of Waterloo

Kluwer Academic / Plenum Publishers
New York, Boston, Dordrecht, London, Moscow

Library of Congress Cataloging-in-Publication Data

Manufacturing of gene therapeutics: methods, processing, regulation, and validation/
edited by G. Subramanian.
        p.   cm.
    Includes bibliographical references and index.
    ISBN 0-306-46680-5
      1. Gene therapy.   2. Genetic vectors.   3. Genetic transformation.   I. Subramanian, G.,
    1935–
      [DNLM: 1. Gene Therapy   2. Drug Approval.   3. Genetic Engineering.   4. Genetic
    Vectors—biosynthesis.   5. Pharmaceutical Preparations.   QZ 50 M294 2001]
    RB155.8 .M36 2001
    616.'042—dc21

                                                                                    2001038587

ISBN: 0-306-46680-5

©2002 Kluwer Academic / Plenum Publishers, New York
233 Spring Street, New York, N.Y. 10013

http://www.wkap.nl/

10 9 8 7 6 5 4 3 2 1

A C.I.P. record for this book is available from the Library of Congress

All rights reserved

No part of this book may be reproduced, stored in a retrieval system, or transmitted
in any form or by any means, electronic, mechanical, photocopying, microfilming,
recording, or otherwise, without written permission from the Publisher.

Printed in the United States of America

Preface

Advances in molecular biology and recombinant DNA technology have accelerated progress in many fields of life science research including gene therapy. A large number of genetic engineering approaches and methods are readily available for gene cloning and therapeutic vector construction. Significant progress is being made in genomic, DNA sequencing, gene expression, gene delivery and cloning. Thus gene therapy has already shown that it holds great promise for the treatment of many diseases and disorders. In general it involves the delivery of recombinant genes or transgenes into somatic cells to replace proteins with a genetic defect or to transfer with the pathological process of an illness. The viral and non-viral delivery systems may hold the potential for future non-invasive, cost effective oral therapy of genetically based disorders.

Recent years have seen considerable progress in the discovery and early clinical development of a variety of gene therapeutic products. The availability, validation and implementation have enabled success but also for testing and evaluation. New challenges will need to be overcome to ensure that products will also be successful in later clinical development and ultimately for marketing authorisation. These new challenges will include improvements in delivery systems, better control of in-vivo targeting, increased level transduction and duration of expression of the gene, and manufacturing process efficiencies that enable reduction in production costs. Perhaps profound understanding of regulated gene design may result in innovative bioproducts exhibiting safety and efficacy profiles that are significantly superior to those achieved by the use of naturally occurring genes. This procedure may considerably contribute to fulfil standards set by regulatory authorities.

This book aims to project an overview of the current advances in the field of gene therapy and the methods that are being successfully applied in the manufacture of gene therapeutic products.

I am indebted to the international group of contributors who have shared their practical knowledge and experience. Each chapter represents an overview of its chosen topic. Chapters one and two provide an overview of gene therapy and gene therapy for cancer. Gene self-assembly and gene expression are discussed in chapters three and four. Genotype and response to cytotoxic gene therapy is reviewed in chapter five. Vector assembly and gene transfer is discussed in chapters six and seven. Plasmid manufacturing is reviewed in chapter eight. The importance of quality control and assurances and the analytical methods are discussed in chapters nine and ten. Chapter eleven provides an insight into validation aspects in gene therapy and gene delivery is reviewed in chapter twelve. The importance of regulatory issues and guidelines are reviewed for the American market in chapter thirteen and the European market in chapter fourteen, and chapter fifteen discusses the regulatory compliance in contract manufacturing environment. Finally, chapter sixteen discusses the risk assessment in gene therapy.

My thanks to the contributors for the extensive diligence and their patience and goodwill during the production of the book; they deserve the full credit for the source of the book.

It is hoped that this book will be of great value to all those who are engaged in the field of gene therapy and that it will stimulate further progress and advancement in this field to meet the ever increasing demands. I should be most grateful for any suggestion, which could serve to improve future editions of this book.

My deep appreciation to Jo Lawrence of Kluwer Academic/Plenum Publishers for her continuous patience, encouragement and help in guiding all of us through the preparation and the completion of this book.

G. Subramanian

Contributors

Akshay Anand
Department of Immunopathology
Post Graduate Institute of Medical Education and Research
Chandigarh, India

Sunil K. Arora
Department of Immunopathology
Post Graduate Institute of Medical Education and Research
Chandigarh, India

Joy A. Cavagnaro
President, Access Bio LC
Leesburg
VA 20177-1400, USA

Nancy Chew
President, Regulatory Affairs, North America LLC
P.O.Box 72375
Durham
NC 2772, USA

Odile Cohen-Haguenauer
Laboratorie TGOM & Service d'Oncologie Medicale
Hopital Saint-Louis
1, avenue Claude Vellefaux
75475 Paris CEDEX 10, France

Peter Daniel
Department of Haematology, Oncology and Tumour Immunology
University Medical Centre Charité
Campus Berlin-Buch
Humboldt University of Berlin
Lindenberger Weg
13125 Berlin, Germany

Linh Do
Berlex Biosciences
15049 San Pablo Avenue
P.O.Box 4099
Richmond
CA 94804-4089, USA

Vladimir I. Evtushenko
Laboratory of Genetic Engineering
Research Institute of Roentgenology and Radiology
St. Petersburg 189646, Russia

James G. Files
Berlex Biosciences
15049 San Pablo Avenue
P.O.Box 4099
Richmond
CA 94804-4089, USA

Erwin Flaschel
PlasmidFactory GmbH & Co. KG
Meisenstrasse 96
D-33607 Bielefeld, Germany

Karl Friehs
University of Bielefeld
Postfach 10 01 31
D-33501 Bielefeld, Germany

Bernhard Gillissen
Department of Haematology, Oncology and Tumour Immunology
University Medical Centre Charité
Campus Berlin-Buch
Humboldt University of Berlin

Lindenberger Weg
D-13125 Berlin, Germany

Clague P. Hodgson
Nature Technology Corporation
4701 Innovation Drive
Lincoln
Nebraska 68521, USA

John Irving
Berlex Biosciences
15049 San Pablo Avenue
P.O.Box 4099
Richmond
CA 94804-4089, USA

Juan Irwin
Berlex Biosciences
15049 San Pablo Avenue
P.O.Box 4099
Richmond
CA 94804-4089, USA

John Jenco
Dow Biopharmaceutical Contract Manufacturing Services
50 East Loop Road
Stony Brook
NY 11790, USA

Jaspreet Kaur
Department of Biochemistry
Postgraduate Institute of Medical Education and Research
Chandigarh, India

Steven S. Kuwahara
Kuwahara Consulting
PMB #506
1669-2 Hollenbeck Avenue
Sunnyvale
CA 94087-5042, USA

Mark Lawler
Department of Haematology and Oncology
St Patrick Dun Research Labs
James's Street
Dublin 8, Ireland

Elisabeth Lehmberg
Berlex Biosciences
15049 San Pablo Avenue
P.O.Box 4099
Richmond
CA 94804-4089, USA

Bruce Mann
Berlex Biosciences
15049 San Pablo Avenue
P.O.Box 4099
Richmond
CA 94804-4089, USA

Michael T. McCaman
Berlex Biosciences
15049 San Pablo Avenue
P.O.Box 4099
Richmond
CA 94804-4089, USA

Stephen Morris
BioReliance
14920 Broschart Road
Rockville
MD 20850-3349, USA

Munishi Mukesh
National Bureau of Animal Genetics Resources (ICAR)
Karnal, India

Peter K. Murakami
Berlex Biosciences
15049 San Pablo Avenue

P.O.Box 4099
Richmond
CA 94804-4089, USA

Chris Murphy
BioReliance
14920 Broschart Road
Rockville
MD 20850-3349, USA

Jeffrey W. Nelson
Berlex Biosciences
15049 San Pablo Avenue
P.O.Box 4099
Richmond
CA 94804-4089, USA

Eirik Nestaas
Berlex Biosciences
15049 San Pablo Avenue
P.O.Box 4099
Richmond
CA 94804-4089, USA

Erno Pungor, Jr.
Berlex Biosciences
15049 San Pablo Avenue
P.O.Box 4099
Richmond
CA 94804-4089, USA

Martin Schleef
PlasmidFactory GmbH & Co. KG
Meisenstrasse 96
D-33607 Bielefeld, Germany

Torsten Schmidt
PlasmidFactory GmbH & Co. KG
Meisenstrasse 96
D-33607 Bielefeld, Germany

Gail Sofer
BioReliance
14920 Broschart Road
Rockville
MD 20850-3349, USA

Isrid Sturm
Department of Haematology, Oncology and Tumour Immunology
University Medical Centre Charité
Campus Berlin-Buch
Humboldt University of Berlin
Lindenberger Weg
D-13125 Berlin, Germany

Mei P. Tan
Berlex Biosciences
15049 San Pablo Avenue
P.O.Box 4099
Richmond
CA 94804-4089, USA

Joseph A. Trai-Na
Berlex Biosciences
15049 San Pablo Avenue
P.O.Box 4099
Richmond
CA 94804-4089, USA

Spencer Tse
Berlex Biosciences
15049 San Pablo Avenue
P.O.Box 4099
Richmond
CA 94804-4089, USA

Dominic K. Vacante
BioReliance
14920 Broschart Road
Rockville
MD 20850-3349, USA

Martin Weber
Group Manager R&D
Qiagen GmbH
Max-Volmer Strasse 4
D-40724 Hilden, Germany

Tao Yu
Berlex Biosciences
15049 San Pablo Avenue
P.O. Box 4099
Richmond
CA 94804-4089, USA

Contents

Manufacturing of
Gene Therapeutics

Somatic Gene Therapy, Paradigm Shift or Pandora's Box
A perspective on gene therapy

MARK LAWLER

Department of Haematology, St Patrick Dun Research Labs, St James's Hospital and Trinity College Dublin, James's Street, Dublin 8, Ireland

1.   DEFINITIONS

Gene therapy may be defined as the introduction of genetic material into the cells of a patient in an effort to help cure the disease either by producing a gene product which is missing or in reduced amounts in the patient due to a genetic mutation in the individual (*eg* Factor VIII protein for haemophilia) or by introduction of new genetic material which either directly or indirectly will help to combat the disease (*eg* genetic vaccination). Therapeutic genes are delivered using a carrier (called a vector) which may be a non functional viral vector or by using non viral vector approaches such as liposomes or other carrier molecules. All gene therapy protocols involve the introduction of genetic material into cells that have a finite life span such as blood cells, liver cells *etc* (termed somatic tissue), thus the introduced gene is not passed on to the next generation. This type of gene therapy is known as somatic gene therapy and is in contrast to the concept of germ line gene therapy (which would involve a gene being introduced into sperm or ova so that the gene could be inherited by the children of the patient). Germ line gene therapy is subject to an international moratorium.

Somatic gene therapy

Introduction of a gene into a specific tissue or tissues to provide therapeutic benefit to the patient.

1

Germ line gene therapy
Introduction of genetic material into the egg cells or sperm cells of an individual such that the gene will also be passed on to the next generation.

Ex vivo **gene therapy**
Collection of the patient's cells, introduction of therapeutic genetic material into these cells and reintroduction of these cells into the patient.

In vivo **gene therapy**
Direct injection of therapeutic gene to the relevant tissue via a vector.

The promise of gene therapy lies in its proposed ability to treat the causes of disease rather than the symptoms. The first decade of gene therapy has been somewhat of a "roller coaster" ride, with early excitement of the potential of this approach being tempered somewhat by disappointing clinical results. Recently, however, improvements in vector construction and vector delivery to the appropriate tissue has led to better pre clinical and clinical results.

2.   INTRODUCTION

Somatic gene therapy involves the amelioration of a disease by introduction of genetic material with therapeutic potential into a somatic tissue.   It is only in the last few years that gene transfer could be contemplated in the clinic. Advances have been made which allow the transfer and stable existence of genetic material in a foreign host. Gene therapy has benefited from (A) the gene transfer and expression techniques of molecular genetics; (B) the natural ability of retroviruses to infect foreign replicating cells and stably integrate their genetic material into the host genome and (C) the fact that Bone Marrow Transplantation (BMT) provides a straightforward delivery of *in vitro* manipulated material into the blood stream.   Thus a retrovirus can be engineered to contain an appropriate segment of DNA and *ex vivo* transduction of haemopoietic cells allows subsequent introduction of foreign material into the host by the BMT route.
There are a variety of gene delivery systems currently available and some systems may be more useful in targeting particular tissues. Retroviruses were the initial vehicles of choice but many studies have made use of adenoviruses or herpes viruses and new developments such as the use of adeno associated virus and non viral delivery systems have great potential. These systems together with the early work in animal studies have been crucial in bringing gene therapy protocols to the clinic.   While much of the

earlier preclinical gene transfer work has concentrated on murine models, it has become clear that use of large animal models including sheep, dogs, pigs and monkeys may be more relevant as the final preclinical step before proceeding to human trials.

While a large number of single gene defects are candidates for a gene therapy protocol and several, including Adenosine Deaminanse (ADA)[1] and Cystic Fibrosis (CF)[2,3] have been treated with such protocols it is the realisation that gene therapy has potential in acquired disease which is perhaps the most interesting. The ability to tackle malignant disease either directly through the introduction of genes such as Tumour Necrosis Factor (TNF) or indirectly by introducing immune stimulatory genes may have enormous clinical applications. Gene therapy for malignant diseases will be discussed in Chapter 5. The use of immunostimulatory molecules may also yield dividends in the fight against AIDS. Protocols for cardiovascular disease using either vector or antisense based approaches have shown promise[4,5] while gene therapy protocols for neurodegenerative disorders are also being developed and applied[6,7].

2.1   Vectors and target cells

Retroviruses have been the most extensively used vectors to date particularly due to the fact that the therapeutic gene is integrated into the host genome. Haemopoietic cells are probably the easiest to target due to the well characterised hierarchical structure of the haemopoietic system. Thus many of the clinical protocols have involved the introduction of a new gene into haemopoetic cells. A second important advantage is that *ex vivo* gene transfer can be performed - haemopoietic cells can be taken from the individual, infected and then re-introduced to the individual by BMT. *Ex vivo* gene transfer is very efficient as cells can be externally stimulated to allow higher infection rates with retroviral constructs. The other important advantage of the haemopoietic system is the ability to target stem cells as well as lineage specific cells.

Other targets for retroviral vectors include skin fibroblasts where retroviral infection rates of 50% have been reported and several clinically important genes including ADA, purine nucleoside phosphorylase (PNP) and Factor IX have been successfully transferred. The skin may be a highly important target for delivery of therapeutic genes - collagen beads with genetically modified fibroblasts secreting Factor IX have been produced and using skin keratinocytes, therapeutic genes could be delivered directly to the circulation.

The major disadvantage of retroviral vectors is their inability to infect non dividing cells. The other major disadvantages relate to the safety aspects of retroviruses. It is necessary to do stringent testing of cell lines for potential replication competent retroviruses. Clearly insertional mutagenesis and the worry that retroviral activation of cellular oncogenes has occurred in murine systems must also be considered. Finally the fact that the maximum size of DNA which can be efficiently packaged and transduced is approximately 7 kb limits their potential for certain diseases. Specialised packaging cell lines capable of producing high titres of replication deficient recombinant virus free of any wild type viral contaminants have been produced. However, outbreaks have been reported presumably due to contaminating replication competent viruses. The prevention of such outbreaks must of course be avoided. This has been aided by the development of better modifications to packaging lines and by more stringent monitoring of the products of such lines.

2.2   Adenoviruses

Adenoviruses are potentially more useful in *in vivo* gene therapy. They can infect non dividing cells, larger sizes of exogenous DNA (> 30 kbs) can be incorporated and high titres of the virus can be produced which is of great importance for *in vivo* applications. The virus particle itself is relatively stable. One worry is the presence of similar sequences in the human genome which could combine with the inserted sequences leading to development of malignancy. Replication deficient adenovirus can be generated and propagated by growth in cells engineered to express the E1 replication region, thus allowing the development of adenovirus vectors expressing large amounts of a foreign gene product *in vitro*. Pre clinical studies have indicated that efficient transduction *in vivo* occurs and gene expression can be seen for significant periods post transduction. Adenoviruses have been used to deliver the Cystic Fibrosis gene to airway epithelium and the ability of adenoviral vectors to target brain, liver and muscle cells have indicated that adenoviruses may become major vectors in clinical protocols. However adenovirus vectors tend to be recognised by the host's immune system and so need to be modified greatly to avoid immunisation and clearance of the therapeutic vector.

2.3   Other viral vectors

The most important of these include adeno-associated virus (where their ability to integrate at a particular locus on chromosome 19 might allow controlled precise expression of any inserted gene), and herpes simplex virus which could be highly important in neurodegenerative disorders[8]. Adeno associated viruses are very simple viruses to produce and their broad host range in conjunction with adenovirus make them useful vector systems. The scope of such systems is enormous and will mean that even vector systems that we might consider enemies (such as the Human Immunodeficiency virus) may prove to be friends at the gene therapy level due to their target cell specificity[9].

2.4 Non viral delivery systems

The majority of the work currently reported in the literature has focused on the use of viral vectors to deliver the desired gene product to its target cell or tissue; however there is a growing body of evidence that non viral methods may be useful in the potential treatment of several single or polygenic disorders. While a variety of methods including the direct injection of naked DNA into muscle cells, arterial walls or the heart itself have been shown to be feasible, an approach which would have general implications for a variety of diseases has involved the delivery of DNA protein complexes to a specific cell via a receptor molecule intermediate. The main advantages associated with non viral methods are (A) their ability to deliver large transgenes (up to 50 kb) to their target; (B) their ability to target different receptors; (C) a safer approach since viral integration is avoided. This receptor mediated delivery system has been used to deliver the low density lipoprotein receptor (LDLr) to the circulation of Watanebe rats and the lowering of subsequent total serum cholesterol by this treatment provide a good animal model for a gene therapy protocol for hypercholesteremia[10]. Recent studies are also indicating that it may be much easier than we at first believed to get DNA into the cell through the use of anitisense approaches or "naked DNA " injection[11].

2.5 Tissue specific gene delivery

While the majority of pre-clinical work has focused on delivery of therapeutic genes to the haemopoietic system, there have been many attempts to target other tissues also. In the liver, the primary candidates would be the hepatocyte but while transducing hepatocytes is relatively straightforward, the transplantation of such hepatocytes has proven problematic. This may be overcome by using methods for ectopically grafting hepatocytes or by *in vivo* transduction by retrovirus. Delivery of gene products to the circulation will also be important for a wide variety of diseases and so a method for efficient delivery of transgenes is necessary. Retroviral transduction of keratinocytes has been achieved but expression of exogenous genes was short lived. Both factor IX and growth hormone can be expressed in myoblast cell lines using vectors which contained non viral control sequences and these cells can be successfully re-implanted into animals[12], while correction of the lysosomal storage defect in β glucuronidase deficient mice (the animal model for Sly syndrome, a human mucopolysaccharide disorder) has been achieved by autologous transplantation following retroviral infection of fibroblasts[13].

3. WHERE DID GENE THERAPY BEGIN?

The development of gene therapy as we know it today has resulted from two significant advances in science and medicine in the 1960s and 70s - the advances in cellular and transplantation biology leading to effective bone marrow transplant treatment for leukaemia and the advances in molecular biology and genetic engineering leading to the cloning of therapeutic proteins for the treatment of human disease.

3.1 Tumour infiltrating lymphocytes

The Neo[r] TIL protocol, the first approved protocol for gene transfer in humans, was initiated at NIH in 1988 and provided the springboard for the subsequent ADA protocol which will be discussed in section 2. The basis of this work was the finding that there is a class of cell called a tumour infiltrating lymphocyte (TIL) which has potential activity against a patient's own cancer. These cells can be isolated from a patient and expanded using interleukin 2. These TILs are classic cytotoxic T cells and injection of IL2 expanded TILs into cancer bearing mice leads to tumour regression. The next step involved establishing their efficacy in patients. TILs from end stage melanoma patients were collected from tumour nodules, expanded with IL2 and reintroduced intravenously into the patient. 55% of the group responded well to this therapy. Subsequently it was decided to look at the feasibility of marking TILs prior to re-introduction into the patient. The neophosphotransferase resistance (neo[r]) gene was chosen as a marker gene and a clinical trial of end stage melanoma patients was initiated after many reviews and delays to ensure the safety of the technique. On May 22 1989, the first patient was treated and a series of 5 patients were reported on in the *New England Journal of Medicine* in 1990. While this protocol was not a therapeutic protocol but a simple "marking" protocol, it provided the impetus for the second stage in gene transfer in humans, the transfer of a gene with therapeutic potential[14,15].

3.2   ADA deficiency

Initially β thalassemia was thought to be the disease of choice for genetic manipulation but by 1984 it became clear that the haemoglobinopathies were too complex for early attempts at gene therapy. One of the first criteria for a suitable gene therapy model is for the corrected cells to have a growth advantage in the patient. Obvious candidates would be mutations in DNA metabolism as corrected cells would show more efficient cell division. Possibilities focused on the sequential steps in purine metabolism involving the genes Adenosine Deaminase (ADA), Purine Nucleosyl Phosphorylase (PNP) and Hypoxanthine guanine PhosphoRibosyl Transferase (HPRT) which are all known to cause diseases due to enzyme deficiency. ADA deficiency leads to an immune deficiency syndrome and makes these patients very susceptible to infection, thus necessitating that they live in a carefully controlled environment. Evidence from BMT of ADA patients indicated that T cells from the donor easily outgrow ADA deficient cells - therefore ADA was chosen as the initial model system for gene therapy. In ADA deficiency there is a severe depletion of the number and activity of T cells while there is also a debilitating effect on B cells. Thus cellular and humoral activity is severely compromised and death usually ensues from infection in the first 2 years of life. As matched BMT has been shown to cure ADA it is clear that replacement of the abnormal T cells with normal T cells is sufficient to cure the disease. The lack of success with mismatched BMT means that alternative strategies need to be attempted. The finding that individuals can have as little as 5% or as much as 50 times the standard levels of ADA are also important factors for a preliminary gene therapy protocol as the level of expression need not be under stringent control.

Thus on September 14th 1990, the first child was treated for this disease by gene therapy. Blood was taken from the patient, a four year old girl with no immune function[16]. Red cells were given back by leukophoresis and mononuclear cells isolated by Ficoll centrifugation. These cells were grown in tissue culture, stimulated with IL2 and infected with a third generation retrovirus containing the ADA gene and a *neo* marker gene. The girl received 8 infusions in an 11 month period of the transduced cells and was also on weekly injections of Polyethylene glycol (PEG) ADA. PCR showed gene corrected T cells (20-25%) in the mononuclear cell population. Clinical condition had improved so subsequently she received maintenance gene therapy infusions at 6 month intervals. A second patient (a 9 year old girl) received 11 infusions of gene corrected autologous T cells from January 1991. Results in both patients were encouraging with both attending school and showing no side effects and average number of infections for children of their age. ADA levels were at 25% of normal and it was estimated that the half life of gene corrected T cells could be as high as 2-3 years whereas abnormal T cells in the patient have a half life of approximately 2-6 months. Subsequently a second study was performed in Europe which also indicated long term expression of the gene in transfected T cells reintroduced into the patient[17]. However these studies were complicated by the fact that these patients were receiving PEG ADA also and it was unclear if sufficient long term expression of ADA in gene corrected T cells could significantly alter the phenotype. Removal of the PEG - ADA led to a reduction in ADA expression, clearly indicating that while the potential for gene therapy was there, improvements in efficient delivery and gene expression were required[18].

Current work is focusing on the delivery of gene-corrected stem cells so that educable T cells might be achieved with continual ADA production. Here CD 34 cells (early progenitor cells) are used in combination with "adult cells" to preserve potential stem cell repopulation. In order to monitor stem cell repopulation, different vectors are being used for stem cells and differentiated cells and pre treatment with granulocyte colony stimulating factor (G-CSF) would allow stimulation of the circulation of stem cells normally present in bone marrow. Introduction of the ADA gene into primitive cells is being attempted by either manipulation of the microenvironment through the use of molecules such as fibronectin or by using other sources of stem cells such as cord blood. In this regard recent work has indicated that the retroviral infection efficiency may be superior in cord blood stem cells than in adult bone marrow. A number of children have been treated with constructs that have been introduced into umbilical cord cells. Despite an improvement in clinical symptoms and treatment of other children using these protocols, there is no evidence to date of a patient with ADA deficiency having long term cure of their disease by gene therapy[1].

3.3   Other enzyme deficiencies

Other metabolic disorders are also good candidates for gene therapy based approaches, as 5-25% of normal enzyme activity will normally suffice for protection from clinical disease in disorders such as Haemophilia B. Several metabolic disorders are caused by absence of specific lysosomal enzymes that degrade specific compounds. The inability to degrade these compounds can lead to organ dysfunction, both visceral and in the CNS. Gaucher disease, a deficiency of β glucocerebrosidase is treatable with enzyme replacement therapy and BMT, thus showing the potential for a gene therapy directed approach. Clinical trials involving the introduction of retroviral vectors containing the β glucocerebrosidase cDNA into BM or stem cells are underway[19]. Animal studies in Sly syndrome where β glucorinidase deficiency results in accumulation of sulphated glycoaminoglycans has indicated that autologous fibroblasts transfected with appropriate vectors and transplanted in mice can correct the lysosomal storage problems and serve as a model for the human situation[13,20].

Recent studies in the haemophilia B dog model also suggest that gene therapy approaches may be useful in the treatment of this disorder in humans. A single intraportal vein injection of a recombinant adeno-associated virus (rAAV) vector encoding canine factor IX (cFIX) cDNA under the control of a liver-specific enhancer/promoter led to long-term correction of the bleeding disorder in haemophilia B dogs. Both whole-blood clotting time (WBCT) and activated partial thromboplastin time (aPTT) of the treated dogs have been greatly decreased since the treatment supporting the feasibility of using AAV-based vectors for liver-targeted gene therapy of genetic diseases[21].

3.4   Cystic fibrosis

While adenosine deaminase deficiency has perhaps proved the most amenable disease to gene therapy, it is a rare genetic disorder. Cystic fibrosis (CF) however is a lethal inherited disorder which affects approximately 1 in 2,000 caucasians. Since the cloning of the cystic fibrosis transmembrane conductance regulator (CFTR) gene in 1989, effective and safe treatment of the underlying defect by gene therapy has become one of the principle aims of researchers in this area. Several important breakthroughs at the molecular level have meant that this possibility may be imminent, even though a fuller understanding of the molecular defect causing CF is probably required for long term amelioration of the genetic defect.

The first important discovery was that the cystic fibrosis could be corrected *in vitro* by retrovirus mediated gene transfer. Subsequent work indicated that the human CFTR gene could be directed to the lung epithelium in cotton rats using a replication deficient adenovirus vector, a vector known to infect the respiratory epithelium and capable of transferring recombinant genes into non proliferating cells. A second important development in this area was the production of homozygous CFTR deficient mice. These mice were produced by "knock out" homologous recombination and represent the first authentic animal model for CF. These mice allowed the re-introduction and expression of CFTR in a variety of cell types and thus provide vital knowledge for the application of somatic gene therapy for this disease.

The central question for CF gene therapy is what level of CFTR expression is required to achieve and maintain normal function. It may be as little as 10%. This would be important as it may indicate that 100% correction of epithelial cells is not necessary to repair the chloride transport defect that is crucial to the pathogenesis of the disease. A second question relates to the type of cell which must be corrected and at what stage in development this should be performed. Although the epithelia of the lung is the major site of pathology and morbidity associated with the disease, very low expression levels of CFTR are present in adults, whereas high levels of CFTR mRNA can be detected in foetal tissue. This may have implications for early intervention in this disorder. Currently, a variety of protocols have been approved for gene therapy in CF using adenovirus delivery systems to the pulmonary epithelium. While initial results proved promising a worrying aspect occurred during dose escalation studies associated with the trial. A woman who had received a high titre of gene manipulated adenovirus developed fever and lung inflammation which prompted re-evaluation and lowering of the dose of modified virus. A second trial looked at the efficacy of gene transfer to CF patients using a liposomal DNA complex spray administered through the nose. This non viral delivery route may be a more suitable route for administration of normal CFTR to the appropriate tissue. Recently a third generation adenoviral vector containing recombinant human cystic fibrosis transmembrane conductance regulator (CFTR) gene was delivered by bronchoscope in escalating doses to the conducting airway of 11 volunteers with cystic fibrosis. These results demonstrate that gene transfer to epithelium of the lower respiratory tract can be achieved in humans with adenoviral vectors but that efficiency is low and of short duration in the native CF airway[22].

3.5   Neurological disease

While gene therapy for the disorders already discussed usually involves the delivery of the ameliorating gene to a specific cell type, it may appear initially that gene therapy for the nervous system would present more complications due to the wide range of highly differentiated cell types in the central nervous system. Thus any gene therapy protocol would involve a gene targeting system that was suitable for infection of these different cell types. Since most mature neurons do not replicate, the use of retroviral vectors may not be the most suitable in this area. However, the ability to target other cells and use them as surrogate carriers of the corrective gene to the brain may be a more effective method of treating neurological disorders, such as Alzheimer's or Parkinson's disease.

The earliest work in this area concentrated on the modification and transplantation of autologous fibroblasts. Animal studies on Alzheimer's disease indicated that intracerebral grafting of fibroblasts that had been genetically modified to produce nerve growth factor prevented the death of cholinergic neurons in the basal forebrain which had been associated with profound cognitive impairment in this animal model[23]. A similar method may also be useful for Parkinson's disease where non neural cells could be genetically modified to produce dopamine or the dopamine precursor L - DOPA. While these approaches of indirect gene transfer may prove useful it is also clear that for some disorders the genetic material must be transferred directly to the resident brain cell. This would require the development of neurotropic vectors - the herpes simplex virus is an excellent candidate due to its high infection efficiency in neuronal cells. Herpes simplex viral vectors have been developed by a number of groups and in studying Lesch Nyhan disease it has been shown that functional human HPRT could be introduced into the brains of normal rats by replication competent viral vectors[24].

While these results are promising, problems of cytotoxicity and poor gene expression with these systems have indicated that these vectors must be optimised for more routine use. Although heterologous genes can be expressed in the brain using different technologies, further studies are now necessary to characterise the long term viability of transferred cells/vectors and their expressed transgenes and their potential for inducing direct or indirect neuropathological changes.

3.6   Gene therapy for infectious disease

Two strategies are currently in vogue in the potential treatment of diseases such as AIDS. One involves the use of intracellular immunisation and is designed to render cells resistant to viral replication whereas the second involves the use of genetically modified cells which express viral gene products, thus inducing antiviral cellular immune responses[25,26]. RNA based inhibitors include antisense RNA, ribozymes and RNA decoys have also been used in preclinical studies. Antisense technology involves the production of an artificial molecule which is complementary to the normal sense strand and thus blocks the messenger RNA from subsequent translation. Initial antisense studies showed limited efficacy with transcripts directed against *tat, rev, gag* or the primer binding site. *Tat* and *Rev* are key regulatory gene products which bind to specific regions of the viral RNA termed TAR and RRE and activate transcription of viral genes. Subsequent use of an adeno associated virus construct allowed introduction of antisense to the HIV 1 LTR including the TAR sequence. This antisense resulted in specific down regulation of LTR TAR expression *in vitro*[27]. Further studies are underway using this approach.

4.   SAFETY ISSUES ASSOCIATED WITH GENE THERAPY

The US Recombinant DNA Advisory Committee (RAC) was established in 1975 and in the area of gene therapy has advocated open and public discussion of advanced therapeutic products and protocols. Stringent vetting of proposals is performed and they stress the need for full disclosure of positive and negative results and potential side effects of gene therapy. The US Food and Drug Administration (FDA) have issued a Note for Guidance on the use of human somatic cell therapy and gene therapy in March 1998. The European Commission communication (OJ EC C229/4 issued on 22/07/1998) provides details on human gene therapy and regulations but is currently being revised. The European Agency for the Evaluation of Medicinal Products (EMEA) has recognised the need for consistent regulations in relation to gene therapy. In February 1999 it published a concept paper (CPMP/BWP/2257/98) of its Biotechnology Working Party entitled "Concept paper on the development of a committee for proprietary medicinal products (CPMP) points to consider on human somatic cellular therapy". This has led to the release of 2 discussion papers CPMP/BWP/41450/98 and CPMP/BWP/3088/99 for discussion and consultation which form the basis of current regulation of gene therapy and gene therapy products in Europe.

Nevertheless despite these regulations and proposed regulations in force in Europe and North America, in September 1999, the first death directly attributable to gene therapy occurred. The patient was being treated for an enzyme deficiency called ornithine transcarbamylase deficiency (OTC) as part of a phase 1 clinical study. The patient was in the group that received the highest dose in the trial protocol of an adenoviral vector containing the OTC gene. He developed acute respiratory distress syndrome (ARDS) shortly after the gene therapy infusion and died 2 days later from organ failure. Measurement of cytokine levels indicated that he had systemic inflammatory response syndrome; all erythroid precursor cells were wiped out from his marrow and the vector had gone to other organs besides the liver[28]. Subsequently the FDA found procedural problems and shut down all 7 clinical trails at Penn's Institute for Human Gene Therapy. Problems related to consent, and the death of two animals in similar preclinical procedure indicates that as in all other therapeutic approaches, full and frank disclosure of any problems should be performed.

5.   WHERE DO WE GO FROM HERE?

It is important to realise at this stage that gene therapy is no longer a dream; it has become a reality. Protocols have currently been approved in the US and Europe for a variety of genetic and acquired diseases involving over 300 patients. The initial excitement and preliminary work has of course focused on the genetic disorders. However diseases such as ADA deficiency probably only affect 100-150 people in the US and even the more common genetic disorders such as the haemoglobinopathies would affect less than 100,000 individuals in the US although they are far more widespread in the Third World. It is the ability of gene therapy to be a viable modality of therapy for acquired genetic disease or diseases with a genetic component that is potentially the most exciting. If gene therapy strategies can be tailored as outlined to treat neuro-degenerative disorders, cardiovascular disease and cancer, then the potential for cure of disease in millions of patients world-wide is within reach. This is why many biotechnology companies which specialised in recombinant DNA technologies in the 80's are now developing active gene therapy programmes in the new millennium. In addition most large pharmaceutical drug companies have gene discovery and gene therapy programmes.

But how close are we in reality? While the approaches outlined are very exciting, there are still major problems to be addressed. The treatments described thus far have in general concentrated on *ex vivo* gene therapy using viral vectors. The main problem with expanding gene therapy to treat large numbers of patients is the complexity of the current protocols. *Ex vivo* protocols require the extraction and manipulation of patient specific cells and their re-introduction into the patient. This is probably only feasible in highly specialised institutes although several companies will engineer the gene of interest into patient cells and then return them to the physician for subsequent therapy. Also while initial ADA gene therapy infusions were performed in intensive care, recent treatments have involved more routine operations. Gene therapy is perhaps a natural progression from blood transfusion and bone marrow transplantation, indeed the blood transfusion centres of today may be the gene therapy centres of the future as the need for conventional blood products decreases with the availability of synthetic substitutes. Some transfusion centres already collect stem cells for treatment of leukaemia so their metamorphosis to gene therapy vector centres is not too far-fetched. The recent advances in antisense technology also look promising.

It is necessary of course to temper our excitement not only with a degree of realism but also with a high regard for the safety aspects of this new therapy. Gene therapy is at a very early stage so we have no ideas of what long term side effects may ensue. The development of tumours in primates due to replication competent retroviral contamination provided a timely warning of the potential hazards of gene therapy and has led to much more rigorous testing of vectors and the development of 3rd and 4th generation vector systems with added safety features. If we think about it, the long term aim of any protocol would be a type of 'genes in a bottle " approach where the gene of interest could be injected just like a pharmaceutical drug. Thus it could be injected intravenously much as insulin is delivered in diabetics. Perhaps the "gene-drug" is on the horizon.

REFERENCES

1. Parkman R, Weinberg K, Crooks G, Nolta J, Kapoor N, Kohn D, Gene therapy for adenosine deaminase deficiency. *Annu Rev Med.* 2000;**51**:33-47

2. Zeitlin PL Cystic fibrosis gene therapy trials and tribulations. *Mol Ther.* 2000;**1**(1):5-6.

3. Alton E, Geddes D Cystic fibrosis clinical trials. *Adv Drug Deliv Rev.* 1998 2;30(1-3):205-217.

4. Mann MJ Gene therapy for peripheral arterial disease. *Mol Med Today.* 2000;**6**(7):285-91.

5. Raizada MK, Francis SC, Wang H, Gelband CH, Reaves PY, Katovich MJ. Targeting of the renin-angiotensin system by antisense gene therapy: a possible strategy for the long-term control of hypertension. *Hypertens.* 2000 Apr;**18**(4):353-62.

6. Stedman H, Wilson JM, Finke R, Kleckner AL, Mendell J Phase I clinical trial utilizing gene therapy for limb girdle muscular dystrophy: alpha-, beta-, gamma-, or delta-sarcoglycan gene delivered with intramuscular instillations of adeno-associated vectors. *Hum Gene Ther.* 2000 **20**;11(5):777-90.

7. Bjorklund A The use of neural stem cells for gene therapy in the central nervous system. *Gene Med.* 1999;**1**(3):223-6

8. Latchman DS. Herpes virus vectors for gene therapy in the nervous system. *Biochem Soc Trans.* 1999;**27**(6):847-51

9. Costello E, Munoz M, Buetti E, Meylan PR, Diggelmann H, Thali M. Gene transfer into stimulated and unstimulated T lymphocytes by HIV-1-derived lentiviral vectors. *Gene Ther.* 2000 **7**(7):596-604

10. Stein CS, Martins I, Davidson BL Long-term reversal of hypercholesterolemia in low density lipoprotein receptor (LDLR)-deficient mice by adenovirus-mediated LDLR gene transfer combined with CD154 blockade. *J Gene Med.* 2000;2(1):41-51

11. Restifo NP, Ying H, Hwang L, Leitner WW. The promise of nucleic acid vaccines. *Gene Ther.* 2000 ;7:89-92.

12. Thompson AR. Gene therapy for the haemophilias. *Haemophilia.* 2000;**6** Suppl 1:115-9

13. Ohashi T, Yokoo T, Iizuka S, Kobayashi H, Sly WS, Eto Y. Reduction of lysosomal storage in murine mucopolysaccharidosis type VII by transplantation of normal and genetically modified macrophages. *Blood.* 2000;**95**(11):3631-3

14. Rosenberg SA, Aebersold P, Cornetta K, Kasid A, Morgan RA, Moen R, Karson EM, Lotze MT, Yang JC, Topalian SL, *et al* Gene transfer into humans--immunotherapy of patients with advanced melanoma, using tumour-infiltrating lymphocytes modified by retroviral gene transduction. *N Engl J Med.* 1990;**323**(9):570-8

15. Rosenberg SA. Immunotherapy and gene therapy of cancer. *Cancer Res.* 1991;**51**(18 Suppl):5074s-5079

16. Blaese RM, Culver KW, Chang L, Anderson WF, Mullen C, Nienhuis A, Carter C, Dunbar C, Leitman S, Berger M, *et al*. Treatment of severe combined immunodeficiency disease (SCID) due to adenosine deaminase deficiency with CD34+ selected autologous peripheral blood cells transduced with a human ADA gene. Amendment to clinical research project, Project 90-C-195, January 10, 1992. *Hum Gene Ther.* 1993;**4**(4):521-7

17. Bordignon C, Notarangelo LD, Nobili N, Ferrari G, Casorati G, Panina P, Mazzolari E, Maggioni D, Rossi C, Servida P, *et al*. Gene therapy in peripheral blood lymphocytes and bone marrow for ADA- immunodeficient patients. *Science.* 1995;**270**(5235):470-5

18. Anderson WF. Human gene therapy. *Nature.* 1998;**392**(6679 Suppl):25-30

19. Barranger JA, Rice EO, Swaney WP Gene transfer approaches to the lysosomal storage disorders. *Neurochem Res.* 1999;**24**(4):601-15

20. Sferra TJ, Qu G, McNeely D, Rennard R, Clark KR, Lo WD, Johnson PR. Recombinant adeno-associated virus-mediated correction of lysosomal storage within the central nervous system of the adult mucopolysaccharidosis type VII mouse. *Hum Gene Ther.* 2000 Mar;**11**(4):507-19

21. Wang L, Nichols TC, Read MS, Bellinger DA, Verma IM. Sustained expression of therapeutic level of factor IX in hemophilia B dogs by AAV-mediated gene therapy in liver. *Mol Ther.* 2000;**1**(2):154-8

22. Zuckerman JB, Robinson CB, McCoy KS, Shell R, Sferra TJ, Chirmule N, Magosin SA, Propert KJ, Brown-Parr EC, Hughes JV, Tazelaar J, Baker C, Goldman MJ, Wilson JM. A phase I study of adenovirus-mediated transfer of the human cystic fibrosis transmembrane conductance regulator gene to a lung segment of individuals with cystic fibrosis. *Hum Gene Ther.* 1999;**10**(18):2973-85

23. Blesch A, Grill RJ, Tuszynski MH Neurotrophin gene therapy in CNS models of trauma and degeneration. *Prog Brain Res.* 1998;**117**:473-84

24. Lowenstein PR, Southgate TD, Smith-Arica JR, Smith J, Castro MG. Gene therapy for inherited neurological disorders: towards therapeutic intervention in the Lesch-Nyhan syndrome. *Prog Brain Res.* 1998;**117**:485-501

25. Kohn DB, Bauer G, Rice CR, Rothschild JC, Carbonaro DA, Valdez P, Hao Ql, Zhou C, Bahner I, Kearns K, Brody K, Fox S, Haden E, Wilson K, Salata C, Dolan C, Wetter C, Aguilar-Cordova E, Church J. A clinical trial of retroviral-mediated transfer of a rev-responsive element decoy gene into CD34(+) cells from the bone marrow of human immunodeficiency virus 1-infected children. *Blood.* 1999;**94** (1):368-71.

26. Amado RG, Mitsuyasu RT, Zack JA. Gene therapy for the treatment of AIDS: animal models and human clinical experience. *Front Biosci.* 1999;**4**:D468-75

27. Lamothe B, Joshi S. Current developments and future prospects for HIV gene therapy using interfering RNA-based strategies. *Front Biosci.* 2000;**5**:D527-55

28. Gene Therapy - a loss of innocence. *Nature Medicine* 2000; **6** (1) 1

Gene Therapy for Cancer
Deceiving the malignant cell

MARK LAWLER
Department of Haematology, St Patrick Dun Research Labs, St James's Hospital and Trinity College Dublin, James's Street, Dublin 8, Ireland

1. INTRODUCTION

Acquired genetic changes are fundamental to tumourigenesis, tumour progression and the development of resistance to chemotherapy induced apoptosis. Thus approaches which manipulate the genetic material of a cancer cell may prove fruitful in preventing or reversing the malignant phenotype. Gene therapy protocols in cancer have thus far outnumbered those for any other disease due to the fact that retroviral vectors, which were the principle "first generation" gene delivery systems, could deliver genes to cancer cells due to their higher proliferative capacity, as retroviral vectors can more readily integrate their therapeutic gene into the genomes of actively dividing cells. Thus over 60% of gene therapy protocols are cancer gene therapy protocols[1]. Successful cancer gene therapy depends on two principal elements (i) the selection of an appropriate gene and (ii) an effective gene transfer system. As many molecular changes in cancer involve a perturbation of the signal transduction pathway, there are numerous potential targets for a gene mediated treatment, either by introduction of a gene that will halt or stall cancer development or by prevention of "malignant" gene expression through antisense or antigene approaches. Cancer gene therapy protocols can involve a number of different strategies including (i) developing an immune response against the tumour by cancer vaccination strategies, (ii) causing the malignant tumour to be killed by introduction of a suicide gene into cancer cells, (iii) re-introduction of a functional tumour suppressor gene or (iv) downregulation of expression of

17

an activated oncogene[1-3]. In this chapter some of these strategies will be elaborated on and the prospects for cancer gene will be addressed.

2.   CANCER GENE THERAPY STRATEGIES

2.1   Tumour infiltrating lymphocytes and cancer vaccination; adoptive immunotherapeutic approaches against malignancy

Gene therapy strategies for cancer trace their beginnings to the use of marker genes in the Tumour Infiltrating Lymphocyte (TIL) mediated gene transfer trial[4,5] (see Chapter 1). Having established that a variety of marker genes could be successfully introduced without any side effects, it was observed that the TILs had the ability to home back to the tumour site in patients with end stage melanoma. This prompted the introduction of cytotoxic genes into the TILs before re-infusion back into the patient. In the first of these studies, the gene for tumour necrosis factor (TNF) was inserted into TILs in an effort to increase their therapeutic effectiveness[6,7]. If TNF is to be effective in killing tumour cells preferentially, there is a need to target large doses of TNF to the tumour site without build up of TNF in the liver or other organs where it could be detrimental to the patient. TNF interferes with the tumour's own blood supply so it was hypothesised that vectors expressing TNF in TILs should home back to the tumour site and kill the malignant cells. In Jan 1991 the first patient was treated. Tumour cells were taken from a 46 year old man with metastatic melanoma and engineered to produce excess TNF to make the tumour more susceptible to attack from the body's immune system. Fifteen patients were subsequently entered in this clinical trial. This Phase I trial used TILs that secrete up to 100 times the normal level of TNF[7]. However there were problems associated with the TNF protocol which caused some rethinking on this approach of adoptive immunotherapy. Indications were that the homing mechanism of TILs to tumours might not be as specific as was initially expected as TNF was also detected in the liver in some of these patients. Re-evaluation studies and more *in vitro* and animal studies were performed and while successful TNF-TIL protocols have not been reported, the experience gained with this approach allowed similar protocols to be developed and applied with a number of cytokine genes introduced for therapeutic benefit. Animal experiments revealed that transduction of tumour cells with cytokine genes can enhance tumour immunogenicity and, thus, increase the recognition of the tumour as foreign by the host. Clinical trials based on these observations in which patients are immunized against their own autologous tumours that were transduced with the genes for TNF or interleukin-2 thus became the forerunner for the cancer vaccination approach, as we know it today.

This type of approach can also be used to develop a gene-modified vaccine in an attempt to induce specific antitumour cytotoxicity. Activated CD-8 T lymphocytes are the primary antitumour effector cells present in the immune system. Tumour vaccine gene therapy attempts to augment this system by engineering tumour cells to secrete cytokines, Studies in animal tumour models have demonstrated that gene-modified vaccines can produce potent, specific, and long-lasting antitumour immunity[8]. Several cytokines, including the interleukins (IL-2, IL-4, IL-7), granulocyte-macrophage colony-stimulating factor (GM-CSF), tumour growth factor-beta 2 and interferon-gamma are currently undergoing phase I and II clinical trials. Adoptive transfer of autologous tumour-infiltrating lymphocytes (TIL) with IL-2 into patients with cancer resulted in tumour regression, indicating that T cells play an important role in this process[9]. Preclinical studies have demonstrated the feasibility of this approach both in prostate cancer and in renal cell carcinoma[10]. Tumour vaccine approaches using IL-2 gene-modified tumour cells has also induced tumour regression and increased survival in a mouse bladder cancer model. Animals that became disease-free were resistant to subsequent tumour challenge suggesting that long-term immunological memory develops in these animals[11] and this approach may have benefit for the management of human bladder cancer.

Another approach that has been used to initiate an immune response to the tumour is by injection of a gene therapy vector directly into the tumour mass that will express a foreign HLA antigen. HLA-B7 is the most common gene used in current protocols. A proportion of the tumour cells will take up the vector and begin to express the foreign antigen. An immune response develops against the tumour cells that express the foreign antigen. Simultaneously an immune reaction generated to other foreign surface antigens expressed by the tumour cells would generate a subsequent systemic response to tumour cells elsewhere in the body. Preliminary reports suggest that this approach is associated with low toxicity. Partial responses have been seen in patients with metastatic melanoma[12], and recent results also suggest that this approach may have relevance for patients with head and neck cancer[13].

2.2 Gene marking studies

While initial work concentrated on TILs as described in section 1, the use of gene marking type approaches has also been shown to be useful in the assessment of cancer therapy. In September 1991, two children with AML

were given autologous marrow which had been marked with the **neo**
resistance gene. The objective was to see if the re-emerging leukaemia
originated from the autologous marrow infused or if relapse in the patient
was due to residual leukaemia in the patient's own marrow. The first patient
had a 3 way translocation involving chromosomes 1,8 and 21. The 8:21
translocation is characterised by a fusion product which can be detected by
Reverse Transcriptase PCR. Day 180 marrow had 30% blasts and the
molecular marker of relapse and neo[r] were both detected in blast cells. The
second patient relapsed at day 67 and the blast cells carried the neo[r] gene.
The marker gene is lost when transduced cells die - thus the detection of
marked malignant cells after transplantation indicates that the reinfused
marrow contributed to relapse[14]. This type of study has important
ramifications for a purging protocol (where the aim is to deplete the
autologous marrow of malignant cells prior to reinfusion into the patient),
and subsequent trials have begun for other leukaemias and neuroblastoma
where the effects of purging have been analysed in a randomised clinical
trial using two different retroviral vectors[15]. Thus a knowledge of the biology
of malignant relapse can help guide the development of new protocols
leading to higher success rates for this type of clinical approach.

2.3   Genetic prodrug activation therapy (GPAT) strategies

While gene marking studies will provide important information
concerning the pathogenesis of the disease and provide good information for
clinical intervention, strategies have been developed which allow for a more
directed gene therapy approach in cancer. The protocols outlined above in
Section 2.1 indicate the possibilities for gene directed approaches, although
there are problems as outlined above.   A different approach involves the
insertion of drug activating genes directly into tumour cells. In this way the
cytotoxic effects of the prodrug may be targeted to the relevant tissue.
Currently several systems have been developed and a number of clinical
trials have been completed. The rationale in this approach, which is also
referred to as "suicide gene therapy" is to express genes in tumour cells that
will selectively activate nontoxic prodrugs to their active cytotoxic form.
Toxic metabolites will be generated from these prodrugs and will cause
tumour cell death. The most commonly used gene is that coding for the
herpes simplex virus thymidine kinase (HSV-tk). Mammalian cells contain a
thymidine kinase gene that is unable to phosphorylate a thymidine
nucleoside base. However, the HSV-tk gene can phosphorylate a
nucleosidase base. Therefore, any cell into which the HSV-tk gene is
transferred can phosphorylate the prodrug gancyclovir. This will generate

abnormal nucleoside bases that enter and block the DNA synthesis pathway and result in cell death. A similar system using the enzyme cytosine deaminase, which selectively converts the antifungal 5-fluorocytosine to the antimetabolite 5-fluorouracil, also has been used in a number of cancer gene therapy protocols including gene therapy for breast and pancreatic cancer[16]. In addition other prodrug activating strategies are currently being tested for transduction efficiency and cytotoxic efficacy[17].

2.3.1 Brain tumours

Suicide gene therapy approaches are particularly suited to the treatment of brain tumours as the apparent disadvantage of retroviral vectors in their inability to transfect non dividing cells is converted into an advantage – the only actively dividing cells in patients with gliomas will be in the tumour itself and thus specific delivery of HSV-Tk vectors can be achieved while sparing non dividing brain/neural tissue. Initial studies in rat glioma models proved successful, with significant reduction of tumour burden[18-21]. The GPAT strategy has been applied extensively as an approach to treat malignant brain tumours[22,23,25-27]. Ongoing gene therapy trials consist of direct injection of the vectors into the tumour mass or into residual tumour after incomplete resection followed by gancyclovir (GCV) therapy. Incorporation of the gene into every tumour cell is not necessary for effective therapy with the suicide gene system; complete tumour responses have been reported in animals when less than 20% of the cells express the transferred gene. This phenomenon is known as the "bystander effect"[24] and appears to occur through a mechanism that apparently involves connections between cells through gap junctions (Fig. 1). Thus targeting of as few as 50% of tumour cells may have a therapeutic effect. Clinical studies have revealed that the HSV-tk/GCV approach is safe, but also that responses are observed only in very small brain tumours, indicating insufficient vector distribution and very low transduction efficiency with replication-deficient vector systems[25]. A phase III, multicenter, randomized, open-label, parallel-group, controlled trial of the technique in the treatment of 248 patients with newly diagnosed, previously untreated glioblastoma multiforme (GBM) has recently been reported. Patients received either standard therapy (surgical resection and radiotherapy) or standard therapy plus adjuvant gene therapy during surgery. Progression-free median survival in the gene therapy group was 180 days compared with 183 days in control subjects. Median survival was 365 versus 354 days, and 12-month survival rates were 50 versus 55% in the gene therapy and control groups, respectively. These differences were not significant. Therefore, the adjuvant treatment improved neither time to tumour progression nor overall survival time, although the feasibility and good biosafety profile of this gene therapy strategy were further supported. The failure of this specific protocol may be due mainly to the poor rate of delivery of the HSV-tk gene to tumour cells[26]. To date, no convincing clinical trial has emerged that provides objective proof of the superiority of gene therapy strategies as compared to conventional treatment[27].

Mechanism of GCV cell killing of HSV-Tk transduced cells

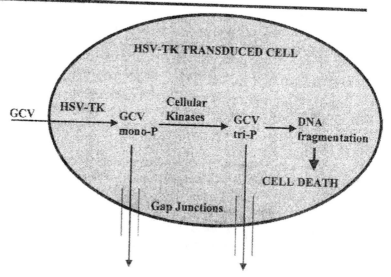

Figure 1. Mechanism of GCV killing of HSV-TK transduced cells. Following treatment with GCV, the HSV Tk gene in combination with cellular kinases converts the non toxic prodrug into its toxic byproducts which kill the cell by apoptosis. The presence of gap junctions between cells allows these toxic byproducts to promote a "bystander effect".

2.3.2 GPAT for other tumours

In order for GPAT to be applicable in other solid tumours, accurate targeting is required since gene expression in normal cells followed by exposure to GCV will result in normal cell death. This problem has been addressed by placing the viral thymidine kinase gene under control of a tissue-specific (or preferentially a tumour-specific) promoter so that the gene will be expressed only in a select population of targeted cells. In liver cancer, delivery of the TK gene into cells can be put under the regulation of a liver tumour specific promoter such as the promoter for the alpha-fetoprotein promoter. Thus while the mode of delivery may be random, the TK gene will only be expressed in the cells that express alpha-feto protein, ie the malignantly transformed hepatocytes[28,29]. In addition a tissue specific tumour gene therapy approach has recently entered stage 1 clinical trials for breast cancer using the CD suicide gene approach under the control of the erbB2 promoter which is over expressed in 20-30% of breast cancers. Twelve breast cancer patients received direct intratumoural injection of a plasmid construct containing the *Escherichia coli* cytosine deaminase gene driven by the tumour-specific erbB-2 promoter, thus allowing activation of fluorocytosine to the active cytotoxic fluorouracil only within tumour cells that express the oncogene. Targeted gene expression in up to 90% of cases and significant levels of expression of the suicide gene were specifically restricted to erbB-2-positive tumour cells, confirming the selectivity of the approach. These results are encouraging for the development of genetic prodrug activation therapies that exploit the transcriptional profile of cancer cells[16]. Other tissue specific gene therapy approaches include the use of the CD system under the control of the Prostate Specific Antigen promoter to direct expression of the suicide gene in malignant prostate cells. We and a number of other groups are investigating this approach[30,31], particularly in efforts to synergise suicide gene therapy and radiotherapy.

2.3.3 GPAT for management of graft versus host disease in allogeneic stem cell transplantation for malignancy

Another application of the suicide gene therapy approach has been in the control of graft versus host disease (GVHD), a source of significant morbidity and mortality following allogeneic stem cell transplantation (SCT). GVHD is mediated at least in part by immunocompetent T cells from the donor. In this approach, the suicide gene is inserted into donor T cells as opposed to tumour cells. These gene modified cells are then introduced into the patient either simultaneously with or following an allogeneic SCT. Donor T cells mediate a powerful anti-leukaemic effect, termed the graft versus leukaemia (GvL) effect, which is crucial to the success of allogeneic SCT approaches. However graft versus host disease (GvHD) is a major complication of allo SCT, mediated by donor T cells. The GPAT strategy in stem cell transplantation seeks to preserve the GvL effect while introduction of the prodrug eg Gancyclovir following development of GVHD has allowed control of GVHD in this setting (Fig. 2). The initial study involved a group of 8 patients who relapsed or developed Epstein-Barr virus-induced lymphoma after T cell-depleted BMT for leukaemia or other haematological malignancies[32]. All patients received donor lymphocytes transduced with the herpes simplex virus thymidine kinase (HSV-TK) suicide gene. The transduced lymphocytes survived for up to 2.5 years, resulting in antitumour activity in five patients. Three patients developed

GvHD, which could be effectively controlled by ganciclovir-induced elimination of the transduced cells. Thus genetic manipulation of donor lymphocytes may increase the efficacy and safety of allo-BMT and expand its application to a larger number of patients. Currently a number of groups are using this approach in Phase I and Phase 2 studies and a European Biomed II funded consortium involving partners in Italy, Ireland, the Netherlands, France, UK and Germany are developing *in vitro* and *in vivo* studies with existing and new vectors/approaches to establish this novel application of suicide gene therapy in the context of allogeneic stem cell transplant for malignant disease. While this approach has shown the great promise of GPAT, a number of problems have been identified including the development of a strong immunogenicity against the vector[33] and the development of GCV resistance due to the presence of a cryptic splice donor and acceptor sites within the original vector[34]. Nevertheless development of new vectors and subsequent fine tuning of this approach could have enormous ramifications in the transplant field, widening the number of donors that can be used for allogeneic SCT and providing a less toxic alternative to standard SCT approaches.

SUICIDE GENE THERAPY: MAXIMISING GVL WHILE ABROGATING GVH.

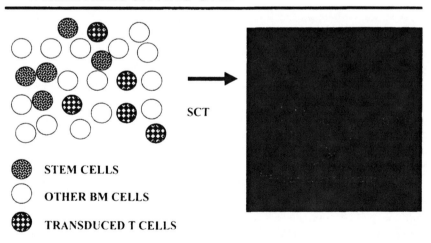

Figure 2. Suicide Gene Therapy: maximising GvL while abrogating GvH. Donor T cells are transduced with the suicide gene construct and the resulting cells are infused into the patients. The donor T cells help mediate a reduction in the leukaemia cell load, promulgating the GvL effect. If GVHD occurs, treatment with gancyclovir will kill the GVHD causing T cells and abrogate GVHD.

2.4 Transfer of multidrug resistance genes into normal haemopoetic progenitor cells

Common chemotherapeutic regimens for the treatment of cancer such as those including cyclophosphamide, anthracyclines and epipodophyllotoxins, while effective against certain forms of cancer, are limited by the effects of these agents on normal stem cells, particularly haemopoietic stem cells, leading to bone marrow toxicity. Myelosuppression can increase the risk of infection and bleeding, thus limiting the usefulness of dose escalation with chemotherapeutic agents. If substantial drug resistant gene transfer into these haemopoietic cells could be optimised, then significant reduction in chemotherapy induced cytopenia and long term marrow toxicity could be achieved. A number of genes involved in drug resistance mechanisms have been identified including the multidrug resistance gene MDR-1, Dihydrofolate reductase (DHFR), alkyltransferase (MGMT), aldehyde dehydrogenase (ALDH), glutathione S Transferase (GST), superoxide dismutase (SOD) and cytosine deaminase but for the purposes of this review we will concentrate on the MDR 1 gene. MDR-1 codes for P-glycoprotein, a 170 k-Da ATP-dependent membrane-bound protein that acts as an energy-dependent drug efflux system. Thus, MDR-1 expression pumps chemotherapeutic drugs out of the cell and decreases intracellular drug concentrations to below the toxic threshold[35]. Transfer of the MDR-1 gene into the normal bone marrow cells of mice has allowed subsequent administration of higher doses of chemotherapy without the associated bone marrow toxicity[36] leading to the development of a number of clinical studies to test the efficacy of bone marrow protection after MDR-1 gene transfer. These studies, while showing proof of principle of the approach, have been limited by low MDR expression in transfected cells[37] but preliminary results indicate that colocalisation of retroviral vector particles and fibronectin can increase the expression levels of MDR in haemopoietic stem cells[38] and could hold promise for this type of approach. Obviously as in all gene therapy approaches that utilise viral vectors, particularly retroviral vectors, there is the a theoretical risk that genes affecting normal functions in normal stem cells could be interrupted by a retroviral construct. Another concern is to guard against tumour cells being infected with an MDR construct.

2.5 Using molecular information to treat cancer: introduction of tumour suppressor genes and downregulation of oncogene expression

Loss of expression of tumour suppressor genes and over expression of oncogenes are important factors in the initiation and progression of certain tumours. Gene therapy has the potential to replace lost tumour suppressor genes or to down-regulate the action of oncogenes. Thus, tumour suppressor genes and oncogenes are attractive targets for gene therapy. Clinical studies have been initiated in a number of malignancies including acute myelogenous leukaemia, non-small cell lung carcinoma, head and neck cancer to test the ability of reintroduction of the *p53* tumour suppressor gene to inhibit tumour progression.

2.5.1   Reintroduction of tumour suppressor genes

The *p53* tumour suppressor gene is a 393-amino acid nuclear phosphoprotein that acts as a transcription factor to control expression of proteins involved in the cell cycle. The *p53* gene maps to the p arm of chromosome 17 with loss of heterozygosity resulting in expression of a mutant allele. *p53* is perhaps the most attractive target for a gene replacement strategy due to its pivotal role as "Guardian of the genome". Loss of function of *p53* either through deletion or point mutation contributes significantly to tumour development and chemotherapeutic resistance. *p53* may also inhibit angiogenesis. In addition the frequency at which *p53* is mutated in a number of different cancers[39] indicates that a *p53* replacement strategy might have widespread applications in cancer therapy[40].

Preclinical studies have concentrated mainly on the development of an adenoviral vector (Ad-*p53*) to deliver *p53* and direct antitumour responses. Phase I and phase II clinical trials have been initiated and provide support for the use of this approach in advanced head and neck cancers[41]. A phase I clinical trial was recently reported using Ad-*p53* in 28 patients with non-small-cell lung cancer (NSCLC) whose cancers had progressed on conventional treatments. Repeated intratumoural injections of Ad-*p53* appeared to be well tolerated, resulted in transgene expression of wild-type *p53*, and mediated antitumour activity in a subset of patients with advanced NSCLC[42]. Overall these results suggest that administration of Ad-*p53* in combination with chemotherapeutic drugs may have relevance in a wide range of cancers including head and neck, prostate, ovarian and breast cancer[43,44].

In addition to *p53* mediated gene therapy, an attractive approach has been developed based on the use of ONYX15, an adenoviral vector which has a deletion of the E1B55-kDa gene, enabling the vector to selectively replicate in *p53* deficient cancer cells while sparing normal cells with functional *p53*. This "smart bomb" virus as it has been nicknamed, has been shown to have short lived efficacy in a number of phase I clinical trial including head and neck cancer but a recent phase II trial combining ONYX-15 with cisplatin and 5-fluorouracil has shown responses in patients with recurrent squamous cell cancer of the head and neck[45]. The potential of current gene-therapy approaches to restore functional *p53* to tumours as a means of modulating the effects of radiation and chemotherapy should be studied in more detail.

Other tumour suppressor genes that have potential in a gene therapy replacement strategies include *p21* which interacts with *p53* and has been shown to inhibit cancer cell growth in vivo; *Rb* which regulates cell cycle progression by binding to the *E2F* family of transcription factors; *p27* which is a downstream regulator of TGF beta which also restrains cell growth in a number of cancers including gliomas. Indeed a number of preclinical studies have demonstrated the feasibility of *Rb* replacement therapy for the treatment of bladder cancer[46] and lung cancer[47].

2.5.2 Antisense approaches to downregulating gene expression

Antisense approaches are also very attractive in the treatment of cancer, particularly in downregulating expression of genes such as bcl-2 or bcr-abl that make cancer cells resistant to chemotherapy induced apoptosis. While this will be discussed in more detail in the chapter on antisense and antigene approaches, results from an antisense approach to targeting bcl-2 overexpression in Non Hodgkins Lymphoma (NHL) by Prof Finbarr Cotter's group has been very encouraging with reduction of circulating bcl-2 and partial reduction of tumour mass in patients who have been unresponsive to current chemotherapeutic approaches. Bcl-2 protein was reduced in seven of 16 assessable patients; there was one complete response, 2 minor responses, nine cases of stable disease, and nine cases of progressive disease. This study indicated that Bcl-2 antisense therapy is feasible and shows potential for antitumour activity in NHL[48,49].

3. THE FUTURE OF GENE THERAPY

Presently, more than 250 active protocols are evaluating gene therapy in the treatment of cancer. The future of gene therapy approaches in cancer will depend on further development of vector systems at the basic science level as well as a better understanding of the genes involved in tumour induction and proliferation. The ideal cancer therapy would selectively eliminate cancer cells without damaging normal surrounding tissue. To achieve this goal, a gene therapy vector would need to be delivered to the tumour at a sufficiently high concentration, be selectively activated in the tumour, selectively kill tumour cells, have no adverse effect on normal cells, and be deactivated (perhaps by a GPAT type approach once the therapeutic goal is achieved. Identification of tumour suppressor genes lost in specific tumours may allow selective gene reconstitution and cause cell death in tumour cells but not in cells normally expressing the gene. Several inducible promoter systems have been described that allow selective gene activation and inactivation[50]. Inclusion of such promoter systems in the gene therapy vector may allow selective control of the therapeutic effect.

4.   CONCLUSIONS

Gene therapy approaches in cancer have come a long way since the first gene marking protocols. Generation of tumour vaccines show considerable promise in a number of solid tumours. Introduction of multidrug resistance to normal cells may have some benefit but dose escalation can probably only achieve a modest improvement in treatment for cancer. Suicide gene therapy approaches have a certain elegance about them but need new vectors and delivery to surmount the problems of antigenicity and insufficient delivery to the tumour site. Phase I clinical trials now show that p53 gene replacement therapy using both retroviral and adenoviral vectors is feasible and safe. In addition, p53 gene replacement therapy induces tumour regression in patients with advanced NSCLC and in those with recurrent head and neck cancer. Identification of new gene targets and the development of new vector systems will allow accurate and effective gene therapy of cancer. New developments in the control of gene expression will be crucial to the success of some of the approaches outlined in this chapter.

REFERENCES

1. Anderson WF. Gene therapy of cancer. *Hum Gene Ther*. 1994; **5**:1-2.

2. Gutierrez AA, Lemoine NR, Sikora K. Gene therapy of cancer. *Lancet*. 1992; **339**: 715-721.

3. Clinical protocols. *Cancer Gene Ther*. 1996; **3**: 58-68.

4. Rosenberg SA, Aebersold P, Cornetta K, Kasid A, Morgan RA, Moen R, Karson EM, Lotze MT, Yang JC, Topalian SL, *et al*. Gene transfer into humans--immunotherapy of patients with advanced melanoma, using tumour-infiltrating lymphocytes modified by retroviral gene transduction. *N Engl J Med*. 1990; **323**: 570-578.

5. Rosenberg SA. Adoptive immunotherapy for cancer. *Sci Am*. 1990 ; **262**: 62-69.

6. Hwu P, Rosenberg SA. The genetic modification of T cells for cancer therapy: an overview of laboratory and clinical trials. *Cancer Detect Prev*. 1994; **18**:43-50.

7. Rosenberg SA, Anderson WF, Blaese M, Hwu P, Yannelli JR, Yang JC, Topalian SL, Schwartzentruber DJ, Weber JS, Ettinghausen SE, *et al*. The development of gene therapy for the treatment of cancer. *Ann Surg.* 1993 ; **218**:4 55-463;

8. Pardoll D. Immunotherapy with cytokine gene-transduced tumour cells: the next wave in gene therapy for cancer. *Curr Opin Oncol.* 1992; **4**; 1124-1129.

9. Wang RF, Rosenberg SA. Human tumour antigens for cancer vaccine development. *Immunol Rev.* 1999; **170**: 85-100.

10. Sanda MG, Ayyagari SR, Jaffee EM, et al. Demonstration of a rational strategy for human prostate cancer gene therapy. *J Urol.* 1994;**151**:622-628.

11. Connor J, Bannerji R, Saito S, *et al*. Regression of bladder tumours in mice treated with interleukin 2 gene-modified tumour cells. *J Exp Med.* 1993; **177**: 1127-1134.

12. Darrow TL, Abdel-Wahab Z, Seigler HF. Immunotherapy of Human Melanoma With Gene-Modified Tumour Cell Vaccines. *Cancer Control.* 1995; **2**: 415-423.

13. Gleich LL. Gene therapy for head and neck cancer. *Laryngoscope.* 2000; **110**: 708-26.

14. Brenner M, Krance R, Heslop HE, Santana V, Ihle J, Ribeiro R, Roberts WM, Mahmoud H, Boyett J, Moen RC, *et al*. Assessment of the efficacy of purging by using gene marked autologous marrow transplantation for children with AML in first complete remission. *Hum Gene Ther.* 1994; **5**: 481-499.

15. Rosenberg SA, Blaese RM, Brenner MK, Deisseroth AB, Ledley FD, Lotze MT, Wilson JM, Nabel GJ, Cornetta K, Economou JS, Freeman SM, Riddell SR, Oldfield E, Gansbacher B, Dunbar C, Walker RE, Schuening FG, Roth JA, Crystal RG, Welsh MJ, Culver K, Heslop HE, Simons J, Wilmott RW, Aebischer P, *et al*. Human gene marker/ therapy clinical protocols. *Hum Gene Ther.* 1996 **20**; 7:1621-1647.

16. Pandha HS, Martin LA, Rigg A, Hurst HC, Stamp GW, Sikora K, Lemoine NR. Genetic prodrug activation therapy for breast cancer: A phase I clinical trial of erbB-2-directed suicide gene expression. *J Clin Oncol.* 1999; **17**; 2180-2189.

17. Springer CJ, Niculescu-Duvaz I. Prodrug-activating systems in suicide gene therapy. *J Clin Invest.* 2000; **105**: 1161-1167.

18. Culver KW, Ram Z, Wallbridge S, Ishii H, Oldfield EH, Blaese RM. In vivo gene transfer with retroviral vector-producer cells for treatment of experimental brain tumours. *Science.* 1992; **256**: 1550-1552.

19. Vincent AJ, Vogels R, Someren GV, Esandi MC, Noteboom JL, Avezaat CJ, Vecht C, Bekkum DW, Valerio D, Bout A, Hoogerbrugge PM. Herpes simplex virus thymidine kinase gene therapy for rat malignant brain tumours. *Hum Gene Ther.* 1996; 7:197-205.

20. Barba D, Hardin J, Sadelain M, Gage FH. Development of anti-tumour immunity following thymidine kinase-mediated killing of experimental brain tumours. *Proc Natl Acad Sci U S A.* 1994; **91**:4348-4352.

21. Chen SH, Shine HD, Goodman JC, Grossman RG, Woo SL. Gene therapy for brain tumours: regression of experimental gliomas by adenovirus-mediated gene transfer in vivo. *Proc Natl Acad Sci U S A.* 1994; **91**:3054-3057.

22. Niranjan A, Moriuchi S, Lunsford LD, Kondziolka D, Flickinger JC, Fellows W, Rajendiran S, Tamura M, Cohen JB, Glorioso JC. Effective treatment of experimental glioblastoma by HSV vector-mediated TNF alpha and HSV-tk gene transfer in combination with radiosurgery and ganciclovir administration. *Mol Ther.* 2000; **2**: 114-120.

23. Trask TW, Trask RP, Aguilar-Cordova E, Shine HD, Wyde PR, Goodman JC, Hamilton WJ, Rojas-Martinez A, Chen SH, Woo SL, Grossman RG. Phase I study of adenoviral delivery of the HSV-tk gene and ganciclovir administration in patients with current malignant brain tumours. *Mol Ther.* 2000; **1**:195-203.

24. Freeman SM, Abboud CN, Whartenby KA, *et al.* The "bystander effect": tumour regression when a fraction of the tumour mass is genetically modified. *Cancer Res.* 1993; **53**: 5274-5283.

25. Wildner O. In situ use of suicide genes for therapy of brain tumours. *Ann Med.* 1999; 31421-31429.

26. Rainov NG. A phase III clinical evaluation of herpes simplex virus type 1 thymidine kinase and ganciclovir gene therapy as an adjuvant to surgical resection and radiation in adults with previously untreated glioblastoma multiforme. *Hum Gene Ther.* 2000; **11**: 2389-2401.

27. Gupta N. Current status of viral gene therapy for brain tumours. *Expert Opin Investig Drugs.* 2000; **9**: 713-726.

28. Huber BE, Richards CA, Krenitsky TA. Retroviral-mediated gene therapy for the treatment of hepatocellular carcinoma: an innovative approach for cancer therapy. *Proc Natl Acad Sci U S A.* 1991; **88**:8039-8043.

29. Ruiz J, Qian C, Drozdzik M, Prieto J. Gene therapy of viral hepatitis and hepatocellular carcinoma. *J Viral Hepat.* 1999; **6**:17-34.

30. O'Keefe DS, Uchida A, Bacich DJ, Watt FB, Martorana A, Molloy PL, Heston WD. Prostate-specific suicide gene therapy using the prostate-specific membrane antigen promoter and enhancer. *Prostate.* 2000; **45**:149-157.

31. Foley R., Hollywood D., Lawler M. Suicide gene therapy approaches for prostate cancer, potential synergy with radiotherapy. *Cancer Gene Therapy* 2000; **7**: 12 S21.

32. Bonini C, Ferrari G, Verzeletti S, Servida P, Zappone E, Ruggieri L, Ponzoni M, Rossini S, Mavilio F, Traversari C, Bordignon CHSV-TK gene transfer into donor lymphocytes for control of allogeneic graft-versus-leukemia. *Science.* 1997; **276**:1719-1724.

33. Verzeletti S, Bonini C, Marktel S, Nobili N, Ciceri F, Traversari C, Bordignon C. Herpes simplex virus thymidine kinase gene transfer for controlled graft-versus-host disease and

graft-versus-leukemia: clinical follow-up and improved new vectors. *Hum Gene Ther.* 1998; **9**:2243-2251.

34. Garin MI, Garrett E, Tiberghien P, Apperley JF, Chalmers D, Melo JV, Ferrand C. Molecular mechanism for ganciclovir resistance in human T lymphocytes transduced with retroviral vectors carrying the herpes simplex virus thymidine kinase gene. *Blood.* 2001; **97**:122-129.

35. Kartner N, Riordan JR, Ling V. Cell surface P-glycoprotein associated with multidrug resistance in mammalian cell lines. *Science.* 1983; **221**;1085-1088.

36. Richardson C, Bank A. Preselection of transduced murine hematopoietic stem cell populations leads to increased long-term stability and expression of the human multiple drug resistance gene. *Blood.* 1995; **86**:2579-89.

37. Hesdorffer C, Ayello J, Ward M, Kaubisch A, Vahdat L, Balmaceda C, Garrett T, Fetell M, Reiss R, Bank A, Antman K. Phase I trial of retroviral-mediated transfer of the human MDR1 gene as marrow chemoprotection in patients undergoing high-dose chemotherapy and autologous stem-cell transplantation. *J Clin Oncol.* 1998; **16**:165-172.

38. Abonour R, Williams DA, Einhorn L, Hall KM, Chen J, Coffman J, Traycoff CM, Bank A, Kato I, Ward M, Williams SD, Hromas R, Robertson MJ, Smith FO, Woo D, Mills B, Srour EF, Cornetta K. Efficient retrovirus-mediated transfer of the multidrug resistance 1 gene into autologous human long-term repopulating hematopoietic stem cells. *Nat Med.* 2000; **6**: 652-658.

39. Esrig D, Elmajian D, Groshen S, *et al.* Accumulation of nuclear p53 and tumour progression in bladder cancer. *N Engl J Med.* 1994; **331**:1259-1264.

40. Gallagher WM, Brown R. p53-oriented cancer therapies: current progress. *Ann Oncol.* 1999; **10**:139-150.

41. Clayman GL, Frank DK, Bruso PA, Goepfert H. Adenovirus-mediated wild-type p53 gene transfer as a surgical adjuvant in advanced head and neck cancers. *Clin Cancer Res.* 1999; **5**:1715-1722.

42. Swisher SG, Roth JA, Nemunaitis J, Lawrence DD, Kemp BL, Carrasco CH, Connors DG, El-Naggar AK, Fossella F, Glisson BS, Hong WK, Khuri FR, Kurie JM, Lee JJ, Lee JS, Mack M, Merritt JA, Nguyen DM, Nesbitt JC, Perez-Soler R, Pisters KM, Putnam JB Jr, Richli WR, Savin M, Waugh MK, *et al.* Adenovirus-mediated p53 gene transfer in advanced non-small-cell lung cancer. *J Natl Cancer Inst.* 1999; **91**:763-771.

43. Gurnani M, Lipari P, Dell J, Shi B, Nielsen LL. Adenovirus-mediated p53 gene therapy has greater efficacy when combined with chemotherapy against human head and neck, ovarian, prostate, and breast cancer. *Cancer Chemother Pharmacol.* 1999; **44**:143-51.

44. Kigawa J, Terakawa N. Adenovirus-mediated transfer of A p53 gene in ovarian cancer. In: *Cancer Gene Therapy*, (N.A. Habib, ed.) Kluwer Academic/Plenum Publishers, New York, 2000; pp207-214.

45. Khuri FR, Nemunaitis J, Ganly I, Arseneau J, Tannock IF, Romel L, Gore M, Ironside J, MacDougall RH, Heise C, Randlev B, Gillenwater AM, Bruso P, Kaye SB, Hong WK, Kim DH. A controlled trial of intratumoural ONYX-015, a selectively-replicating adenovirus, in combination with cisplatin and 5-fluorouracil in patients with recurrent head and neck cancer. *Nat Med.* 2000; **6**:879-885.

46. Seigne JD, Hu SX, Kong CT, *et al.* Rationale and development of retinoblastoma gene therapy for bladder cancer. *J Urol.* 1996; **155**:320.

47. Claudio PP, Howard CM, Pacilio C, Cinti C, Romano G, Minimo C, Maraldi NM, Minna JD, Gelbert L, Leoncini L, Tosi GM, Hicheli P, Caputi M, Giordano GG, Giordano A. Mutations in the retinoblastoma-related gene RB2/p130 in lung tumours and suppression of tumour growth in vivo by retrovirus-mediated gene transfer. *Cancer Res.* 2000; **60**: 372-382.

48. Webb A, Cunningham D, Cotter F, Clarke PA, di Stefano F, Ross P, Corbo M, Dziewanowska Z. BCL-2 antisense therapy in patients with non-Hodgkin lymphoma. *Lancet.* 1997; **349**:1137-1141.

49. Waters JS, Webb A, Cunningham D, Clarke PA, Raynaud F, di Stefano F, Cotter FE. Phase I clinical and pharmacokinetic study of bcl-2 antisense oligonucleotide therapy in patients with non-Hodgkin's lymphoma. *J Clin Oncol.* 2000; **18**:1812-1823.

50. Gossen M, Bujard H. Tight control of gene expression in mammalian cells by tetracycline-responsive promoters. *Proc Natl Acad Sci U S A.* 1992; **89**:5547-5551.

Gene Self-Assembly (GENSA)

Facilitating the construction of genes and vectors

CLAGUE P. HODGSON

Nature Technology Corporation, 4701 Innovation Drive, Lincoln, Nebraska, 68521, USA

1.   INTRODUCTION

Traditional recombinant DNA technology (rDNA), now almost 30 years old, is based on the use of restriction enzymes and DNA ligase, to 'cut' and 'paste' DNA sequences, respectively. Major improvements in rDNA technology resulted from: 1) the introduction of synthetic DNA (oligonucleotide) chemistry; and 2) gene amplification (PCR). Over the years, however, the limits of rDNA technology have been tested, particularly as a result of demanding applications such as gene therapy. Indeed, some of the most promising new vector systems being used in humans are the adenovirus and herpes simplex virus families, which have 36 kb and 150 kb genomes, respectively. Large DNA molecules are difficult to manipulate by cut and paste technology. Even more extreme are mammalian artificial chromosomes, which are capable of carrying megabase-sized DNA.

The predominant method for moving DNA in and out of vectors of adenovirus size and larger is *in vivo* recombination. This approach uses a shuttle vector, having sequences homologous to the insertion site, with the recombinant genes inserted at a site within the recombination target zone. When the shuttle vector and the virus backbone are introduced into bacteria or mammalian cells at the same time, homologous recombination may lead to the desired recombinant, which can then be rescued by a selection process.

Another problem in gene construction is the making of large synthetic cDNA molecules. These are generally made from oligodeoxynucleotides (ODNs), the maximum size of which is around 200-300 bp. Several overlap

33

and fill-in methods are available to increase the size, but mutations occur, leading to a high probability of error in longer molecules.

In place of complicated processes of gene assembly, it would be desirable to have complete control over the sequence content of a construct. It would also be desirable to decrease the amount of time required to construct complex vectors, which presently can take months or years if many steps are involved. Thus, there is a bottleneck in vector development, and a need for new technology aimed at rapid, precise gene construction. In this article, we will briefly examine the reasons why the old rDNA technology does not work as well as it should, and what is being done to overcome those limits.

2.   THE FALSE PROMISE OF PALINDROMES

For many years, molecular biologists have passed along the idea that restriction enzymes create idealised junctions in the form of palindromes, the symmetrical terminal extensions that are familiar to everyone who has used class II restriction enzymes like *Bam*HI or *Eco*RI. Class II enzymes leave either a blunt end or one in which either the 5'- or 3'-end have a short extension that can easily be paired with a similar extension due to its symmetry. Although this is a useful cloning tool, it is also the source of many problems, because (regardless of whether blunt or palindrome ends are used) the number of permutations and combinations of two fragments (a hypothetical 'vector' and 'insert') are almost infinite. Two *Eco*RI fragments, for example, can ligate together in either orientation, or to themselves, in many copies, as interlinked circles or monomeric circles. Although three fragments or more can be recombined in theory, in practical terms it is usually not feasible to sub-clone more than one fragment at a time. Screening for the correct construct is time-consuming and expensive, and results in no new biological knowledge or utility. The source of these costly problems is the symmetry of the termini, or the inability to distinguish exact addresses of recombination except by probability and screening.

3.   THE CONCEPT OF IDEAL GENE
CONSTRUCTION TECHNOLOGY

Clearly, new technology is needed to facilitate the synthesis of complex genes and vectors *in vitro*. We can begin to search for this technology by defining the endpoint of a hypothetical 'ideal vector construction technology', and then work forward toward that point.

The ideal vector technology should permit a large number of fragments to be joined together in an unambiguous manner, regardless of the size or complexity of the fragments. The assembly process should be seamless in the sense that it should join exact desired bases to each other (independent of the positions of restriction sites and of linkers or adapters). The process should not require sub-cloning or sequential addition, and the resulting constructs should form spontaneously, not requiring a complicated screening process in order to yield the vector with essential purity. Ideally, it should be designed so as to be auto-selecting. Ideally also, vectors constructed *in vitro* from several or many components should not require an intermediate host, such as *E. coli* (for a human vector). Clearly, the ability to do recombinant DNA processes *in vitro*, without cloning or sub-cloning, would be a significant advantage.

4.   THE FUNDAMENTALS OF GENSA

In order for the technology to work, each incoming DNA fragment needs an instruction set specifying its exact position and orientation within the assembly. This is possible if the DNA molecules have at their termini 'unique address labels'. These are similar to the overhanging termini that are used in classical cloning experiments. However, palindromes and blunt-ended fragments cannot be used because their symmetry destroys uniqueness by creating a duplication of addresses. The fundamental principle of GENSA is that each fragment must have unique, non-palindromic termini. Let us now consider some of the ways this can be accomplished.

4.1   Class IIS restriction enzymes and GENSA

Class IIS restriction enzymes (IIS) are a sub-class of the traditional class II enzymes (reviewed in Szybalski *et al.* 1991[1]). These enzymes differ from regular class II enzymes in that they digest the DNA a short distance to one side, away from the restriction enzyme recognition site (see Table 1). Class IIS enzymes generate overhanging termini of 1-5 bases in length, with digestion sites located 1-17 bases away from the recognition site. These enzymes have been used extensively for cloning as well as for site directed mutagenesis, usually in combination with regular class II enzymes.

Class IIS enzymes have been in use for over 30 years, and are readily available (*e.g.*, New England BioLabs, Beverly, MA). Historically, they have proven to be of somewhat limited usefulness for constructing multi-fragment gene assemblies. Some notable early examples of successful multi-fragment cloning procedures and technologies using class IIS enzymes

included: reassembly of a virus digested with *Fok*1; assembly of synthetic
(PCR product) exons into a functional interleukin-1α processed gene
(Lebedenko *et al.*, 1991 [2]); construction of a functional epidermal growth
factor gene (Urdea *et al.*, 1983[3]) and the *Fok*I method of gene synthesis
(Mandecki and Bolling, 1988[4]). In addition, class IIS enzymes led to novel
technologies for: 1) a universal restriction enzyme (Szybalski, W., 1985[5]);
gene building by successive addition of ODNs into a unique *Hph*I cleavage
site (Cohen, S.N., 1981[6]); and 2) site-directed mutagenesis (Tomic *et al.*,
1990[7]).

Table 1. Restriction endonucleases useful in GENSA.

Enzyme:	Site size (bp):	Distance to overlap:	Size of overlap:	Overlap type:
*Alw*26 I	5	1-5bp	4bp	5'-overhang
*Bbs*I	6	2-6bp	4bp	5'-overhang
*Bpm*I	6	16-14bp	2bp	3'-overhang
*Bsm*BI	6	1-5bp	4bp	5'-overhang
*Bsp*MI	6	4-8bp	4bp	5'-overhang
*Bsr*DI	6	0-2bp	2bp	3'-overhang
*Eco*57I	6	16-14bp	2bp	3'-overhang
*Fok*I	5	9-13bp	4bp	5'-overhang
*Hga*I	5	5-10bp	5bp	5'-overhang
*Hph*I	5	8-7bp	1bp	3'-overhang
*Mnl*I	5	7-6bp	1bp	3'-overhang
*Ple*I	5	4-5bp	1bp	5'-overhang
*Sap*I	7	1-4bp	3bp	5'-overhang
*Sfa*NI	5	5-9bp	4bp	5'-overhang

The class IIS enzyme's characteristic of digesting the DNA a short
distance away from the recognition site at first appears to be a negative
factor, but it can be used to great advantage in cloning. For example, if a
PCR primer contains a IIS restriction site aimed to cut inside the primer, it
can be positioned so as to cleave the DNA at any desired base near the inside
of the primer. If the bases naturally found at that site do not agree with those
of the desired joining fragment, the primer can be changed so that the
sequence complementarity matches exactly with that of its mate. This form
of sequence tailoring is seamless in the sense that it does not alter any base
of either construct unless the investigator chooses to do so. Thus, any two
sequences can be joined exactly, at any base.   There are no known
exceptions to this rule, although some may arise in time. In this scenario, the
largest number of fragments that can be joined together in a construct is
limited by the size of the extension, and by whether the extension contains
an odd or an even number of bases.

A single base extension enables only two fragments to be joined, because there are only four bases to choose from (A, C, G and T), and one base must be used at each terminus of each fragment. A two base extension means that sixteen possible dinucleotide combinations exist, sufficient for up to eight fragments to be joined by means of unique address tags. However, four out of 16 dinucleotides (AT, TA, CG, and GC) are palindromes (heresy to GENSA) that could ligate to themselves in a forward or backward direction, and thus cannot be used to specify a unique address label in a complex construct. Thus, two base extensions can be used to unambiguously self-assemble a maximum of six fragments.

For example, we recently used several IIS enzymes (*Bpm*I, *Eco*57I, *Bsr*DI) to self-assemble a retrovirus vector from six individual PCR fragments, using 2-base extensions. Interestingly, it was not necessary to clone the fragments in *E. coli* and then to screen for correct recombinants. Instead, the ligation product was introduced into retroviral helper cells, and the viral supernatant from those cells was subsequently passed to several recipient cell lines, where the vector integrated efficiently and expressed both neomycin resistance and human growth hormone genes according to the design (Hodgson *et al.*, 1998[8],[9]). RNA and DNA blot analysis confirmed that the construct was the expected design. The GENSA procedure thus saved five sub-cloning steps. Interestingly, the construct took only a few days to make.

The enzymes used in the above example generate 2bp 3'-extensions. As discussed earlier, two base pairs can specify a maximum of six fragments, while conserving the ideas of uniqueness and non-palindromicity. However, there are no three base or five base palindromes (because an odd numbered extension lacks an axis of symmetry). Thus, an enzyme such as *Sap*1 (which leaves a 5'-overhang of 3 bases) is ideal for many GENSA purposes. Second, it is a rare cutting enzyme, recognising a 7 bp sequence that occurs approximately every 16,384 bp on average. This is important, because there are few internal *Sap*1 sites that must be factored into the construct. Lastly, there are 64 possible combinations of 3 bases (also referred to as codons), permitting up to 32 fragments to be ligated together at one time. Although palindromes are not an issue when using *Sap1*, it is important to keep track of which extensions have been used once (to avoid repeated use). To do so, an extension reservation table can easily be constructed for any extension length. Uniqueness and non-palindromicity are equally important attributes of successful GENSA. It is possible to use other desirable enzymes (Table 1), such as *Fok*1, that create a longer (in this case, 4 base) overhanging terminus (although the more frequent occurrence of sites with the five base *Fok*1 recognition sequence can be a problem with this enzyme). When using class IIS enzymes with four base overhanging termini, one must remember not to use any palindromic termini (e.g., ACGT). A five base overhang (for example, *Hga*1) permits up to 512 fragments to be specified with unique labels. To do so, however, will require optimal ligation conditions, because of the rarity of each address label in such a complex mixture. Enzymes specifying 32 or 512 fragments thus have potential for building mini-chromosomes, herpes virus vectors, adeno-vectors, *etc.* The more complicated bookkeeping required for unique terminal usage can also be done by a computer. In the future it should thus be possible to create chromosomes from a combination of natural class IIS restriction fragments and synthetic DNAs made in the laboratory. The ability to assemble whole chromosomes is the prerequisite for making synthetic organisms*. GENSA is thus a technology with potential for precise chromosome building.

5.   TOWARD A GENE ASSEMBLY LINE

As illustrated above, it is possible to assemble and generate vectors from entirely synthetic (PCR) DNA, without cloning in *E. coli*. However, as the complexity and number of fragments becomes greater, there is a potentially greater chance of human handling error. For this reason, it is desirable to replace current manual operations with simplified robotic handling. For example, computer aided construct design can be combined with robotic platforms that perform operations such as PCR, restriction digest, fragment purification and ligation.

A major source of human error is associated with construct design. This is because it is very difficult to visualise the overlaps needed for joining all of the pieces together. Although a set of 'paper-dolls' can be used to model a simple construct, it is desirable to use computers to select overlap sequences as a part of construct design. Although research aimed at development of a gene assembly line is intended primarily to reduce human error inherent in repetitive processes, it will also save time in assembling complex recombinant molecules.

---

* In this scenario, one or more synthetic chromosomes (comprising a genome) are injected or otherwise inserted into the cells (such as embryonic stem [ES] cells) or eggs of an organism that has been irradiated or otherwise treated to eliminate the host DNA. The transplanted DNA is selected by growing the ES cells in selective media, or by implanting the eggs into a surrogate parent, respectively, providing a synthetic organism.

6. GENSA VECTORS FOR MULTI-FRAGMENT ASSEMBLY

A potential pitfall of conventional GENSA procedures (as described above) is that one or more of the input fragments may contain mutations introduced as a result of gene amplification or synthetic oligonucleotide chemistry used in synthesis. To prevent this, a GENSA vector system has recently been developed in our laboratory. Using the vector pWIZ-BANG[tm*] (Fig.1) it is not necessary to include class IIS sites in PCR primers. In fact, any blunt restriction fragments, PCR products, or synthetic (oligonucleotide)-based DNA molecules can be used alone or in combination in a GENSA assembly using this vector. The prerequisite for GENSA using pWIZ-BANG is to design and make fragment sets wherein each fragment overlaps its neighbours by 3 bp. All component fragments are ligated into the vector (which is made available as a linear, dephosphorylated DNA molecule [Nature Technology Corporation, Lincoln NE]). The primary purpose of pWIZ-BANG is to remove 3 bases from the 3'-termini at both ends of the insert, leaving the 5'-terminus phosphorylated and the 3'-end recessed by 3bp. This allows up to 32 fragments to be connected by means of their unique 3bp address labels. Blue-white selection is available to indicate whether an insert is present. A secondary purpose of pWIZ-BANG is to permit the sequencing of each cloned fragment (at least at their termini), using standard pUC18 sequencing primers (to determine that inserts are correct, and that termini are intact). This ensures that the overlaps will fit together properly. It also permits investigators to determine in advance whether the final construct is free of mutations. Although the sequencing step is not essential, it is highly desirable, especially if the construct is a synthetic (oligo)-based one (as these are more prone to mutation than restriction fragments or PCR products). All types of DNA molecules (ODNs, PCR products, or regular restriction fragments) can be used in the vector, provided they are introduced as blunt molecules.

Once all fragments have been cloned and validated in pWIZ-BANG, the clones are digested with *Sap*1, and the inserts are combined and ligated together in a single reaction, creating the desired construct.

This vector approach is useful for combining a dozen or more synthetic cDNA molecules into a composite cDNA or vector. Although there is no upper limit to the size of fragments that can be combined by this method, the typical size of *Sap*1 digestion fragments (approximately 16kb) suggests that the approach may be useful for assembling artificial chromosomes.

---

[*] Trade Mark, Patents pending, Nature Technology Corp., Lincoln, NE.

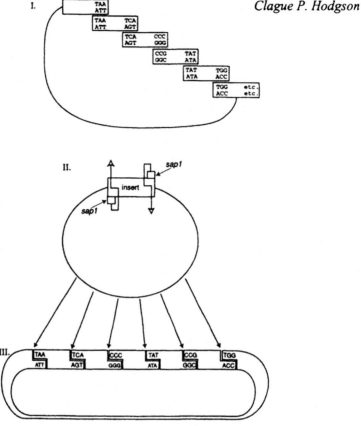

Figure 1. pWiz-Bang. (I) The construct is designed to comprise overlapping sets of fragments. Each fragment overlaps its neighbours by 3 bp. The fragments can be any double stranded DNA molecules (*eg.* ODNs, restriction fragments, or PCR products). Once the fragments have been made, they are cloned into the blunt *Nru*1 site of pWIZ-BANG (II). The clones can be identified by blue-white screening (on X-gal plates), as the clone is a pUC18 derivative. The inserts can be sequenced using pUC18 sequencing primers to determine whether the ends are intact, or whether there are mutations in the inserts. The orientation of the inserts within the vector is not important. Once a correct copy of each insert is identified, the inserts are removed by digestion with *Sap*1, which creates a 3-base, 5'-overhanging terminus at each end. (III) All fragments are combined and are joined together by annealing and ligation, resulting in the expected construct.

One fragment in the mixture is usually the vector (containing an origin of replication of some sort, as well as a selectable marker that is suitable for selection in the desired host cells), although the various vector functions can be divided up on two or more fragments. Notice that the tripartite functions traditionally associated with vectors (origin of replication, selectable marker, and unique cloning site) is reduced to only two functions (origin and selectable marker). The unique cloning site is in GENSA no longer needed. In addition, it may be irrelevant (or not) to include in the construct a prokaryotic origin of replication, as in the previous example of the retrovirus vector. Conversely, it is possible to include more than one origin of replication, if multiple hosts are contemplated. Constructs can be circular or linear, depending upon the design. Linear chromosomes, for example, might contain centromeres, origins of replication, and telomeres.

6.1   Enhanced vectoring capabilities made possible by the approach

The ability to assemble complex DNA molecules from many components has its advantages in gene therapy vectors that may have from three to six or more components. Adenoviruses, particularly, are important and difficult to assemble. The so-called 'gutless' adenovirus vectors have most or all of the 36kb adenovirus genome removed and replaced with foreign DNA sequences. However, although it is relatively easy to remove the genes from the DNA backbone (leaving only the inverted terminal repeats and packaging signal, about 600 bp, total), it is relatively difficult to rebuild a genome of this size. GENSA can be used to make selective additions and deletions in a very precise manner. Possible future uses include the engineering of large viral genomes and mini-chromosomes. In these situations, internal class IIS sites will conflict with the assembly process. These sites can be incorporated into the design (*i.e.*, they can be left alone and their sequence tags factored into the reservation table), or they can be eliminated by using their sites as fragment junctions. Where duplicate overhangs naturally or mistakenly occur within assemblies, it is possible to divide the ligation reactions into separate pools, and the pools are then combined after the ligation has proceeded to completion.

7.   POSSIBLE FUTURE USES OF GENE SELF-ASSEMBLY TECHNOLOGY

An important application of GENSA technology in the future relates to combinatorial chemistry. Allelic sets, for example the domains of immunoglobulin molecules (V, J, and H, for example), can be recombined by copying (PCR, RT-PCR) them *en masse* from genomic DNA, libraries, or mRNA, and then recombining them using GENSA to obtain all possible permutations. For example, three domains each having 100 allelic variants can be combined as GENSA fragment groups, resulting in 1,000,000 possible combinations. A similar approach is being used in our laboratory to create seamless promoter banks (Hodgson *et al.*, 1998, *Supra*). The precision of GENSA assembly is inherently greater than combinatorial methods that are based on random homologous recombination events.

GENSA could also be used to assemble telomeres of controlled length at the tips of chromosomes, for example, by iterative addition of a telomeric repeat with the same address tag at both ends of the repeat. This approach would allow control over the potential lifespan of artificial chromosomes by adjusting telomere length.   Once the assembly of chromosomes is completed, it should be possible to use them in lieu of natural ones, creating artificial organisms by introducing synthetic chromosomes into an organism, cell or egg that has had its natural chromatin ablated, for example by laser or radiation.

Introduction of genes *in vitro* or *in vivo* also requires a selection process. Selection could be performed, for example, in ES cells after injecting the artificial chromosomes, by first identifying viable cell clones containing markers, and then subjecting them to a selection process. The ES cells could then be introduced into an embryo and bred to a homozygous state. In the simplest version of biosynthetic organisms, a single chromosome could be deleted by means of precise laser surgery, followed by insertion of the genetically engineered substitute chromosome. The substitute should ideally be equipped with at least one selectable marker as well as a reporter gene and other genes of interest.

In conclusion, gene self-assembly is a promising alternative to traditional recombinant DNA methods.   Improved technology for gene assembly is badly needed for gene therapy and other complex biological vector systems. Advantages of the process are speed, economy and precision.

ACKNOWLEDGEMENTS

I would like to thank The National Institutes of Health, The National Institute of General Medical Sciences Grant 1R43 GM58361, (to Nature Technology Corporation) for generous support, and Kris Hodgson, for artwork.

REFERENCES

1. Szybalski, W., Kim, S.C., Hasan, N., and Podhajska, A.J., 1991, Class-IIS restriction enzymes – a review. *Gene* **100**:13-26.

2. Lebedenko, E.N., Birikh, K.R., Plutalov, O.V., and Berlin, Y.A., 1991, Method of artificial DNA splicing by directed ligation (SDL). *Nucleic Acids Res.* **19**:6757-6761.

3. Urdea, M.A., Merryweather, J.P., Mullenbach, G.T., Coit, D., Heberlein, U., Valenzuela, P., and Barr, P.J., 1983, Chemical synthesis of a gene for human epidermal growth factor urogastrone and its expression in yeast. *Proc. Natl. Acad. Sci. USA* **80**:7461-7465.

4. Mandecki, W., and Bolling, T.J., 1989, *Fok*I method of gene synthesis. *Gene* **68**:101-107.

5. Szybalski, W., 1985, Universal restriction endonucleases: designing novel cleavage specificities by combining adapter oligodeoxynucleotide and enzyme moieties. *Gene* **40**: 169-173.

6. Cohen, S.N., 1981, Method for synthesizing DNA sequentially. US Patent No. 4,293,652.

7. Tomic, M., Sunjevaric, I., Savtchenko, E.S., and Blumenberg, M., 1990, A rapid and simple method for introducing specific mutations into any position of DNA leaving all other positions unaltered. *Nucleic Acids Res.* **18**:1656.

8. Hodgson, C.P., Xu, G., and Zink, M.A. (1998) Self-Assembling Genes, vectors and uses thereof, World Intellectual Property Organization (WIPO) Patent publication No. WO 98/38326.

9. Solaiman, F., Zink, M.A., Xu, G.-P., Grunkemeyer, J., Cosgrove, D., Saenz, J., and Hodgson, C.P., 2000, Modular retro-vectors for transgenic and therapeutic use. *Mol. Reprod. Dev.* **56**:309-315.

Gene Expression

JASPREET KAUR[*], MANISHI MUKESH[#] AND AKSHAY ANAND[†],
*Department of Biochemistry, †Department of Immunopathology, Postgraduate Institute of
Medical Education and Research, Chandigarh; #National Bureau of Animal Genetics Resources
(ICAR), Karnal, India

1.   INTRODUCTION

'Gene expression' involves RNA synthesis (transcription) and then protein synthesis (translation) of a specific nucleotide sequence to yield the biologically active products. Number and types of RNAs and proteins in different tissues have been found to be different even though each cell of the organism has identical copies of DNA suggesting thereby that these differences are due to differential expression of genes in these cells. The manner in which a cell responds to the environment is determined by its precise combination of expressed and unexpressed genes. The first step to understand the mechanism of gene regulation and cellular functions is identification of the expressed genes.

2.   REGULATION OF GENE EXPRESSION

2.1   Steps involved in gene expression

Various steps involved in expression of a gene are:
1. Activation of chromatin
2. Transcription
3. Processing of the transcript

4. Transport to cytoplasm
5. Translation

The expression of a gene might be regulated at any of these sequential steps.

2.1.1 Activation of chromatin

Coiling of DNA around a histone octamer in the nucleosome establishes repression of genes, and transcription is brought about by uncoiling of these genes by specific positive regulatory mechanisms. The mechanisms by which specific genes are activated in chromatin have been extensively investigated in a variety of biochemical and genetic systems and reviewed by Paranjape *et al.*[1] Nucleosomes repress transcription in different ways by interfering with the interaction of activator and repressor proteins, polymerases, transcription factors etc.[2] Structural basis of repression by nucleosomes is dependent on histone configuration[3,4] and its acetylation status.[5] Acetylation of histone can relieve repression by chromatin, whereas deacetylation can reestablish it and there is a correlation between DNA methylation and histone deacetylation. A protein that binds to methylated DNA recruits a multiprotein histone deacetylase complex[6]. In general eukaryotic gene expression occurs by positive regulation. Positive regulatory elements of DNA known as enhancers, and proteins termed activators stimulate transcription by interacting with chromatin. These enhancers and activators are very specific for genes and respond to various stimuli in and out of the cell. Activators are often latent and become functional during a physiological response. Particularly well studied in this regard are STAT proteins (signal transducer and activator of transcription proteins),[7] NF-*k*B (nuclear factor *k*B)[8] and nuclear hormone receptors.[9] Activator proteins have distinct DNA binding and activation domains. The DNA binding domains fall into several major families whose structures have been elucidated.[10] Activation domains are poorly understood so far.

In short chromatin remodelling and transcription activation are separate processes that can be regulated by distinct proteins or subunits of a given protein. Recently, Armstrong and Emerson[11] have reported that transcription of chromatin packaged genes involve highly regulated changes in nucleosomal structure. Two systems that facilitate these changes are ATP-dependent chromatin remodelling complexes and enzymatic complexes which control histone acetylation and deacetylation.

In recent papers Kornberg and Lorch,[12] Bird[13] and Kornberg[14] have extensively reviewed mechanisms of eukaryotic transcriptional control.

2.1.2   Initiation of transcription

Transcription involves the synthesis of an RNA in 5'-3' direction complementary to the template DNA. The overall transcription cycle involving binding, initiation, elongation and termination is shown in Figure 1.

RNA synthesis is catalysed by 3 classes of enzymes called RNA polymerases, designated I, II and III. The polymerases are large multiprotein complexes which along with other transcription factors, interact with a particular DNA segment called promoter and thus initiate transcription. In eukaryotes, RNA polymerases do not bind to the promoter. RNA polymerase II is the most extensively studied and is responsible for mRNA synthesis, whereas RNA polymerases I and III synthesise ribosomal RNA (rRNA) and transfer-RNA (t-RNA) respectively. Much research is being done to resolve their structural and functional complexity.[15-21] The promoters for RNA polymerase I and II are upstream of the initiation site, but some promoters for RNA polymerase III are downstream. Identification and characterisation of promoters in eukaryotes is considerably more difficult than in prokaryotes. Transcription factors recognise the promoters (promoters are about 25 b.p. upstream from the transcription start site and are called TATA box or Hogness box which have the consensus TATA AAT), which are flanked by high G+C sequences. Many, but certainly not all promoters have another common sequence, 75 b.p. upstream the start site with the consensus GG(T/C) CAATCT in which T & C are equally frequent at the third position from the left. This sequence is called CAAT box. These promoters are the sites of binding for transcription factors but not of RNA polymerase. RNA polymerase II probably does not interact directly with any part of the promoter and this is the major difference between prokaryotic and eukaryotic promoters. TATA box probably determines the base that is first transcribed. Experimental evidence is that deletion or alteration of the TATA box does not prevent initiation: instead initiation occurs at a variety of nearby places. Promoters for RNA polymerase III differ significantly, and they are downstream the transcription start site and within the transcribed DNA. Transcription factors called TFIIIA or the 40 kd protein are responsible for transcription of 5S-rRNA and in regulation of cell division and development. The general transcription factors for polymerase II, designated TFIIB, -D, -E, -F and –H have been isolated from yeast, rat and human cells[22]. The complete set of factors comprising 23 polypeptides, along with the 12 subunits of the polymerase have been identified in yeast and human.[23] Transcription factors for RNA polymerase III have also been described.[24] In some cases, activity of a promoter is enormously increased by the presence of an enhancer, located at a variable distance from promoter

a) Binding of RNA polymerase to duplex DNA : thereby uncoiling the at the site of binding .

b) Initiation : 2-9 bases are synthesized complementary to one strand of DNA.

c)(i) Elongation : RNA polymerase moves, RNA is transcribed complimentary to one strand of DNA.

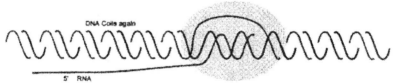

c) (ii) Elongation :RNA polymerase move and reaches the end of gene.

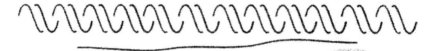

d) Termination : RNA polymerase and newly synthesized RNA are released at terminator

Figure -1 : Overall transcription cycle involving the stages as shown above.

and can function in either orientation. It has been shown that enhancer can stimulate any promoter placed in its vicinity.[25] Enhancers have a modulatory effect like promoters. Some elements are found in both enhancers and promoters. The essential role of the enhancer may be to increase the concentration of transcription factors in the vicinity of the promoter. By experimental manipulation an enhancer can be moved upto a thousand base-pairs upstream or downstream without significant loss of activity. As the distance between an enhancer and a promoter increases, the enhancing effect is gradually reduced. By genetic engineering an enhancer can be removed from its normal site and reinserted in reverse orientation with respect to the promoter without loss of its enhancing activity but molecular basis of this enhancement is not understood. If normal sequence is experimentally inverted, RNA is still made at a high rate. Deletion of the enhancer reduces transcription more than a hundredfold.

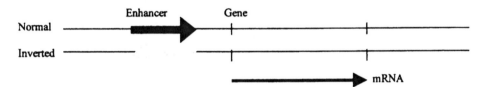

Transcription elongation is a complex process. A number of elongation factors have been characterised and these can function by directly suppressing the transient pausing or arrest of polymerase II at specific sites in RNA. mRNA synthesis takes place in eukaryote at a rate of ~1200-2000 nucleotides/min[26,27] whereas elongation by purified RNA polymerase II in vitro proceeds under optimal conditions at rates of only 100-300 nucleotides/min,[28] suggesting the existence of a class of elongation factors that stimulate the overall rate of elongation of eukaryotic mRNA. The switch from transcriptional initiation to transcriptional elongation is associated with the phosphorylation of the carboxy terminal tail of polymerase II[29] or by specifically facilitating the passage of polymerase II through chromatin templates. Regulation of gene expression at the level of transcription has been studied in detail.[30-33]

2.1.3   Processing the transcript (RNA splicing)

The primary transcripts of higher eukaryotes contain untranslated intervening sequences (introns) that interrupt the coding sequences (exons). In the processing of RNA in higher eukaryote, 50-90% of primary transcript is excised. The remaining segments (exons) are joined together to form the finished mRNA. The excision of the intron and formation of the final mRNA molecule by joining of exons is called RNA splicing.

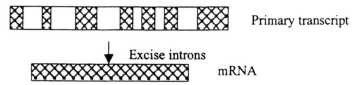

Primary transcript

Excise introns

mRNA

A variety of techniques have been used to measure intron and exon length.[34] In yeasts, introns are not present and the coding portions of the genes may be over 1000 nucleotides in length whereas long exons of over 1000 nucleotides are found very rarely in the vertebrates, but rather more frequently in plants, insects and fungi. Less than 5% of vertebrate genes (notably nearly all histone genes) do not contain introns, whereas 95% of genes have introns varying in length from 300-2000 nucleotides. Since a considerable amount of energy is used in synthesising intronic sequences that are subsequently excised and degraded and never translated, the question is what is the significance of these sequences. Extensive studies have shown that some introns contain potential regulatory sequences such as enhancers.[25] Role of enhancers is already discussed in section 2.1.2.

Fidelity of excision and splicing is extraordinary and achieved by recognition of particular base-sequences at or near the intron-exon junction. Rate of mutation in introns is higher than in exons since there are fewer constraints on the sequences of introns. This is apparent when the sequences of homologous introns and exons in the corresponding protein of different species are compared. Introns can be detected and visualised by electron microscopy of heteroduplexes of mRNA and the genomic DNA from which it was transcribed.

Recently Bentley[35] have reported a connection between transcriptional initiation and mRNA processing. Carboxy terminal domain of polymerase II is phosphorylated that interacts directly with factors that mediate mRNA capping, 3'-end processing and splicing thereby recruiting the various RNA processing machineries to the transcription apparatus and the RNA transcript, concluding that events occurring during transcriptional initiation have regulatory consequences at the level of mRNA processing and hence affect the final products of gene expression.

'Alternative' splicing is a regulatory phenomenon in eukaryotes as different sequence of steps in splicing have been observed in different cells of an organism. This method may be the major mechanism for protein diversity in eukaryotes.[36] Off/on regulation of gene expression at the level of splicing has also been reported by Bingham *et al*.[37]

In addition to splicing, most pre-mRNAs are also processed at their 3'-ends by a coupled process of cleavage and polyadenylation. This process depends on specific signals in the pre-mRNA and is mediated by a large complex of proteins which have been identified and characterised and reviewed by Minviella-Sebastia.[38]

2.1.4   Transport to cytoplasm

After processing is completed the mature RNA is transported to the cytoplasm to be translated; it is not known what prevents the primary transcript and partially processed molecules from being transported. An attractive idea is that splicing occurs on the nuclear membrane and is linked to transport through the nucleopores to the cytoplasm.

2.1.5   Translation of mRNA

Translation does not occur until processing is complete. Thus nuclear RNA contains widely varied sizes and given the name heterogenous nuclear RNA (hnRNA). Translation is a complex process requiring the participation of about 200 distinct macromolecules and involves initiating protein synthesis, linking together the amino acids in correct order, terminating the chain and often post-translational modification of the newly synthesised chain.

Protein synthesis can be divided into three stages: (1) polypeptide chain initiation, (2) chain elongation and (3) chain termination. In prokaryotes, no nuclear membrane separates the DNA and ribosomes and there is coupled transcription – translation i.e. translation begins as soon as the mRNA synthesis starts but it does not occur in eukaryotes because mRNA is synthesised and processed in the nucleus and later transported through the nuclear membrane to the cytoplasm where ribosomes are located. Translation of mRNA molecule occurs in the 5'-3' direction. Peptide bond is formed by an enzyme complex called peptidyl transferase and elongation factors along with GTP are required for further steps. The main cellular components involved in the stages of protein synthesis in prokaryotes and eukaryotes are listed in Table 1.

Table 1. Components required for the three stages of protein synthesis in prokaryotes and eukaryotes

Stage	Essential components	
	Prokaryotes	Eukaryotes
Initiation	mRNA	mRNA
	N-formyl methionyl RNA	Initiating methionyl-t-RNA
	Initiation codon in mRNA (AUG)	Initiation codon in mRNA (AUG)
	30 S subunit	40 S subunit
	50 S subunit	60 S subunit
	Initiation factor (IF-1, IF-2, IF-3)	Eukaryotic initiation factors (eIF2, eIF2B, eIF3, eIF4A, eIF4B, eIF4E, eIF4G, eIF5, eIF6)
		Poly A binding protein (PAB)
	GTP, Mg^{2+}	GTP, Mg^{2+}
Elongation	Functional 70S ribosome (initiation complex)	Functional 80S ribosome (initiation complex)
	'Aminoacyl t RNA's' specified by codons	'Aminoacyl t RNA's' specified by codons
	Elongation factors EF-Tu, EF-Ts, EFG	Eukaryotic elongation factors eEF-1α, eEF-1βγ, eEF-2
	GTP, Mg^{2+}	GTP, Mg^{2+}
Termination	Termination codons in mRNA	Termination codons in mRNA
	Polypeptide release factors (RF1, RF2, RF3)	Only one polypeptide release factor eRF

The main features of initiation step are binding of mRNA to the ribosome, selection of the initiation codon and binding of the charged t-RNA bearing the first amino acid (fMet or Met). In the elongation stage there are two processes – joining together two amino acids by peptide bond formation and moving the mRNA and the ribosome w.r.t. one another in order that each codon can be translated successively, called translocation. In termination stage the completed protein is dissociated from the synthetic machinery and the ribosomes are released to begin another cycle of synthesis.

Translational control means regulation of the number of times a finished mRNA molecule is translated. There are 2 ways in which translation of a particular mRNA may be regulated (1) by the lifetime of mRNA, (2) by regulation of the rate of overall protein synthesis.

Translational control of gene expression has also been found via covalent modification of some of the translation factors such as phosphorylation of eukaryotic initiation factor-2 (eIF-2) and elongation factor-2 (EF-2) by protein kinases.[39]

2.2 Problems in the study of gene regulation

Knowledge of regulation in eukaryote is much less detailed than for prokaryotes. Studies with multicellular eukaroytes have been hampered by two problems – the inability to obtain regulatory mutants and to manipulate genes with ease, and lack of an in vitro RNA polymerase II transcription system. Thus the information available is presented as a collection of observations illustrating the variety of regulatory mechanisms used by different cells. Many simple environmental signals that result in changes in gene expression in prokaryotes also cause changes in single celled eukaryotes. However, the cells of multicellular organisms do not, in general, respond to such signals, because in whole organisms a constant environment is normally maintained for individual cells and the signals that do bring about differential control of genes in multicellular organisms have a different purpose: to cause cells to perform specialised functions leading to formation of organs and whole organisms. In most cases these developmental processes cannot be reversed or repeated within a single cell and lead to cell differentiation. Most of the gene expression studies have been carried out at the single-cell level.[40-42]

Although regulation by hormones, growth factors or other transcriptional inducers such as cAMP or phorbol esters has been studied in detail but regulation of tissue-specific gene expression is still not understood. Also lesser understood is the loss of gene regulation in certain diseases. The human disease that exhibits the most extensive malregulation of gene expression is cancer. Thus, not only does cancer often result from the overexpression of certain cellular genes called oncogenes, due to errors in their regulation, but several of these genes themselves actually encode transcription factors and cause the disease by affecting the expression of other genes.

Dreyfuss and Struhl[43] and Bird and Kouzarides[13] have presented editorial overviews on gene-expression mechanisms and role of nucleus and other multiprotein complexes.

2.3 Gene expression in gene therapy

Gene therapy is a complex series of events relying heavily on new biotechnological techniques. It is truly a multidisciplinary activity involving skills in molecular biology, cell biology, virology and pharmacology.

Expression of introduced genes has been highly variable in animal trials. In many cases the introduced genes were expressed well in culture, then not at all when the cells were transferred to an animal. The big problem is how to deliver the gene to the right cell and get it incorporated into the chromosome such that it is expressed at appropriate time. Transformation of animal cells by introduction of genes is problematic because introduced DNA can disrupt essential genes. Different integration sites can also greatly affect the expression of an integrated gene, because integrated genes are not transcribed equally well everywhere in the genome. New strategies for gene expression are being developed. Various vectors are being tested to alter the genotype of animals such as pUC18, M13+, M13-, bacteriophage λ, cosmid vectors etc. The situation may be changing with the development of new approaches for assembling and studying the effects of mammalian artificial chromosomes. Recently artificial chromosomal vectors and several other vectors have been studied for their use in gene therapy[44] but its use is still in infancy and further research is needed for their appropriate expression of the introduced genes.

REFERENCES

1. Paranjape, S.M., Kamakaka, R.T., Kadonaga, J.T., 1994, Role of chromatin structure in the regulation of transcription by RNA polymerase II. *Ann. Rev. Biochem.* **63**: 265-297

2. Workman, J.L. and Kingston, R.E., 1998, Alteration of nucleosome structure as a mechanism of transcriptional regulation. *Ann. Rev. Biochem.* **67**: 545-579

3. Ramakrishnan, V., 1997, Histone H1 and chromatin higher-order structure. *Crit. Rev. Eukaryotic Gene Expr.* **7**: 215-230

4. Grunstein, M., 1998, Yeast heterochromatin: regulation of its assembly and inheritance by histones. *Cell* **93**: 325-328

5. Struhl, K., 1998, Histone acetylation and transcriptional regulatory mechanisms. *Genes Dev.* **12**: 599-606

6. Nig, H.H. and Bird, A., 1999, DNA methylation and chromatin modification. *Curr. Opin. Genet. Dev.* **9**: 158-163

7. Darnell, J.E.J., 1997, STATs and gene regulation. *Science* **277**: 1630-1635

8. Ghosh, S. *et al.*, 1998, NF-kappa B and Rel proteins: evolutionarily conserved mediators of immune responses. *Ann. Rev. Immunol.* **16**: 225-260

9. Mangestsdorf, D.J. *et al.*, 1995, The nuclear receptor superfamily: the second decade. *Cell* **83**: 835-839

10. Harrison, S.C., 1991, A structural taxonomy of DNA-binding domains. *Nature* **353**: 715-719

11. Armstrong, J.A. and Emerson, B.M., 1998, Transcription of chromatin: these are complex times. *Current Opinion Genet. & Dev.* **8**: 165-172

12. Kornberg, R.D. and Lorch, Y., 1999, Twenty-five years of the nucleosome, fundamental particle of the eukaryote chromosome. *Cell* **98**: 285-294

13. Bird, A. and Kouzarides, T., 2000, Chromosomes and expression mechanisms. *Current Opinion Genet. & Dev.* **10**: 141-143

14. Kornberg, R.D., 2000, Eukaryotic transcriptional control. *Trends in Genetics* **15**: M46-49 (Millennium issue)

15. Roeder, R.G. and Rutter, W.J., 1969, Multiple forms of DNA-dependent RNA polymerase in eukaryotic organisms. *Nature* **224**: 234-237

16. Kedinger, C. *et al.*, 1970, Alpha-amanitin: a specific inhibitor of one of two DNA-dependent RNA polymerase activities from calf thymus. *Biochem. Biophys. Res. Commun.* **38**: 165-171

17. Sklar, V.E.F., *et al.*, 1975, Distinct molecular structures of nuclear class I, II and III DNA dependent RNA polymerase. *Proc. Natl. Acad. Sci., U.S.A.* , 348-352.

18. Flanagan, P.M. *et al.*, 1991, A mediator for activation of RNA polymerase II transcription in vitro. *Nature* **350**: 436-438

19. Asturias, F.J. and Kornberg, R.D., 1999, Protein crystallization on lipid layers and structure determination of the RNA polymerase II transcription initiation complex. *J. Biol. Chem.* **274**: 6813-6817

20. Poglitsch, C.L., Meredith, G., Gnatt, A., Jensen, G.J., Chang, W-H., Fu, J. and Kornberg, R.D., 1999, Electron crystal structure of an RNA polymerase II transcription elongation complex. *Cell* **98**: 791-798

21. Cramer, P., Bushnell, D.A., Fu, J., Gnatt, A.L., Mater-Davis, B., Thompson, N.E., Burgess, R.R., Edwards, A.E., David, P.R. and Kornberg, R.D., 2000, Architecture of RNA polymerase II and implications for the transcription mechanism. *Science* **288**: 640-649

22. Conaway, R. and Conaway, J., 1997, General transcription factors for RNA polymerase II. *Prog. Nucleic Acids Res. Mol. Biol.* **56**: 327-346

23. Feaver, W.J. *et al.*, 1996, Genes for Tfb2, Tfb3, and Tfb4 subunits of yeast transcription/repair factor IIH: Homology to human cyclin-dependent kinase activating kinase and IIH subunits. *J. Biol. Chem.* **272**: 19319-19327

24. Gabrielsen, O.S. and Sentenac, A., 1991, RNA polymerase III(c) and its transcription factors. *Trends Biochem. Sci.* **16**: 412-416

25. Atchison, M.L., 1988, Enhancers: mechanisms of action and cell specificity. *Ann. Rev. Cell Biol.* **4**: 127-153

26. Thummel, C.S., Burtis, K.C. and Hogness, D.S., 1990, Spatial and temporal patterns of E74 transcription during Drosophila development. *Cell* **60**: 101-111

27. Tennyson, C.N., Klamut, H.J., Worton, R.G., 1995, The human dystrophin gene requires 16 hour to be transcribed and is cotranscriptionally spliced. *Nat. Genet.* **9**: 184-190

28. Izban, M.G., Luse, D.S., 1992, Factors stimulated RNA polymerase II transcribes at physiological elongation rates on naked DNA but very poorly on chromatin templates. *J. Biol. Chem.* **267**: 13647-13655

29. Conaway, R. and Conaway, J., 1999, Mechanism and regulation of transcriptional elongation by RNA polymerase II. *Current Opinion Cell Biol.* **11**: 342-346

30. Struhl, K., 1989, Molecular mechanisms of transcriptional regulation in yeast. *Ann. Rev. Biochem.* **58**: 1051-71

31. Mitchell, P.J. and Tjian, R., 1989, Transcriptional regulation in mammalian cell by sequence specific DNA binding proteins. *Science* **245**: 371-78

32. McKinney, J.D. and Heintz, N., 1991, Transcriptional regulation in eukaryotic cell cycle. *Trends Biochem. Sci.* **16**: 430-434

33. Roeder, R.G., 1991, Complexities of eukaryotic transcription initiation: Regulation of preinitiation complex assembly. *Trends Biochem. Sci.* **16**: 402-407

34. Abelson, J., 1979, RNA processing and the intervening sequence problem. *Ann. Rev. Biochem.* **48**: 1035

35. *Bentley, 1999, *Current Opinion Cell Biol.* **11**: 347-351

36. Brietbart, R.E., Andreadis, A. and Nadal-Ginard, B., 1987, Alternative splicing: A ubiquitous mechanism for the generation of multiple protein isoforms from single genes. *Ann. Rev. Biochem.* **56**: 467-495

37. Bingham, P.M., Chou, T., Mims, I. And Zachari, Z., 1988, On/off regulation of gene expression at the level of splicing. *Trends Genet.* **4**: 134

38. *Minviella-Sebastia, 1999, *Current Opinion Cell Biol.* **11**: 352-357

39. Hinnebusch, A.G., 1990, Involvement of an initiation factor and protein phosphorylation in translational control of GCN4 mRNA. *Trends Biochem. Sci.* **15**: 148-152

40. Eberwine, J. *et al.*, 1992, Analysis of gene expression in single live neurons. *Proc. Natl. Acad. Sci. U.S.A.* **89**: 3010-3014

41. Dixon, A.K. *et al.*, 1998, Expression profiling of simple cell using 3 prime end amplification (TPEA) PCR. *Nucleic Acids Res.* **26**: 4426-4431

42. Dixon A.K. , Richardson, P.J., Pinnock, R.D. and Lee, K., 2000, Gene expression analysis at the single cell-level. *TIPS* **21**: 65-70

43. Dreyfuss, G. and Struhl, K., 1999, Nucleus and gene expression: Multiprotein complexes, mechanistic connections and nuclear organization. *Current Opinion Cell Biol.* **11**: 303-306

44. Marshall, D.J. and Leiden, J.M., 1998, Recent advances in skeletal-muscle based gene therapy. *Current Opinion Genet. & Dev.* **8**: 360-365

Tumour Genotype and Response to Cytotoxic Gene Therapy

PETER T. DANIEL, BERNHARD GILLISSEN AND ISRID STURM
Department of Haematology, Oncology, and Tumour Immunology, University Medical Centre Charité, Campus Berlin-Buch, Humboldt University Berlin, Lindenberger Weg 80, 13125 Berlin, Germany

1.   INTRODUCTION

The development of cancer is a multifactorial process and the result of alterations that affect different cellular machineries resulting ultimately in deregulated proliferation. The rapid progress in elucidating the participating signalling pathways involved in tumourigenesis made clear that the direct or indirect disturbance of both cell cycle regulation and the cellular death programmes, i.e. apoptosis, are the major players determining the switch from normal to malignant growth. When the damage is done and a tumour has developed, the further disruption of cell growth and apoptosis control is influenced by a wide variety of factors including the factors regulating genetic and chromosomal stability (e.g. DNA repair pathways and cell cycle checkpoints) and the tumour cell microenvironment, e.g. cell-to-cell and cell-to-matrix interactions and angiogenesis.

Therapeutic strategies aim to eliminate all cancer cells. Virtually all antitumour therapies (except surgery) act via the induction of apoptosis and utilise to this end the endogenous death machinery. It is therefore not surprising that most therapeutic strategies fail to ultimately cure the malignant disease due to the selection of resistant clones. This deregulation of growth, cell death and survival is not only a feature of tumour development but additionally occurs further downstream during the evolution of a tumour and this process is called tumour progression.

59

Promising approaches for the gene therapy of cancer are based on the viral or non-viral transfer of genes mediating cell cycle arrest and apoptosis such as the tumour suppressor genes p53 or p16[INK4a]. Other approaches are based on the transfer of genes for prodrug converting enzymes such as *Herpes simplex* Thymidine Kinase (HS-TK) or Cytosine Deaminase (CD)[1]. The strategy of these approaches is to induce cytotoxicity in the genetically manipulated cell and the surrounding cancer cells by the local conversion of an otherwise non-toxic prodrug, such as the metabolism of ganciclovir to toxic triphosphates which induce apoptosis[2] and mediate the bystander effect (i.e. the induction of apoptotic cell death in neighbouring (tumour) cells)[2, 3].

The initial data obtained from *in vitro* studies, animal experiments and early clinical phase I studies are promising. Nevertheless, an increasing number of reports show that these approaches encounter similar problems as in conventional cytotoxic therapy, i.e. constitutive and acquired resistance to therapy. In this line, the vast majority of studies both *in vitro* and *in vivo* are dealing with the additional problem of the still far from optimal gene transfer efficiency. The genotype of the tumour cell that governs the outcome of cytotoxic therapies is, however, completely neglected in most of these ongoing preclinical and clinical trials. Thus, tumour cells showing defects in the targeted signalling cascades may not respond, even when the gene transfer is successful and the transgene is expressed in the majority or even all cells of the tumour.

Most of the current gene transfer strategies aim at the induction of apoptosis. This may be achieved by the transfer of genes which directly (e.g. p53 or Bax)[4-6] or indirectly (e.g. prodrug converting enzymes HS-TK or CD)[7-9] activate the cell death programme. Others are aimed to induce growth arrest (and apoptosis) of the tumour by manipulation of cell cycle checkpoints (e.g. p16, p21, pRb, E2F-1 or p14ARF)[5, 10-13]. Other approaches are based on the gene transfer of oncolytic viruses, i.e. utilise the apoptotic potential of a lytic viral infection such as by adenovirus. The action of these oncolytic viruses depends, at least to some extent, on the reprogramming of the cell cycle and apoptosis machinery, cellular DNA, RNA and protein synthesis factories by viral proteins[14-19]. The major targets of all these strategies are therefore signalling pathways regulating apoptosis and cell cycle checkpoints.

The right choice of a therapeutic gene is subject of current *in vitro* and *in vivo* investigations, and the knowledge of pathways and interactions is as essential as the question to focus on the relevant target genes. This report therefore aims to improve the understanding of the problems of therapeutic gene transfer for cancer therapy in the context of the current knowledge of the targeted signalling pathways.

2. CELLULAR SIGNALLING CASCADES TARGETED BY CANCER THERAPY

The progress in cell cycle and apoptosis research made in recent years showed the relevance of dysregulation of apoptosis in disease. An excess of apoptosis may result, e.g. in T cell depletion in HIV-infected individuals, neurodegenerative diseases (Alzheimer´s disease) or hepatocellular degeneration (Wilson´s disease)[20-21]. Impaired apoptosis, however, is frequently observed in cancer which contributes not only to tumourigenesis but is also associated with the development of therapy-refractory and metastatic disease. On the other hand, dysregulated cell proliferation leads, if it is not balanced by increased cell death, to the development of malignant disease or diminishes the effectiveness of anticancer therapy[22-]. Furthermore, cell cycle control and cell death pathways are closely interconnected to ensure that cells with deregulated growth are eliminated by activation of apoptosis programmes[20, 23].

The major targets for inactivation of cell cycle control in cancer are the members of the Rb-pathway, i.e. those genes ensuring negative cell cycle control: Rb-family members themselves or inhibitors of cyclin-dependent kinases (CDKI´s), or the genes linking negative cell cycle control to apoptosis, (e.g. p53, mdm-2 or the p14ARF gene). Thus, every cancer displays inactivation of one or multiple of these signal modules[28-36]. In addition, such defects are observed in the signalling cascades of apoptosis which, like in the case of the Rb pathway, may occur at any level of the signalling cascades[20,25,26,37-39]. Animal models show that the targeted disruption of such key regulators leads to cancer, e.g. like in the case of c-myc and p14(p19)ARF[28]. Most of these genetic defects are, however, also involved in the constitutive or the acquired resistance to cytotoxic therapies.

2.1 Cell cycle regulation and the G1-restriction point

The entry of resting cells from the G0 to the G1-phase (gap phase 1) and from G1 to the S-phase is controlled by the G1 restriction point which is active in the late G1-phase (Figure 1)[40]. Once this cell cycle checkpoint is inactivated, the cell is allowed to progress into the S-phase were DNA synthesis takes place. Once the S-phase is completed the cell passes through the G2 phase where the segregation of the replicated DNA strands is prepared to the M-phase (Mitosis) which leads to cell division and the end of the cell cycle and re-entry in G1. Apart from the G1 restriction point, additional checkpoints are active which control the progression through each of the cell cycle phases. The cell is very vulnerable during the cell cycle. Thus, cells may arrest after genotoxic damage not only in the G1-phase by activation of the G1 restriction point but also at other checkpoints active in the S-phase, the G2/M transition and the different steps of the mitotic cell division. The progression of the cell through the cell cycle is controlled by the Rb pathway[41, 42].

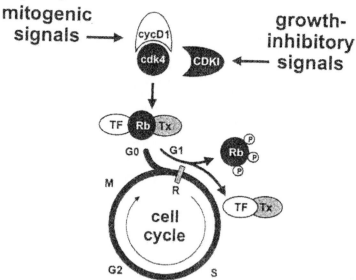

Figure 1. Regulation of the G1 restriction point. Growth factors promote entry of the cell in the cycle and inactivate the G1 restriction point by inducing cyclin D expression and activation of cyclin-dependent kinases such as CDK4 and CDK6. This leads to phosphorylation of pRb-family proteins and thereby promotes entry of the cell into the S-phase. Growth-inhibitory signals inhibit the activity of the CDK's. This results in Rb-dephosphorylation and activation of cell cycle restriction points such as in the late G1-phase. Rb-phosphorylation results in the release of transcription factors (TF) which require a dimerisation partner (Tx), eg. like in the case of E2F proteins which dimerise with Dp-family proteins or c-myc which dimerises e.g. with Max. These dimers constitute the active factors which trigger the transcription of genes required for cell cycle progression. Similar programs are active in the other phases of the cycle.

2.1.1   The retinoblastoma gene (Rb) family

The Rb family consists of the functionally and structurally closely related proteins pRb, p107 and p130[41]. All three proteins may inhibit growth in certain cell types. Inactivation of both copies of the pRb gene is associated with the development of retinoblastomas in humans. Other types of tumours also exhibit inactivation of the retinoblastoma gene product, or they harbour a dysregulation in a closely associated part of the pathway. The growth-controlling activity of the retinoblastoma protein (pRB) relies on its capacity to reversibly associate with E2F and other cellular transcription factors (Figure 1). Rb is a transcriptional repressor that is selectively targeted to promoters through an interaction with the E2F family of cell cycle transcription factors. When Rb is tethered to a promoter through E2F, it not only blocks E2F activity, it also binds surrounding transcription factors, preventing their interaction with the basal transcription complex, thus resulting in a dominant inhibitory effect on transcription of cell cycle genes. The binding region (pocket domain) of these factors led to the term "pocket" proteins which is often used as a synonym for this protein family[43]. The two domains in the Rb pocket, A and B, which are conserved across species and in the Rb-related proteins p107 and p130, are both required for repressor activity. These domains form a repressor motif that can both interact with E2F and have a dominant inhibitory effect on transcription.

The phosphorylation state of pRb determines its ability to bind to and inactivate transcription factors which are required for the transition of the cell through the phases of the cell cycle (Figure 2). In its underphosphorylated state, pRb acts as a repressor of the E2F/DP family of transcription factors[44,45] that control the expression of genes required for DNA synthesis (see below). Passage through the restriction point requires the phosphorylation and functional inactivation of pRb by cyclin-dependent kinases (CDK's). In growing cells, the phosphorylation state of pRb changes periodically and in accordance with the phase of the cellular division cycle. As well as being positively regulated by the cyclins, CDK activity can be negatively regulated by inhibitory proteins, termed CDKI's[35,36]. Two distinct types of CDKI's have been identified in mammalian cells, the CIP/KIP family which inhibit multiple cyclin-CDK combinations and the INK4 family which are specific inhibitors of CDK4 and CDK6.

2.1.2 Cyclins, CDKs and their inhibitors

The cell cycle transition points are in general regulated by cyclin dependent kinases (CDKs). These enzymes form a complex together with their respective cyclins. This complex is able to phosphorylate substrates, e.g. the pocket proteins of the retinoblastoma family[41]. Cyclin-CDK complexes regulate many of these important steps in the cell cycle, and when quiescent cells are stimulated with growth factors, the initial steps are attributed to the D-type cyclins (D1, D2 and D3) and their catalytic partners, CDK4 and CDK6. These enzymes initiate the phosphorylation of pRb but in subsequent parts of the cycle, complexes of cyclin E with CDK2 and cyclin A with CDK2 may contribute to or maintain the phosphorylated state. Thus, different cyclins and cyclin dependent kinases are involved at the different stages of the cell cycle (cyclin D family and CDK4 and CDK6 in G1 (see below), cyclin E/CDK2 in the late G1/early S phase, cyclin A/CDK2 in the S

Peter T. Daniel, Bernhard Gillissen and Isrid Sturm

phase, cyclin A/cdc2 in the late S and S/G2 transition, and cyclin B/cdc2 in the G2 phase and G2/M (Figure 2).

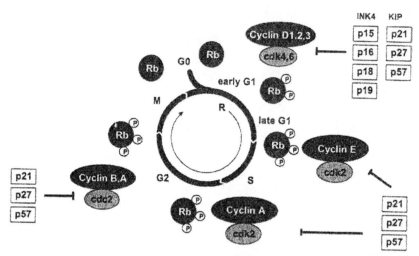

Figure 2. Cyclin and cyclin-dependent kinase activity in the cell cycle. The expression of cyclins oscillates through the phases of the cell cycle. Specific cyclins are upregulated in the different phases of the cell cycle and thereby activate specific cyclin-dependent kinases (CDKs). This maintains pRb family members in a phosphorylated (inactive) state and ensures progression through the cycle. CDKs activate the expression of phase-specific genes and thereby orchestrate these complex regulatory processes. CDK inhibitors of the INK4 and the Cip/Kip families show also cell cycle specific activities: the members of the INK4 family act on G1-phase CDK/cyclin complexes. The members of the Cip/Kip family can inhibit CDKs during all phases of the cell cycle. The decrease of CDK activity in the late G2 and the M-phase restores the hypo-phosphorylated status of pRb family members and ends the cell cycle after the cell division. R: restriction point; such R-points exist in every phase of the cell cycle and ensure under normal circumstances that cells may progress only when all requirements are fulfilled.

The CDKs show a relatively long half-life throughout the cell cycle. In contrast, the expression levels of their cyclin partners which determine activation of the CDKs oscillate and the cyclins are rapidly degraded once the cell progresses to the next phase of the cell cycle and the next cyclin is upregulated. The continuous activity of different cyclin/CDK complexes ensures that Rb proteins are maintained in a phosphorylated state[46]. Thus, the downregulation of cyclin B levels/cdc2 activity in the M-phase restores the dephosphorylated status of Rb and brings the cell cycle to an end[36, 41].

The D-type cyclins (cyclin D1, D2 and D3) are important for the G1-restriction point regulation: The retinoblastoma gene product (pRb) is phosphorylated by these cyclin D/kinase-complexes during mid – to late G1-phase[47]. The pRb-phosphorylation disrupts its association with histone deacetylase and E2F transcription factors, allowing the transcription of genes whose activities promote the transition into the late G1-phase (mainly cyclin E). The progression from late G1 to S phase is triggered by a second phosphorylation step of Rb which is mediated by the cyclin E/cdk4 complex[48,49]. This promotes release of E2F factors and transcription of genes promoting entry into and execution of the S-Phase (see below).

The E2F family of transcription factors is composed of 6 members (E2F-1 to -6)[44,45]. The formation of the active transcription factor requires, like in the case of other transcription factors such as c-myc, AP-1 or NF-kB, the binding of dimerisation partners, i.e. the DP1 proteins in the case of E2F (Figure 1). The Rb pocket protein-family members associate differentially with these transcription factors (i.e. pRb with E2F-1 to -4, p130 and 107 with E2F-4 and -5)[41]. Hyperphosphorylation of Rb releases these transcription factors, and allows their activity. The genes regulated by E2F transcription factors include genes coding for further cell cycle regulatory proteins (cyclin E, cyclin A, cdc2, the cyclin dependent kinase inhibitor p21), the protooncogene c-myc and proteins involved in DNA metabolism, like dihydrofolate reductase, thymidine kinase, thymidilate synthase[41]. Among the E2F-responsive genes are cyclins E and A, which combine with and activate CDK2 to facilitate S phase entry and progression. Accumulation of cyclin D-dependent kinases during G1-phase sequesters CDK2 inhibitors of the Cip/Kip family, complementing the effects of the E2F transcriptional programme by facilitating cyclin E-CDK2 activation at the G1-S transition.

The cell cycle checkpoints, i.e. the arrest of cell cycle progression at specific phases of the cycle are controlled by cyclin-dependent-kinase inhibitors, termed CDKI's[35,36]. Two distinct types of CDKI have been identified in mammalian cells, the CIP/KIP family which inhibit multiple cyclin-CDK combinations and the INK4 family which are specific inhibitors of CDK4 and CDK6 (Figure 2). The activity of cyclin-dependent kinase 2 (CDK2 associates with Cyclins A and E) can be blocked by the Cip/Kip family of CDK inhibitors ($p21^{Wafl/Cip1}$, $p27^{Kip1}$ and $p57^{Kip2}$). These Cip/Kip proteins share a homologous inhibitory domain, which is both necessary and sufficient for binding and inhibition of CDK4- and CDK2-containing complexes. These proteins act as stoichiometric inhibitors, and, although *in vitro*, they inhibit all G1 complexes, they preferentially act on CDK2 complexes.

The INK4a family (inhibit CDK4) consists of 4 closely related members ($p16^{INK4a}$, $p15^{INK4b}$, $p18^{INK4c}$ and $p14^{INK4d}$ ($p19^{INK4d}$ in mice)). The INK4a family members block cells from exiting G1-phase by inhibiting the phosphorylation of Rb. This occurs by inhibiting the D-type-dependent CDK4 and CDK6. Thus, the hypophosphorylated Rb inactivates the transcription factors necessary for the G1 transition, and thereby prevents the synthesis and activity of cyclin E and prevents the cell from entering the S-phase. This also explains why the INK4 CDKIs depend on the presence of Rb. In cells with defective pRb the INK4a CDKIs fail to work[49].

2.2   Molecular ordering of cell death pathways

As briefly mentioned before, the cell cycle and apoptosis are closely linked. This ensures, under normal conditions, that cells entering the cell cycle by deregulated activity, e.g. of oncogenes such as E2Fs or c-myc, rapidly undergo cell death[34,42]. Such an abortive cell cycle activates cellular suicide programmes which thereby maintain tissue homeostasis and prevent deregulated cellular growth. This occurs via activation of the p53/Bax pathway of apoptosis (Figure 3).

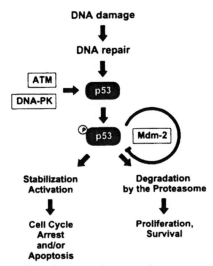

Figure 3. The p53 pathway. DNA damage activates repair programmes and a cellular stress response. This is mediated in large parts via the p53 pathway. Phosphorylation of p53 triggers p53 activity and stabilisation. Active p53 may induce cell cycle arrest, e.g. by transcriptional activation of p21$^{Cip/WAF}$ or cell death via the apoptosis-promoting Bax. P53 also induces transcription of mdm-2 (hdm-2 in humans) which exerts a negative feedback control and mediates p53-inactivation by mediating degradation of p53 via the proteasome.

2.2.1   The cell cycle and apoptosis

It is established knowledge that the p53 gene is one of the most commonly inactivated genes in cancer which may result in an impaired response to therapy and poor clinical prognosis[38]. DNA damage, hypoxia, viral infection or oncogene activation stabilise the p53 protein. While p53 acts as a DNA-binding protein which mediates transcriptional activation of target genes (e.g. Bax, p21$^{Cip/WAF-1}$, 14-3-3, Gadd45, IGF-BP-3, PIGs (p53-

induced **genes**)), the expression of other genes may be transcriptionally repressed. The events triggered by p53 such as cell cycle arrest or induction of apoptosis are to some extent cell type specific which may explain the diversity of observed p53 responses (see below and Figure 3). The inactivation of the homologues of p53 such as the p73 protein is less well investigated but there is evidence that these genes may also be inactivated, either by mutation or deletion of the corresponding chromosomal region.

The p53-triggered cell cycle arrest and initiation of DNA-repair upon DNA-damage by irradiation or chemical mutagenesis may permit the cell to recover from the damage and to survive. Upon failure, the apoptosis-pathways also initiated by p53 are executed and the cell dies. Experimental evidence for this model came from the group of Vogelstein[50] who showed that the knock-out of p21$^{Cip/WAF-1}$ impaired the ability of such cells to arrest in the cell cycle at the G1 and G2/M checkpoints upon treatment with cytostatic drugs. In consequence of the disruption of p21-mediated cell cycle arrest, these cells continued through the cell cycle and became polyploid without undergoing mitosis and subsequently died by apoptosis. The DNA-synthesis occurred independently from cell division, presumably due to intact M-phase checkpoint control. In contrast to the p21 proficient cells, these polyploid p21 k.o. cells died by apoptosis. This may explain why p53-inactivated tumour cells are genetically unstable which in turn may facilitate accumulation of further genetic alterations related to therapy resistance. Such a genetic instability may be due to a loss of Gadd45 activity in p53-inactivated tumours. Evidence therefore comes from experiments in Gadd45 k.o. mice which are highly aneuploid and carry many chromosomal aberrations[51]. Cell cycle arrest at the G1 restriction point is dependent on the induction of the cyclin-dependent kinase inhibitor p21 while the arrest at the G2/M restriction point occurs due to both p21 and a 14-3-3-dependent sequestration of cyclin B2/cdc2 complexes.

Accumulation of p53 is not due to induced p53 gene transcription or mRNA stabilisation but is mediated by stabilisation of the p53 protein. P53 protein levels are under the control of Mdm-2 (Hdm-2 in humans). Mdm-2 binds to p53 and mediates p53 degradation by the proteasome. Phosphorylation of the amino terminus of p53 disrupts binding of Mdm-2 and reduces p53 destabilisation by Mdm-2. Such phosphorylation events are assumed to be mediated via the ATM (Ataxia teleangiectasia mutated) and the DNA-PK kinases (Figure 3).

Prolongation of p53 half life and activity is also the mechanism how oncogenes such as E2F-1 and c-myc induce apoptosis. Under normal circumstances these genes promote cell cycle progression from the G1-phase to the S-phase. Their uncoordinated activation may result in cell death instead of cell cycle progression (i.e. when concomitant survival signals are missing). This signalling pathway was recently shown to be mediated via the second product of the INK4a gene locus, p14ARF (ARF stands for alternative reading frame, Figure 4)[34]. The first gene product discovered to be transcribed from the INK4a gene locus was the cell cycle inhibitory cyclin-dependent kinase inhibitor p16$^{CDKN2/INK4A}$. P16 is known to be induced by DNA damage via p53-independent mechanisms and mediates an arrest at the G1 restriction point of the cell cycle (Figures 2 and 5). The second gene product, p14ARF binds to Mdm-2 and promotes the degradation of Mdm-2 via the proteasome. This in turn stabilises the p53 protein and activates the p53-dependent events of cell cycle arrest and apoptosis (Figure 3 and 5). Both genes are regulated by independent promoters and have been shown to be inactivated independently from each other by promoter methylation or concomitantly by chromosomal 9p deletions.

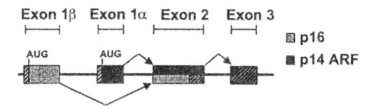

Figure 4. Intron/Exon structure of the INK4a gene locus. Alternative splicing and exon usage leads to expression of 2 genes from the same gene locus. Both genes are regulated independently by individual promoters. The p16^{INK4a} CDKI is produced from exon 1α, exon 2 and exon 3. In contrast, the completely non-homologous p14ARF (p19ARF in mice) is produced from exon 1β and a part of exon 2. Exon 2 is translated in p14ARF by the use of an alternative reading frame (ARF). The two genes are regulated by two independent promoters which may also be differentially inactivated in tumours.

P53 may then activate transcription-dependent and -independent mechanisms of apoptosis. Transcription-dependent events include induction of apoptosis-mediating genes, such as Bax, PIG3, and IGF-BP3 resulting in enhanced protein expression of these factors (Figure 5). Synergistic effects between the apoptosis promoting transcription factor E2F-1 and p53 may be required to induce p53-dependent apoptosis. Transcription-independent events are still poorly understood but may involve the 53BP2 protein which interacts competitively with p53 and Bcl-2.

These data also support the hypothesis that impairment of this pathway either by loss of function of p14ARF, p53, or Bax, or a gain of function of Mdm-2 may result in resistance of the tumour cell to cytotoxic therapies.

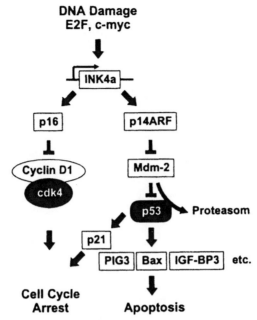

Figure 5. The p14ARF/p53 signalling pathway. DNA damage or dysregulated activation of cell cycle promoting factors leads to activation of p14ARF. The expression of the p16$^{CDKN2/INK4A}$ protein leads to cell cycle arrest in G1 via the inhibition of D-type cyclins/CDK complexes. The p14ARF protein mediates degradation of Mdm-2 by the proteasome and thereby initiates p53-dependent apoptosis and cell cycle arrest. Mdm-2 is an inhibitor of p53 which mediates degradation of p53 protein by the proteasome. Inactivation of Mdm-2 therefore increases p53 protein levels which triggers the spectrum of p53 signalling events: transcriptional activation of death pathways, cell cycle arrest in G1 and G2/M, DNA repair and transcriptional repression of survival genes.

2.2.2 The cell death programmes

Recent discoveries established that multiple distinct signalling pathways regulate apoptosis. Such pathways are activated in general by the formation of a death-inducing signalling complex (DISC). This event is triggered by an apoptosis promoter such as a death ligand. Activation of the DISC results in the early recruitment of inducer cysteinyl aspartases (caspases) (e.g. caspases-2, -8-, -9, -12). These inducer caspases then amplify the apoptosis signal by cleavage and activation of effector caspases (caspases-3, -6, -7) which execute apoptosis by degrading hundreds of regulatory proteins and activation of endonucleases and other non-caspase proteases. A similar complex (of non-homologous) proteins is observed in mitochondrial apoptosis where a DISC is formed by release of mitochondrial factors (cytochrome c, and probably also caspase-9) to the cytoplasm and formation of a complex with cytoplasmic proteins (APAF-1 and other CARD

(Caspase-Activation and Recruitment Domain) proteins such as ARC (Apoptosis Repressor with a CARD) and CIPER (CED-3/ICH-1 Prodomain homologous, E10-like Regulator; identical to bcl-10, mE10). This "mitochondrial" DISC is also designated as "apoptosome" (Figure 6). Inhibitors of such apoptosis activators may work at any step of these apoptosis signalling cascades. The basic principle of apoptosis promoters acting in a cascade leading to activation of executing enzymes, e.g. the caspases, is evolutionary conserved and can be traced back to archaic organisms such as *Cenorhabditis elegans* and possibly even simpler organisms such as slime moulds (*Dictyostelium*) and even bacteria.

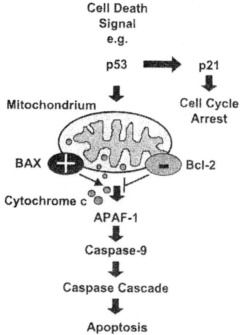

Figure 6. Activation of the mitochondrial apoptosis signalling cascade. Cytoplasmic death signals may activate the mitochondrial apoptosis signalling cascade. Such signals can be generated, e.g. upon genotoxic stress which induces p53-family protein expression. These signals activate the Bax protein (in the case of p53 this may occur via the transcriptional activation of the gene resulting in elevated Bax protein levels). Bax (and its homologues) is a direct activator of the mitochondrial signalling cascade and induces release of cytochrome c, possibly by regulating channels in the mitochondrial membrane. This is a decisive step in the activation of the apoptosome, i.e. the mitochondrial DISC (death inducing signalling complex): the cytoplasmic apoptosis regulator APAF-1 binds the cytochrome c in the WD40-repeat region and (d)ATP in the ced4-homologous loop domain of the molecule. This leads to recruitment of procaspase-9 to the complex and autocatalysis and cleavage of the procaspase-9 zymogen to the active caspase-9 enzyme. Caspase-9 then activates the downstream apoptosis cascade by activation of effector caspases.

The most prominent of these pathways are activated either by mitochondria or the death receptors belonging to the superfamily of TNF receptors.

2.2.3   Death receptors

The analysis of such distinct apoptosis pathways led to the definition of death receptor-induced signalling complexes (DISC) and the mitochondrial apoptosome. In the case of the death receptors, binding of a trimeric ligand such as TNF, TRAIL, or the Fas ligand triggers recruitment of adaptor proteins like FADD (TNF and Fas pathway[52]) or RIP and RAIDD (TNF receptor 1) to the death domains of the receptor[53] (Figure 7). These adaptor proteins then recruit procaspases which are thereby activated to form enzymatically active heterotetramers. Caspase-8 is the dominant death receptor-activated caspase and is recruited to the DISC, e.g. by the TRAIL receptors (DR3 and DR4), CD95/Fas, and TNF-R1. Caspase-10, which was mainly implicated in TRAIL-mediated death may contribute but is apparently not decisive in TRAIL and Fas signalling. In addition, TNF-R1 may recruit Caspase-2 via the RIP and RAIDD adaptor proteins. While the exact mechanisms how these inducer caspases become cleaved and activated are so far unknown, autocatalysis like in the case of procaspase-9 (which becomes activated by autocleavage upon multimerisation in a complex with APAF-1, see below) has been suggested.

The interest for the death receptors is supported by experimental evidence that death receptors may be involved in the cellular response to stress, e.g. upon DNA damage. Expression of death receptors such as CD95/Fas or TRAIL receptors was observed upon DNA damage and the expression of these receptors was shown to be induced via a p53-dependent pathway. Nevertheless, accumulating evidence suggests that the apoptosis upon chemo- or radiotherapy or p53 gene transfer is independent from such receptor/ligand interactions[54]. Thus, the dominant pathway upon DNA damage or deregulated cell cycle activation appears to be the mitochondria-dependent activation of apoptosis execution via caspase-dependent and caspase-independent mechanisms.

2.2.4   Mitochondria and apoptosis

The mitochondrial activation during apoptosis is inhibited by Bcl-2, Bcl-xL and other apoptosis-preventing members of the Bcl-2 gene family. In contrast, apoptosis-promoting members of this gene family, such as of the Bax-subfamily (Bax, Bak and Bok) have been demonstrated to be direct activators of the mitochondria[55-57]. Both upregulation of Bcl-2 or loss of Bax

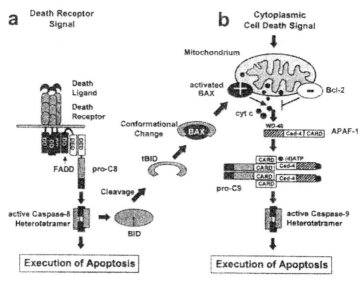

Figure 7. Interconnection of death receptor and mitochondrial apoptosis signalling. Death receptor signalling may contribute to the activation of the mitochondrial apoptosis cascade. Activation of death receptors leads to formation of a DISC. In the case of the CD95/Fas receptor, a trimeric Fas ligand binds to the trimerised receptor. This initiates the formation of the DISC: the adaptor molecule FADD is recruited via interaction of the death domains in Fas and FADD. This mediates the recruitment of procaspase-8 to the DISC by interaction of the DED in the caspase-8 pro-domain and FADD. This results (probably by autocatalysis) in cleavage of the pro-enzyme to the active caspase-8 heterotetramer which consists of 2 small and 2 large subunits. Caspase-8 cleaves the apoptosis-promoting Bcl-2 homologue Bid to form tBid (truncated Bid). tBid has been shown to physically interact with Bax and to mediate a conformational change in the N-terminal portion of the protein. This activation of Bax coincides with cytochrome c release. This is a decisive step in the activation of the apoptosome, i.e. the mitochondrial DISC. tBid connects the receptor-mediated apoptosis which is induced by death ligands to the mitochondrial apoptosis signalling pathway which is activated by cytoplasmic signals, e.g. upon DNA damage. DD: death domain, DED: death effector domain, DISC: death-inducing signalling complex, FADD: Fas-associated death domain, tBid: truncated Bid, CARD: caspase-activation and recruitment domain.

is associated with development of therapy resistance in many types of cancer[22,25,26,39]. Members of the Bcl-2 family, the mitochondrial protein cytochrome c and the cytosolic APAF-1 form the so-called mitochondrial apoptosome, also known as mitochondrial DISC (Figure 6). Binding of cytochrome c and (d)ATP to APAF-1 has been shown to trigger conformational changes in APAF-1 leading to the activation of the caspase-activation and recruitment domain (CARD) of APAF-1 which then binds procaspase-9[55-57]. Complexing of a minimum of 2 APAF-1 and 2 procaspase-9 molecules has been suggested to mediate autocatalysis of procaspase-9 to the active enzyme. Previous data suggested that Bcl-xL and possibly Bcl-2 act by keeping APAF-1 in an inactive form. Bax was shown in immunoprecipitation experiments to disrupt the interaction between APAF-1 (or the *C. elegans* homologue ced-4) and Bcl-xL. This, together with cytochrome c release, was suggested to trigger recruitment of procaspase-9. This physical interaction of Bcl-xL with APAF-1, however, appears to be much weaker as could be expected from previous experiments. Thus, the role of Bcl-xL and its homologues appears to be the regulation of cytochrome c release from mitochondria and not the direct regulation of APAF-1 conformational changes.

The mitochondrial apoptosis promoting events can therefore be dissected into distinct steps: The release of cytochrome c from the intimate space between the inner and outer mitochondrial membrane precedes the mitochondrial permeability shift transition, i.e. the opening of channels, ion fluxes, the acidification of the cytoplasm by the release of protons produced by the respiratory chain[56]. Thus, in neuronal cells the pro-apoptotic Bax protein can induce cytochrome c release which is associated (i) with shrinking of the mitochondria, (ii) clustering of apoptotic mitochondria around the nucleus which (iii) cannot be prevented by cyclosporin A or other inhibitors of the mitochondrial permeability transition. These data are in line with the finding that opening of the permeability transition pores of the mitochondria occurs after the cytochrome c release[58].

The apoptosis promoter Bax can form such channels in model systems where recombinant proteins are inserted into artificial membranes. Interestingly, preliminary evidence suggests that these channels might be large enough to permeabilise the outer mitochondrial membrane for the release of cytochrome c which has a molecular weight of 15 kDa. In intact cells, the release of cytochrome c coincides with a conformational change of Bax which has been revealed by the use of conformation-specific N-terminal anti-Bax antibodies[58]. It is, however, not yet clear whether this cytochrome c releasing channel is formed by Bax itself or whether Bax is regulating another channel.

The mitochondrial apoptosis inducing factor, AIF[55,59],was described to be released from apoptotic mitochondria and to translocate to the nucleus where it mediates fragmentation of the DNA into high molecular fragments. The action of AIF is caspase-independent. The contribution of AIF in apoptosis induction is not very clear. It is evident that the complete inhibition of caspase-activation can inhibit some forms of apoptosis, e.g. in neurons, while cell death induced by cytostatic drugs or irradiation is only delayed and not completely inhibited in most assay systems. In cells treated with caspase-inhibitors, the morphology changes from the typical signs of apoptosis (i.e. cell shrinking, chromatin condensation, DNA fragmentation into small oligonucleosomal fragments, membrane blebbing, formation of apoptotic bodies etc.) to a necrotic phenotype. It was suggested in several presentations that AIF may account for these changes, i.e. mediates caspase-independent pathways of apoptosis as originally published.

2.2.5   Interconnecting the cell death signalling pathways

The recent discovery of Bid-cleavage by caspase-8 to tBid (truncated Bid) connects the death-receptor to the mitochondrial apoptosis pathway [60, 61]. This mechanism allows a cross-talk between death receptor signalling and pathways leading to mitochondrial activation (Figure 7). The activation of death receptor signalling and DISC recruitment leads to caspase-8-dependent cleavage of Bid which then binds to Bax. This activates Bax to a conformational change of NH_2-terminal epitopes which coincides with the release of cytochrome c from the mitochondria[58].

In this line, there is evidence that tumour necrosis factor (TNF) also activates Bid-cleavage which in turn mediates a conformational change in Bax. Again, the occurrence of this conformationally altered Bax was shown to coincide with cytochrome c release and caspase-9 activation. TNF-α induces an interaction between the adenoviral E1b 19kDa protein (which shares sequence homology and functional similarities with Bcl-2) and Bax leading to the translocation of E1b to mitochondria and binding of E1b19kDa to Bax upon treatment with TNF-α. This coincides with the conformational change of Bax and the cleavage of Bid to tBid. E1b 19kDa inhibited cytochrome c release, caspase-9 and partially caspase-3 processing while the upstream events, i.e. caspase-8 cleavage, remained unaffected[62].

2.3   The tumour cell microenvironment and epigenetic changes in the response to therapy

The response to cytotoxic therapies, such as chemotherapy or irradiation, may be also influenced by the tumour cell microenvironment. This may not only be true for these "conventional" therapies but also for novel therapeutic strategies such as gene transfer of apoptosis promoters, disruption of survival signals by kinase inhibitors as in the case of bcr-abl positive leukaemia, or surface receptor directed apoptosis-inducing therapies, e.g. by anti-CD20 antibodies in B cell lymphoma.

Thus, there is evidence that the B cell lymphoma response to etoposide-induced apoptosis can be inhibited by cross-linking of the CD40 antigen which leads to an increased expression of the apoptosis inhibitory Bcl-xL and NF-kB[63]. Thus, the CD40 receptor/ligand interaction may counteract the effects of tumour therapy in situations where T/B cell collaboration takes place or where infiltrating T cells or other cells express the CD40 ligand.

A similar effect is observed in the case of hypoxia. Aggressive tumours often outgrow their blood supply and the resulting hypoxia may be a selective factor for genetic alterations which protect from apoptosis[64]. Hypoxia can cause cell death either by necrosis or by apoptosis. Hypoxic cells cover their energy demand by glycolysis which leads to the production of lactic acid and may generate an acidic extracellular microenvironment. This is of potential interest since the provision of enough nutrients and prevention of medium acidification may inhibit hypoxia-induced cell death *in vitro*. Thus, pathways protecting from low pH may be important for tumour cell survival. Hypoxia leads to the expression of the hypoxia-inducible factor 1α, HIF-1α: This occurs by prolongation of HIF-1α half life: under normoxic conditions, HIF-1α is degraded by the proteasome via the Von-Hippel-Lindau (VHL) oncoprotein. In hypoxia this is prevented and HIF-1α translocates to the nucleus where it interacts with the aryl hydrocarbon nuclear translocator (ARNT) to bind to specific DNA sequences which are designated as hypoxia response elements (HREs). Such target genes are glycolytic enzymes (i.e. GLUT-1), vascular endothelial growth factor (VGEF), urokinase receptor, and inducible nitric oxide (NO) synthase which shows that many pathways involved in tumour angiogenesis and invasion are regulated by hypoxia. Interestingly, the effects of HIF-1α on apoptosis of the tumour cells varies and some groups could not observe an apoptosis promoting effect of HIF-1α suggesting a role for the tumour type and the genetic background of the tumour cell with regard to hypoxia-induced apoptosis. Thus, it is not surprising that inactivation of p53 confers resistance to hypoxia-induced apoptosis. In this line, HIF-1α has been shown to physically interact with wild type p53 and to stabilise it thereby contributing to the induction of p53 levels upon hypoxic stress. Thus, it is not surprising that dominant negative HIF-1α mutants can inhibit hypoxia-induced death in neurons. Nevertheless, hypoxia may also induce apoptosis by p53-independent mechanism and HIF-1α-independent mechanisms are involved in the hypoxia-induced p53 response. Hypoxia sensitises as well for other apoptosis pathways such as c-myc and Fas-mediated apoptosis, and can be counteracted by survival signals, e.g. the Bcl-2, IGF-1 and PI3-kinase pathways.

Thus, the tumour microenvironment may exert a similar effect on the inactivation of relevant signalling cascades as compared to the above described genetic alterations and such phenomena therefore need to be considered when the effects of a therapeutic gene transfer are assessed.

3.   INACTIVATION OF PATHWAYS AND RESISTANCE TO CONVENTIONAL CYTOTOXIC THERAPY

Cytotoxic therapies such as irradiation and chemotherapy induce active death of the tumour cell through the activation of endogenous cell death pathways. Experimental therapies such as heat shock (hyperthermia), monoclonal antibodies or novel agents like kinase inhibitors such as flavopiridol or STI571 may also induce cell death either via the activation of cell death programmes or by inactivating anti-apoptotic pathways[20,65].

The inactivation of the death programmes is a common phenomenon both in tumourigenesis and progression of malignant tumours from a therapy-sensitive to a therapy-resistant phenotype. This is associated in many situations with e.g. the inactivation of p14ARF, p53, Bax or caspases, the executioners of apoptotic cell death. Furthermore, this is often associated with the loss of cell cycle checkpoint control which then results in dysregulated proliferation (see below). Similarly, resistance may be caused by dysregulation of survival pathways, e.g. by the overexpression of Bcl-2 or Inhibitor of Apoptosis Proteins (IAPs), or the constitutive activation of anti-apoptotic transcription factors like NF-kB, oncogenes like Ras or kinases in the PI3/Akt kinase pathway[20].

3.1   The p14ARF/mdm-2/p53/Bax pathway

All tumours exhibit such a deregulation of apoptosis programmes which may occur at any level of the signalling cascade. The analysis of such defects may identify patients with a poor prognosis, like in the case of Bax inactivation[26,39]. There is accumulating evidence that loss of p14ARF is involved in the generation of an apoptosis resistant phenotype. Similar findings have been published in the case of overexpression of mdm-2 in a wide variety of cancer types. Thus, with regard to the role of these programmes in gene therapy approaches, we investigated cell cycle dysregulation and inactivation of the p14ARF/p53/Bax apoptosis pathway in clinical samples. Prediction of tumour aggressiveness by means of analysis of such novel, molecularly defined prognostic markers might therefore yield strategies for individual escalation or de-escalation of anti-tumour therapy. To this end, there is increasing evidence that resistance towards apoptosis is not only involved in tumourigenesis, but also confers resistance to anti-tumour therapy[66,67].

In this line, we previously observed a defect in expression of the pro-apoptotic Bax protein, a key promoter of apoptosis in breast cancer[22]. Restoration of Bax expression in breast cancer cell lines inhibited tumourigenicity[24] and increased sensitivity to cytotoxic drug therapy[27]. We previously also showed that overexpression of the Bax-related pro-apoptotic Bik/Nbk could sensitise resistant tumour cells for drug-induced apoptosis[68]. In breast cancer patients, a reduced Bax expression correlates with a poor response to chemotherapy and shorter overall survival[69]. In diffuse aggressive Non-Hodgkin-Lymphoma[70], in ovarian cancer[71] and in pancreatic cancer[72], reduced Bax expression was shown to be a negative prognostic factor. In metastatic colorectal cancer, we recently found that the loss of Bax expression is most deleterious in those patients carrying the wild type p53 gene (Figure 8[26]). This is a further example that inactivation of the apoptotic programme may occur at any level of the signalling cascade (Figure 9).

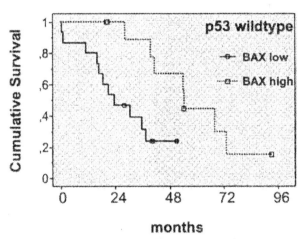

months

Figure 8. Survival analysis for BAX expression in p53 wild type tumours. The effect of Bax inactivation was analysed in patients with hepatic metastases of colorectal cancer. The patients underwent multimodal therapy including surgery of the hepatic metastases and adjuvant chemotherapy (26). Kaplan-Meier analysis for p53 wild type tumours only, stratified for BAX low expressing tumours ($\leq$ 10% stained cells, n=15) and for BAX high expressing tumours (>10% stained cells, n=11) (Logrank-Mantel-Cox p=0.006, Breslow-Gehan-Wilcoxon p=0.004). Censor times are indicated (squares, circles). Dashed lines, squares: BAX high expression; no dash, circles: BAX low expression.

Thus, disruption of the mitochondrial pathway of apoptosis, either by over-expression of apoptosis-preventing or loss of apoptosis-promoting Bcl-2 family members (i.e. Bax) appears to be a common feature in cancer, especially in those refractory to conventional therapies. Experimental evidence suggests that additional, yet to be defined defects may inactivate this pathway up- or downstream of Bax.

Activation of the p53/Bax regulatory pathway of apoptosis is one of the most important signals to induce apoptosis. It is characterised by transcriptional activation of the proapoptotic Bax gene through p53[73]. Since p53 mutations represent one of the significant genetic defects to prevent apoptosis, this pathway is nonfunctional in the majority of malignancies. Consequently, a novel strategy of cancer therapy by re-establishing of p53 function via gene transfer of the corresponding wild-type cDNA emerged. The underlying concept of this strategy is to induce apoptosis in cancer cells carrying p53 mutations or deletions or to make them sensitive to chemotherapy and radiotherapy. However, defects in the p53/Bax apoptosis pathway are not restricted to p53. As demonstrated by frameshift mutations of the Bax gene, genetic defects blocking apoptosis can also impair downstream effectors of p53[67,74].

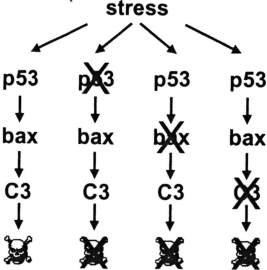

Figure 9. Role of apoptosis defects on tumour cell apoptosis. Genotoxic stress or direct triggering of the apoptosis signalling cascade induces apoptosis. In tumours with defects in these signalling cascades, the cell death programme is impaired and the tumour is resistant to therapy. Thus, there is evidence that inactivation of Bax impairs p53-induced cell death. In parallel, inactivation of caspase-3 renders cells refractory to p53, Bax, or drug-induced apoptosis. Thus, in analogy to the Rb pathway, inactivation of the apoptosis signalling cascade components may occur at any level of the signalling cascade and renders the tumour cell refractory to signals generated upstream in the cascade. Such an inactivation may be complete or incomplete, depending on the existence of redundant pathways.

3.2 The p16^INK4a/cyclin D/CDK4/Rb pathway

Similar findings with regard to resistance to therapy in cancer were obtained from analyses of the Rb pathway of cell cycle regulation. In many human cancers, the p16[INK4a]/cyclin D/CDK4/Rb pathway is disrupted (Figure 10). P16[INK4a] exhibits frequent loss-of function in different tumours, either by promotor hypermethylation, point mutations with or without loss-of-heterozygosity. The inactivation of p16[INK4a] has been correlated with bad prognosis in cancer patients with malignant melanoma[75], pancreatic adenocarcinoma[76], leukemia[77,78], non-small cell lung cancer[79], squamous cell carcinoma of the lung[80], squamous cell carcinoma of esophagus[127], and osteosarcoma (Figure 11). In some cases of familial predisposition to familial pancreatic adenocarcinoma or melanoma, germ line mutations of the p16[INK4a] could be identified[81,82].

For some cancer types the pathway has been carefully investigated, and a disruption at some level of the p16[INK4a]/cyclin D/CDK4/Rb pathway can be found in literally 100% of glioblastomas: 10-25% of tumours sustain Rb loss, around 50-65% have p16[INK4a] loss-of-expression, and CDK4 is amplified in about 15%, whereas cyclin D amplification is rather rare[83,84].

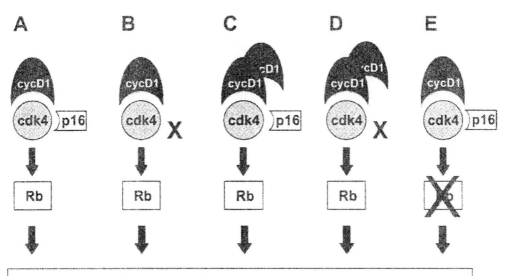

Figure 10. Disruption of the Rb pathway. Different mechanisms may lead to the disruption of the Rb-pathway and impair G1-restriction point control. In non-malignant cells, CDK/Cyclin D complexes promote and CDK inhibitors (like p16[INK4a]) and pRb family members inhibit progression from G1 to the S-phase (A). The following changes may lead to loss of G1 restriction point control: (B) loss of p16 (or other CDKI's such as the Cip/Kip family members p27 and p21[Cip/WAF-1]), (C) overexpression of Cyclins such as Cyclin D1 or Cyclin E, (D) combined deregulation of several genes such as loss of p16 *and* deregulated cyclin expression, (E) loss of Rb.

Peter T. Daniel, Bernhard Gillissen and Isrid Sturm

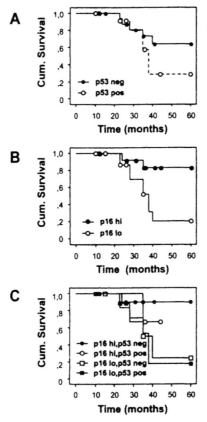

Figure 11. Inactivation of the Rb pathway in osteosarcoma. This graph shows survival curves of young adults with osteosarcoma (n=32). The patients were treated prior and after tumour resection according to the COSS86 polychemotherapy regimen. The treatment had a curative intention. The Kaplan-Meier plots show cumulative overall survival after diagnosis. Patients with disruption of the Rb pathway as determined by analysis of dysregulation of p53 (protein overexpression), p16 (loss of protein expression), and pRb (loss of protein expression). Patients with disturbed p53 (A) or p16^{INK4a} (B) or pRb (data not shown, the curves look similar as in the case of p53) have a much worse prognosis as compared to patients with tumours without these alterations. The combined analysis of these alterations identifies patients with a very poor (loss of two regulators, p53 and p16) or a very good prognosis (i.e. wild-type for both regulators, p53 and p16) and response to this aggressive treatment regimen.

4.   THE TUMOUR GENOTYPE AND THE RESPONSE TO THERAPEUTIC GENE TRANSFER

Up to now, the majority of experience in clinical therapy of human cancer with gene therapeutic approach has been gathered with the overexpression of p53. This approach was undertaken with different kinds of delivery systems, such as non-replicating adenoviruses or the replicating dl1520 (Onyx-015) mutant adenovirus (this virus targets the p53 binding of E1b and is supposed to replicate only in p53 defect cells), retroviral systems, and non-viral delivery systems like cationic liposomes. Such delivery systems were tested in a wide range of human cancers (i.e. head-and-neck squamous cell carcinomas, non-small-cell lung cancers, ovarian cancers, etc.). The approach to transfer the p53 gene alone has, however, not been totally successful. In many of these studies, not even the genotype for p53 of the tumours was determined. This is rather surprising since there is evidence from a wide variety of experimental systems that therapeutic p53 gene transfer is most efficient in tumour cells which carry an inactivated (i.e. loss-of-function) p53. Nevertheless, recent data for the adenoviral gene transfer of p53 in combination with chemotherapy or radiotherapy show a synergistic effect for this combined modality treatment of local p53 gene transfer plus conventional therapy in the tumours, irrespective of the p53 status.

On the other hand, the mere overexpression of p53 will not restore full chemosensitivity in tumours that are resistant to therapy-induced cell death due to defects in the downstream apoptosis signalling cascade (for example, colon cancers with an intact p53 gene, but a defective Bax protein expression)[67]. This is supported by our finding that adenoviral gene transfer of p53 into Bax negative carcinoma cells was unable to induce tumour cell death whereas Bax wild-type but p53 deficient tumour cells responded well, i.e. underwent apoptosis upon transduction with a non-replicating p53 adenovirus (Figure 12). Data from the literature support the finding that p53 acts preferentially (but not exclusively) via a Bax-dependent pathway of apoptosis[67].

There is abundant information in the literature that the gene transfer of cell cycle regulating genes into tumour cells also depends on an intact downstream signalling cascade. Thus, gene transfer of p16, p21, or p53 will not be able to arrest the cell in the G1-phase of the cell cycle when the Rb-pathway is disrupted, e.g. by mutation of CDK's, or loss of pRb itself, either by mutation, deletion, or inactivation by viral proteins interfering with the pathway.

In this line, the effects of adenovirus-mediated p16^{INK4a} expression on cell cycle arrest are clearly dependent on an intact endogenous Rb pathway and the p16 status (Figure 10). Tumour cells that are p16 null and have an intact Rb, are more sensitive to the cytotoxic effects of exogenous p16 expression than are cells that are Rb null, but p16 wildtype[85]. This implies that Ad-p16 gene therapy for human cancers is only highly effective in case the pathway co-player, here Rb, is intact. To a similar extent, inactivation of the CDK's by mutation or overexpression of these CDK's targeted by p16 may impair the effectiveness of such a therapeutic strategy[29,30,86-89]. In addition, the effectiveness of p16 gene transfer depends on an intact mitochondrial apoptosis pathway. Disruption of mitochondrial apoptosis, e.g. by high levels of Bcl-2, also impair induction of cell death by adenoviral p16 gene transfer[90].

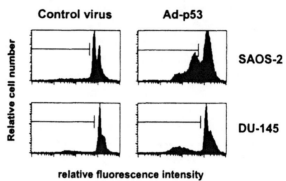

Figure 12. Adenoviral gene transfer of p53 (Ad-p53) in p53-mutated and Bax-deficient DU145 prostate carcinoma cells and p53-deficient/Bax-proficient SAOS-2 osteosarcoma cells. The induction of apoptosis was determined by assessing apoptotic DNA fragmentation on the single cell level by flow cytometric measurement of the cellular DNA content after infection of the cells with the p53 or a control adenovirus. The right side of the four histograms shows the DNA-content peaks of intact cells (G1, S, and G2/M phase; due to the tumour cell aneuploidy, these peaks are not as distinct as in non-malignant cells). The area under the horizontal marker bars depicts cells with DNA fragmentation, i.e. hypodiploid DNA content and which are therefore apoptotic. The Ad-p53 gene transfer induced strong apoptosis in the Bax-positive SAOS-2 but not in the Bax-deficient DU145 cells (which carry a loss-of-function frameshift mutation in the BAX gene). This indicates that loss of the downstream effector Bax reduces sensitivity for p53-triggered cell death.

Further examples for the relevance of the cellular genotype with respect to the response to (over-)expression of genes as a therapeutic approach can easily be found in the literature: For pancreatic tumours, gene therapy approaches involving p53 replacement are promising due to the central role of p53 in the cellular response to DNA damage and the high incidence of p53 mutations in pancreatic tumours. Adenoviruses containing wild-type (wt) p53 cDNA (Ad5CMV-p53) were introduced into four human pancreatic cell lines to examine the impact caused by exogenous wt p53 on these cells. Introduction of wt p53 in mutant p53 cells caused a marked decrease in cell proliferation and induced apoptosis. In contrast, overexpression of p53 did not induce apoptosis in wild-type p53 cells. The presence of p16 contributed to the induction of apoptosis, as demonstrated by introduction of the wt p16 gene (Ad5RSV-p16). Taken together, these results indicate that the effects elicited by exogenous p53 protein depend upon the molecular alterations related to p53 actions on cell cycle and apoptosis[66].

For the Rb, p53 and p16-interaction, a similar observation was made: Analysis of the genomic state of non-small cell lung cancer (NSCLC) cell lines expressing different combinations of normal and mutated p16, p53, and Rb genes with respect to their response to introduction of functional p16 cDNA showed that, as expected, p16-deficient cells infected with Ad-p16 identified growth arrest of the cells in the G0 - G1-phase early on. Apoptosis was identified to occur by the 5th day after infection, but only occurred in cells expressing functional p53. Cells that are -p16, -p53, and +Rb infected with Ad-p16 only exhibit apoptosis by an additional infection with Ad-p53.These studies suggest that p16 is capable of mediating apoptosis in NSCLC cell lines expressing wild-type p53[90].

The same is true for disturbances of the p14(p19)ARF/mdm-2/p53 pathway in cancer: p53-mediated checkpoint control is unperturbed in ARF-null fibroblast strains, whereas p53-negative cell lines are resistant to murine p19ARF-induced growth arrest[91,92]. Thus, the effectiveness of a therapeutic activation of the ARF-pathway, e.g. to overcome a drug-resistant phenotype[13], depends on an intact p53. Alternatively, the overexpression of mdm-2, which is a common feature of many types of cancer, may also impair the effect of the p14ARF on cell cycle arrest and activation of apoptosis pathways. In addition, since p53 gene transfer depends at least in part on an intact Bax and mitochondrial apoptosome, the loss of Bax may further impair such a therapeutic strategy. Therefore, knowledge of the genetic background of tumour cells is crucial to the development of efficient therapies based on the introduction of tumour suppressor genes.

One possibility to circumvent such limitations caused by defects in signalling pathways is to utilise genes which act downstream of such defects. Thus, gene transfer vectors mediating high E2F-1 protein expression may induce both entry into the S-phase and apoptosis[23,44,93,94] and may be used to circumvent defects in the Rb-pathway[95]. A considerable body of evidence points to a role for both cyclin E/cyclin-dependent kinase (CDK)2 activity and E2F transcription activity in the induction of the S-phase. Therefore, overexpression of cyclin E/cdk2 in quiescent cells induces S-phase which coincides with an induction of E2F activity. Similarly, coexpression of E2F enhances the cyclin E/cdk2-mediated induction of S-phase. Likewise, E2F overexpression could induce S-phase progression and does so in the apparent absence of cyclin E/cdk2 activity. In addition, although the inhibition of cyclin E/cdk2 activity blocks the induction of S-phase after growth stimulation, inhibition of cyclin E/cdk2 does not block S-phase induction in cells where E2F is deregulated, i.e. highly expressed, upon adenoviral gene transfer. Adenoviral transfer of E2F-1 is therefore sufficient to drive cells into the S-phase, independent from defects in pRb and p16. Nevertheless, E2F-mediated cell death depends on the presence of the DP family dimerisation partners[23,45] and an intact p14ARF/mdm-2/p53/Bax pathway which is responsible for the apoptosis induced by deregulated oncogenes such as Ras, myc, and the E2Fs[26,34,38,39,67]. Thus, defects in these signal components may result in accelerated growth instead of execution of cell death.

Similarly, we have shown that acquired drug resistance in breast cancer cells may be circumvented when downstream effectors of the apoptosis signalling cascade are introduced. This was shown in human breast carcinoma cells for acquired drug resistance caused by either overexpression of the P-glycoprotein Mdr-1, an ATP-dependent drug export system, and in cells showing a defect in activation of the mitochondrial permeability shift. In these cellular systems, the overexpression of apoptosis-promoting Bik/Nbk[68], Bak (unpublished observation), Bax[27] or Procaspase-3[128] could overcome the resistant phenotype. In these cells, the overexpression of such apoptosis downstream effectors could convert a suboptimal and abortive death signal (i.e. the drug resistant phenotype) to an effective signal by lowering the threshold for apoptosis execution.

5.   THE APPLICATION OF GENETIC DEFECTS IN TUMOURS FOR GENE THERAPY

In addition to the above described implications for the tumour biology and the response to conventional cancer therapy, genetic profiles of tumour cells are important for the efficiency of oncolytic viral vectors. First, the mechanisms of cell death induction depend, at least to some extent, on the integrity of the cellular death programme. Second, the most efficient oncolytic viruses are replicating and utilise the cell cycle machinery both for killing and replication. Thus, defects in cellular genotypes may impair the efficiency of this kind of anti-cancer treatment strategy.

5.1   Utilising the cancer cell genotype for viral replication

On the other hand, these defects can be employed to achieve more or less tumour-specific viral replication. DNA tumour viruses, such as adenoviruses, drive infected cells into the S-phase of the cell cycle to facilitate viral DNA replication. Such replication-selective oncolytic viruses constitute a rapidly evolving and new treatment platform for cancer. Adenoviruses encode the E1b-55kDa protein, which binds and inactivates p53[96]. The human group C adenovirus mutant dl1520 (Onyx-015) does not express the E1b 55KDa protein[96]. Consequently, this virus is believed to replicate selectively in cells that lack p53. Nevertheless, a correlation between the p53 status of tumour cells and dl1520 replication has not been established, thus, dl1520 also replicates in some tumour cells harbouring wild-type p53[97-100].

A recent report showed that the replication of such an oncolytic virus may also be facilitated when the p14ARF pathway is disrupted[101]. This tumour-selective viral replication may also be achieved by manipulation of the adenoviral E1a region. A variety of gene-deleted viruses has been engineered for tumour selectivity, but these gene deletions also reduce the anti-cancer potency of the viruses. Recently, an E1a mutant adenovirus, dl922-947, was described that replicates in and lyses a broad range of cancer cells with abnormalities in cell-cycle checkpoints. This deletion mutant is inactivated in the CR-2 region of E1a which mediates pRB-family binding (as compared to other mutants: dl1520 targets the p53 binding of E1b – see above (Onyx-015 vector), dl1101 targets the p300 binding domain of the CR-1 region). This mutant demonstrated reduced S-phase induction and replication in non-proliferating normal cells. Interestingly, the tumour-killing effect *in vivo* was superior relative to the other gene-deleted adenoviruses tested[102]. This vector showed anti-tumour activity regardless of the specific genetic defect leading to loss of the G1/S-phase checkpoint or the p53 status of the cancer cell line tested. Potent antitumoural efficacy was demonstrated against tumour cells irrespective of, e.g. p53 mutation, pRb mutation or cyclin D1 gene amplification. This is a good example how tumour phenotypes and viruses may be tricked to exploit signalling defects thereby achieving a tumour-selective therapeutic gene transfer and killing of the cancer cell[17].

5.2   Results of clinical trials for cytotoxic gene therapy in cancer

Up to date, only a small number of gene therapy protocols for cancer treatment has been completed and only few of these results are published. For the gene transfer of wild type p53, the intratumoural injection of a retroviral vector in lung cancer lesions has been reported in a feasibility study in 1996 which showed limited local control of the tumour[4]. In more recent reports, the transfer of wild type p53 is achieved by the use of non replicating or replicating adenoviral vectors, e.g. as adjuvant in surgery for advanced head and neck tumours[5,6] or as local therapy for advanced non-small cell lung cancer[103-105]. Further protocols have been published, for example for wildtype p53 local gene therapy in bladder cancer[106] or for the hepatic artery perfusion with an Ad-CMV-P53 in patients with advanced colorectal cancer[107]. The last and most promising report is a phase II study with a replication competent adenoviral vector which is supposed to replicate selectively in p53-defective cells (Onyx-015, see above) in patients with recurrent head and neck tumours. This study demonstrates a moderate response and local control of tumour growth[129]. Phase I studies indicate that better results may be seen in patients treated with a combination of Onyx-015 and chemo-/radiotherapy.

Other approaches in phase I studies for cytotoxic gene therapies are based on the viral transfer of pro-drug converting enzymes, like adenoviral vectors for HS-TK injected directly into the prostate, followed by intravenous administration of the prodrug ganciclovir (GCV)[108]. Other studies applied a similar approach in malignant brain tumours (mostly glioblastomas) either by the use of non-replicating adenoviral vectors[109], or injection of the HS-TK retrovirus producing cells[110,111]. Additional reports were published for malignant melanoma[112], or for mesothelioma[113], and for advanced ovarian cancer with the oral administration of valacyclovir[114]. The results of these studies are disappointing to some extent with regard to the rather limited clinical effect on tumour control and survival of the patients. Another system aims to transfer the enzyme cytosine deaminase, thus allowing activation of fluorocytosine to the active cytotoxic fluorouracil in transduced cells. The clinical evaluation in phase I studies was performed in colon cancer[115] and in breast cancer (here with control of the transgene under the erb2-promoter, thereby aiming for expression preferentially in tumour cells)[116]. No results showing clinical efficiency are yet available. All studies are, however, phase I/II studies, aiming to investigate the tolerable toxicity of the approach and the general feasibility. An estimate of the real antitumoural activity of such a therapeutic approach can hardly be derived from these results.

All these studies have in common that they show the feasibility of such a therapeutic approach. Most of them demonstrated expression of the transduced gene in treated lesions. Nevertheless, none of these studies addressed the genotype of the treated cancer in further detail than by the immunhistochemical staining for p53 expression in the treated tumours prior to treatment. Thus, it is not known whether the treated tumours were p53 wild type (which would render them refractory to such a therapeutic p53 gene transfer) or carried a mutated p53 gene. In addition, no information is available whether the downstream signalling cascades for apoptosis or cell cycle arrest were intact. Therefore, a thorough genotyping approach should be included in future protocols[117]. This is also supported by the fact that most cytotoxic gene therapy trials in the near future will be applied in combination with conventional cytotoxic therapies such as chemo- or radiotherapy. The reason is not only to improve efficiency like in the case of p53 but also to facilitate the application in clinical trials within "standard" therapeutic regimen.

The outcome and the anti-tumour effect of these strategies will, as outlined above, depend not only on the quality of the viral gene transfer but to a major degree on the genetic profile of the tumour. This may also be true for alternative strategies based on the activation of the immune system against the tumour, e.g. by local expression of cytokines[118-121] or costimulatory molecules such as B7.1 (CD80)[122]. Apart from the problem to overcome tolerance and immune evasion mechanisms which prevent activation of the immune system[123-125], the killer attack of immune effector cells depends also on an intact death programme of the targeted tumour cells. Thus, genetic defects of tumour cells may impair the ability of killer cells and their effector molecules such as granzymes and death ligands such as the CD95/Fas ligand to attack and eliminate cancer cells[20,126]. These points may explain why all these strategies failed so far to show a significant therapeutic effect.

6. CONCLUSIONS

Given the impact of inactivation of the above described signalling cascades for the response not only to conventional therapies but also to a therapeutic gene transfer, we believe that an adequate genotype and phenotype analysis of tumours is mandatory prior to therapy. Only the exact characterisation of the response-relevant effector genes for the cell cycle and the cell death regulation will help to identify those patients who will really benefit from a specific cancer therapy.

Conventional therapies, e.g. by the use of radiation or cytotoxic drugs, often activate multiple death pathways. This may explain their initial efficiency in the treatment of many types of cancer. Nevertheless, most patients with disseminated tumour growth ultimately succumb due the development of resistant phenotypes, i.e. tumour progression. In contrast, gene therapy strategies are often focussed on single genes which depend often on a single or only a few effector signalling pathways. Thus, the inactivation of a single but essential signalling component such as CDK4, Rb, p53, Bax, or others, may have deleterious consequences for the efficiency of a therapeutic gene transfer. Thus, the development of adequate diagnostic tools appears to be one of the most promising approaches to increase the therapeutic effects and benefit of gene therapy. Together with more efficient gene transfer systems, such molecular diagnostics will help to adapt the most adequate therapy to an individual cancer patient.

ACKNOWLEDGEMENTS

We wish to thank Drs. Philipp Hemmati and Thomas Wieder for helpful comments to the manuscript. The continuous support of Prof. Bernd Dörken is gratefully acknowledged.

REFERENCES

1. Denning, C., and J. D. Pitts. 1997. Bystander effects of different enzyme-prodrug systems for cancer gene therapy depend on different pathways for intercellular transfer of toxic metabolites, a factor that will govern clinical choice of appropriate regimes. *Hum Gene Ther* 8:*1825.*

2. Freeman, S. M., C. N. Abboud, K. A. Whartenby, C. H. Packman, D. S. Koeplin, F. L. Moolten, and G. N. Abraham. 1993. The "bystander effect": tumour regression when a fraction of the tumour mass is genetically modified. *Cancer Res* 53:*5274.*

3. Andrade-Rozental, A. F., R. Rozental, M. G. Hopperstad, J. K. Wu, F. D. Vrionis, and D. C. Spray. 2000. Gap junctions: the "kiss of death" and the "kiss of life". *Brain Res Brain Res Rev* 32:*308.*

4. Roth, J. A., D. Nguyen, D. D. Lawrence, B. L. Kemp, C. H. Carrasco, D. Z. Ferson, W. K. Hong, R. Komaki, J. J. Lee, J. C. Nesbitt, K. M. Pisters, J. B. Putnam, R. Schea, D. M. Shin, G. L. Walsh, M. M. Dolormente, C. I. Han, F. D. Martin, N. Yen, K. Xu, L. C. Stephens, T. J. McDonnell, T. Mukhopadhyay, and D. Cai. 1996. Retrovirus-mediated wild-type p53 gene transfer to tumours of patients with lung cancer. *Nat Med* 2:*985.*

5. Clayman, G. L., A. K. el-Naggar, S. M. Lippman, Y. C. Henderson, M. Frederick, J. A. Merritt, L. A. Zumstein, T. M. Timmons, T. J. Liu, L. Ginsberg, J. A. Roth, W. K. Hong, P. Bruso, and H. Goepfert. 1998. Adenovirus-mediated p53 gene transfer in patients with advanced recurrent head and neck squamous cell carcinoma. *J Clin Oncol* 16:*2221.*

6. Clayman, G. L., D. K. Frank, P. A. Bruso, and H. Goepfert. 1999. Adenovirus-mediated wild-type p53 gene transfer as a surgical adjuvant in advanced head and neck cancers. *Clin Cancer Res* 5:*1715.*

7. Zwacka, R. M., and M. G. Dunlop. 1998. Gene therapy for colon cancer. *Hematol Oncol Clin North Am* 12:*595.*

8. Wildner, O. 1999. In situ use of suicide genes for therapy of brain tumours. *Ann Med* 31:*421.*

9. Wildner, O., J. C. Morris, N. N. Vahanian, H. Ford, Jr., W. J. Ramsey, and R. M. Blaese. 1999. Adenoviral vectors capable of replication improve the efficacy of HSVtk/GCV suicide gene therapy of cancer. *Gene Ther* 6:*57.*

10. Allay, J. A., M. S. Steiner, Y. Zhang, C. P. Reed, J. Cockroft, and Y. Lu. 2000. Adenovirus p16 gene therapy for prostate cancer. *World J Urol* **18**:*111*.

11. Fueyo, J., C. Gomez-Manzano, V. K. Puduvalli, P. Martin-Duque, R. Perez-Soler, V. A. Levin, W. K. Yung, and A. P. Kyritsis. 1998. Adenovirus-mediated p16 transfer to glioma cells induces G1 arrest and protects from paclitaxel and topotecan: implications for therapy. *Int J Oncol* **12**:*665*.

12. Fueyo, J., C. Gomez-Manzano, W. K. Yung, T. J. Liu, R. Alemany, T. J. McDonnell, X. Shi, J. S. Rao, V. A. Levin, and A. P. Kyritsis. 1998. Overexpression of E2F-1 in glioma triggers apoptosis and suppresses tumour growth in vitro and in vivo. *Nat Med* **4**:*685*.

13. Guo-Chang, F., and W. Chu-Tse. 2000. Transfer of p14ARF gene in drug-resistant human breast cancer MCF-7/Adr cells inhibits proliferation and reduces doxorubicin resistance. *Cancer Lett* **158**:*203*.

14. Alemany, R., C. Balague, and D. T. Curiel. 2000. Replicative adenoviruses for cancer therapy. *Nat Biotechnol* **18**:*723*.

15. Alemany, R., C. Gomez-Manzano, C. Balague, W. K. Yung, D. T. Curiel, A. P. Kyritsis, and J. Fueyo. 1999. Gene therapy for gliomas: molecular targets, adenoviral vectors, and oncolytic adenoviruses. *Exp Cell Res* **252**:*1*.

16. Fueyo, J., C. Gomez-Manzano, R. Alemany, P. S. Lee, T. J. McDonnell, P. Mitlianga, Y. X. Shi, V. A. Levin, W. K. Yung, and A. P. Kyritsis. 2000. A mutant oncolytic adenovirus targeting the Rb pathway produces anti-glioma effect in vivo. *Oncogene* **19**:*2*.

17. Heise, C., and D. H. Kim. 2000. Replication-selective adenoviruses as oncolytic agents. *J Clin Invest* **105**:*847*.

18. Wildner, O., R. M. Blaese, and J. C. Morris. 1999. Therapy of colon cancer with oncolytic adenovirus is enhanced by the addition of herpes simplex virus-thymidine kinase. *Cancer Res* **59**:*410*.

19. Yohn, D. S., W. M. Hammon, R. W. Atchison, and B. C. Casto. 1968. Oncolytic potentials of nonhuman viruses for human cancer. II. Effects of five viruses on heterotransplantable human tumours. *J Natl Cancer Inst* **41**:*523*.

20. Daniel, P. T. 2000. Dissecting the pathways to death. *Leukemia* **14**:*2035*.

21. Krammer, P. H., J. Dhein, H. Walczak, I. Behrmann, S. Mariani, B. Matiba, M. Fath, P. T. Daniel, E. Knipping, M. O. Westendorp, K. Stricker, C. Bäumler, S. Hellbardt, M. Germer, M. E. Peter, and K.-M. Debatin. 1994. The role of APO-1-mediated apoptosis in the immune system. *Immunol Rev* **142**:*175*.

22. Bargou, R. C., P. T. Daniel, M. Y. Mapara, K. Bommert, C. Wagener, B. Kallinich, H. D. Royer, and B. Dörken. 1995. Expression of the bcl-2 gene family in normal and malignant breast tissue: low bax-alpha expression in tumour cells correlates with resistance towards apoptosis. *Int J Cancer* **60**:*854*.

23. Bargou, R. C., C. Wagener, K. Bommert, W. Arnold, P. T. Daniel, M. Y. Mapara, E. Grinstein, H. D. Royer, and B. Dörken. 1996. Blocking the transcription factor E2F/DP by dominant-negative mutants in a normal breast epithelial cell line efficiently inhibits apoptosis and induces tumour growth in SCID mice. *J Exp Med* **183**:*1205.*

24. Bargou, R. C., C. Wagener, K. Bommert, M. Y. Mapara, P. T. Daniel, W. Arnold, M. Dietel, H. Guski, A. Feller, H. D. Royer, and B. Dörken. 1996. Overexpression of the death-promoting gene bax-alpha which is downregulated in breast cancer restores sensitivity to different apoptotic stimuli and reduces tumour growth in SCID mice. *J Clin Invest* **97**:*2651.*

25. Prokop, A., T. Wieder, I. Sturm, F. Essmann, K. Seeger, C. Wuchter, W.-D. Ludwig, G. Henze, B. Dörken, and P. T. Daniel. 2000. Relapse in childhood acute lymphoblastic leukemia is associated with decrease of Bax/Bcl-2- ratio and loss of spontaneous caspase-3 processing in vivo. *Leukemia* **14**:*1606.*

26. Sturm, I., C. H. Kohne, G. Wolff, H. Petrowsky, T. Hillebrand, S. Hauptmann, M. Lorenz, B. Dörken, and P. T. Daniel. 1999. Analysis of the p53/BAX pathway in colorectal cancer: low BAX is a negative prognostic factor in patients with resected liver metastases. *J Clin Oncol* **17**:*1364.*

27. Wagener, C., R. C. Bargou, P. T. Daniel, K. Bommert, M. Y. Mapara, H. D. Royer, and B. Dörken. 1996. Induction of the death-promoting gene bax-alpha sensitizes cultured breast-cancer cells to drug-induced apoptosis. *Int J Cancer* **67**:*138.*

28. Eischen, C. M., J. D. Weber, M. F. Roussel, C. J. Sherr, and J. L. Cleveland. 1999. Disruption of the ARF-Mdm2-p53 tumour suppressor pathway in Myc-induced lymphomagenesis. *Genes Dev* **13**:*2658.*

29. Maelandsmo, G. M., V. A. Florenes, E. Hovig, T. Oyjord, O. Engebraaten, R. Holm, A. L. Borresen, and O. Fodstad. 1996. Involvement of the pRb/p16/cdk4/cyclin D1 pathway in the tumourigenesis of sporadic malignant melanomas. *Br J Cancer* **73**:*909.*

30. Palmero, I., and G. Peters. 1996. Perturbation of cell cycle regulators in human cancer. *Cancer Surv* **27**:*351.*

31. Schutte, M., R. H. Hruban, J. Geradts, R. Maynard, W. Hilgers, S. K. Rabindran, C. A. Moskaluk, S. A. Hahn, I. Schwarte-Waldhoff, W. Schmiegel, S. B. Baylin, S. E. Kern, and J. G. Herman. 1997. Abrogation of the Rb/p16 tumour-suppressive pathway in virtually all pancreatic carcinomas. *Cancer Res* **57**:*3126.*

32. Shapiro, G. I., C. D. Edwards, L. Kobzik, J. Godleski, W. Richards, D. J. Sugarbaker, and B. J. Rollins. 1995. Reciprocal Rb inactivation and p16INK4 expression in primary lung cancers and cell lines. *Cancer Res* **55**:*505.*

33. Shapiro, G. I., J. E. Park, C. D. Edwards, L. Mao, A. Merlo, D. Sidransky, M. E. Ewen, and B. J. Rollins. 1995. Multiple mechanisms of p16INK4A inactivation in non-small cell lung cancer cell lines. *Cancer Res* **55**:*6200.*

34. Sherr, C., and J. Weber. 2000. The ARF/p53 pathway. *Curr Opin Gen Develop* **10**:*94.*

35. Tsihlias, J., L. Kapusta, and J. Slingerland. 1999. The prognostic significance of altered cyclin-dependent kinase inhibitors in human cancer. *Annu Rev Med* **50**:401.

36. Vidal, A., and A. Koff. 2000. Cell-cycle inhibitors: three families united by a common cause. *Gene* **247**:*1*.

37. Petrowsky, H., I. Sturm, O. Graubitz, D. A. Kooby, E. Staib-Sebler, C. Gog, C.-H. Köhne, T. Hillebrand, P. T. Daniel, Y. Fong, and M. Lorenz. 2001. Relevance of proliferation and K-ras mutation in colorectal liver metastasis. *Eur J Surg Oncol* **27**:*80*.

38. Soussi, T., K. Debouche, and C. Béroud. 2000. P53 website and analysis of p53 gene mutations in cancer: forging a link between epidemiology and carcinogenesis. *Hum Mutat* **15**:*105*.

39. Sturm, I., S. Papadopoulos, T. Hillebrand, T. Benter, H.-J. Lück, G. Wolff, B. Dörken, and P. T. Daniel. 2000. Impaired BAX protein expression in breast cancer: mutational analysis of the BAX and the p53 gene. *Int J Cancer* **87**:*517*.

40. Pardee, A. B. 1989. G1 events and regulation of cell proliferation. *Science* **246**:*603*.

41. Grana, X., J. Garriga, and X. Mayol. 1998. Role of the retinoblastoma protein family, pRB, p107 and p130 in the negative control of cell growth. *Oncogene* **17**:*3365*.

42. King, K. L., and J. A. Cidlwoski. 1998. Cell cycle regulation and apoptosis. *Annu Rev Physiol* **60**:*601*.

43. Chow, K. N., and D. C. Dean. 1996. Domains A and B in the Rb pocket interact to form a transcriptional repressor motif. *Mol Cell Biol* **16**:*4862*.

44. Müller, H., and K. Helin. 2000. The E2F transcription factors: key regulators of cell proliferation. *Biochim Biophys Acta* **1470**:*M1*.

45. Yamasaki, L. 1999. Balancing proliferation and apoptosis in vivo: the Goldilocks theory of E2F/DP action. *Biochim Biophys Acta* **1423**:*M9*.

46. Knudsen, E. S., C. Buckmaster, T. T. Chen, J. R. Feramisco, and J. Y. Wang. 1998. Inhibition of DNA synthesis by RB: effects on G1/S transition and S-phase progression. *Genes Dev* **12**:*2278*.

7. Kato, J. Y. 1997. Control of G1 progression by D-type cyclins: key event for cell proliferation. *Leukemia* **11**:*347*.

48. Ohtani, K., J. DeGregori, and J. R. Nevins. 1995. Regulation of the cyclin E gene by transcription factor E2F1. *Proc Natl Acad Sci U S A* **92**:*12146*.

49. Sherr, C. J., and J. M. Roberts. 1999. CDK inhibitors: positive and negative regulators of G1-phase progression. *Genes Dev* **13**:*1501*.

50. Waldman, T., C. Lengauer, K. W. Kinzler, and B. Vogelstein. 1996. Uncoupling of S phase and mitosis induced by anticancer agents in cells lacking p21. *Nature* **381**:*713*.

51. Hollander, M., M. Sheikh, D. Bulavin, K. Lundgren, L. Augeri-Henmueller, R. Shehee, T. Molinaro, K. Kim, E. Tolosa, J. Ashwell, M. Rosenberg, Q. Zhan, P. Fernandez-Salguero, W. Morgan, C. Deng, and A. Fornace. 1999. Genomic instability in Gadd45a-deficient mice. *Nat Genet* **23**:*176*.

52. Chinnaiyan, A. M., K. O'Rourke, M. Tewari, and V. M. Dixit. 1995. FADD, a novel death domain-containing protein, interacts with the death domain of Fas and initiates apoptosis. *Cell* **81**:*505*.

53. Blagosklonny, M. V. 2000. Cell death beyond apoptosis. *Leukemia* *14:1502*.

54. Wieder, T., F. Essmann, A. Prokop, K. Schmelz, K. Schulze-Osthoff, R. Beyaert, B. Dörken, and P. T. Daniel. 2001. Activation of Caspase-8 in drug-induced apoptosis of B-lymphoid cells is independent of CD95/Fas receptor ligand interaction and occurs downstream of Caspase-3. *Blood* **97**:*1378*.

55. Kroemer, G., and J. C. Reed. 2000. Mitochondrial control of death. *Nat Med* **6**:*513*.

56. Shimizu, S., and Y. Tsujimoto. 2000. Proapoptotic BH3-only Bcl-2 family members induce cytochrome c release, but not mitochondrial membrane potential loss, and do not directly modulate voltage-dependent anion channel activity. *Proc Natl Acad Sci USA* **97**:*577*.

57. Van der Heiden, M. G., and C. B. Thompson. 1999. Bcl-2 proteins: regulators of apoptosis or of mitochondrial homeostasis? *Nat Cell Biol* **1**:*E209*.

58. Desagher, S., A. Osen-Sand, A. Nichols, R. Eskes, S. Montessuit, S. Lauper, K. Maundrell, B. Antonsson, and J. Martinou. 1999. Bid-induced Conformational Change of Bax Is Responsible for Mitochondrial Cytochrome c Release during Apoptosis. *J Cell Biol* **144**:*891*.

59. Lorenzo, H. K., S. A. Susin, J. Penninger, and G. Kroemer. 1999. Apoptosis inducing factor (AIF): a phylogenetically old, caspase-independent effector of cell death. *Cell Death Differ* **6**:*516*.

60. Li, H., H. Zhu, C. Xu, and J. Yuan. 1998. Cleavage of BID by caspase 8 mediates the mitochondrial damage in the Fas pathway of apoptosis. *Cell* **94**:*491*.

61. Luo, X., I. Budihardjo, H. Zou, C. Slaughter, and X. Wang. 1998. Bid, a Bcl-2 interacting protein, mediates cytochrome c release in response to activation of cell surface death receptors. *Cell* **94**:*481*.

62. Perez, D., and E. White. 2000. TNF-alpha signals apoptosis through a bid-dependent conformational change in Bax that is inhibited by E1B 19K. *Mol Cell* **6**:*53*.

63. Walker, A., S. T. Taylor, J. A. Hickman, and C. Dive. 1997. Germinal center-derived signals act with Bcl-2 to decrease apoptosis and increase clonogenicity of drug-treated human B lymphoma cells. *Cancer Res* **57**:*1939*.

64. Graeber, T., C. Osmanian, T. Jacks, D. Housman, C. Koch, S. Lowe, and A. Giaccia. 1996. Hypoxia-mediated selection of cells with diminished apoptotic potential in solid tumours. *Nature* **379**:*88*.

65. Daniel, P. T., A. Pezzutto, and B. Dörken. 1999. Humoral immunotherapy and the use of monoclonal antibodies. In: *Textbook of Malignant Haematology*, ed. by Degos, I., Linch, D. C., and Lowenberg, B. , Martin Dunitz Ltd., London:425.

66. Cascallo, M., E. Mercade, G. Capella, F. Lluis, C. Fillat, A. M. Gomez-Foix, and A. Mazo. 1999. Genetic background determines the response to adenovirus-mediated wild- type p53 expression in pancreatic tumour cells. *Cancer Gene Ther* **6**:*428*.

67. McCurrach, M. E., T. M. Connor, C. M. Knudson, S. J. Korsmeyer, and S. W. Lowe. 1997. bax-deficiency promotes drug resistance and oncogenic transformation by attenuating p53-dependent apoptosis. *Proc Natl Acad Sci U S A* **94**:*2345*.

68. Daniel, P. T., K. T. Pun, S. Ritschel, I. Sturm, J. Holler, B. Dörken, and R. Brown. 1999. Expression of the death gene Bik/Nbk promotes sensitivity to drug-induced apoptosis in corticosteroid-resistant T-cell lymphoma and prevents tumour growth in severe combined immunodeficient mice. *Blood* **94**:*1100*.

69. Krajewski, S., C. Blomqvist, K. Franssila, M. Krajewska, V. M. Wasenius, E. Niskanen, S. Nordling, and J. C. Reed. 1995. Reduced expression of proapoptotic gene BAX is associated with poor response rates to combination chemotherapy and shorter survival in women with metastatic breast adenocarcinoma. *Cancer Res* **55**:*4471*.

70. Gascoyne, R. D., M. Krajewska, S. Krajewsky, J. M. Connors, and J. C. Reed. 1997. Prognostic significance of BAX protein expression in diffuse aggressive Non-Hodgkin's lymphoma. *Blood* **90**:*3173*.

71. Tai, Y. T., S. Lee, E. Niloff, C. Weisman, T. Strobel, and S. A. Cannistra. 1998. BAX protein expression and clinical outcome in epithelial ovarian cancer. *J Clin Oncol* **16**:*2583*.

72. Friess, H., Z. Lu, H. U. Graber, A. Zimmermann, G. Adler, M. Korc, R. M. Schmid, and M. W. Büchler. 1998. bax, but not bcl-2, influences the prognosis of human pancreatic cancer. *Gut* **43**:*414*.

73. Miyashita, T., and J. C. Reed. 1995. Tumour suppressor p53 is a direct transcriptional activator of the human bax gene. *Cell* **80**:*293*.

74. Rampino, N., H. Yamamoto, Y. Ionov, Y. Li, H. Sawai, J. C. Reed, and M. Perucho. 1997. Somatic frameshift mutations in the BAX gene in colon cancers of the microsatellite mutator phenotype. *Science* **275**:*967*.

75. Reed, J. A., F. Loganzo, Jr., C. R. Shea, G. J. Walker, J. F. Flores, J. M. Glendening, J. K. Bogdany, M. J. Shiel, F. G. Haluska, J. W. Fountain, and *et al.* 1995. Loss of expression of the p16/cyclin-dependent kinase inhibitor 2 tumour suppressor gene in melanocytic lesions correlates with invasive stage of tumour progression. *Cancer Res* **55**:*2713*.

76. Hu, Y. X., H. Watanabe, K. Ohtsubo, Y. Yamaguchi, A. Ha, T. Okai, and N. Sawabu. 1997. Frequent loss of p16 expression and its correlation with clinicopathological parameters in pancreatic carcinoma. *Clin Cancer Res* **3**:*1473*.

77. Fizzotti, M., G. Cimino, S. Pisegna, G. Alimena, C. Quartarone, F. Mandelli, P. G. Pelicci, and F. Lo Coco. 1995. Detection of homozygous deletions of the cyclin-dependent kinase 4 inhibitor (p16) gene in acute lymphoblastic leukemia and association with adverse prognostic features. *Blood* **85**:*2685*.

78. Heyman, M., O. Rasool, L. Borgonovo Brandter, Y. Liu, D. Grander, S. Soderhall, G. Gustavsson, and S. Einhorn. 1996. Prognostic importance of p15INK4B and p16INK4 gene inactivation in childhood acute lymphocytic leukemia. *J Clin Oncol* **14**:*1512*.

79. Kratzke, R. A., T. M. Greatens, J. B. Rubins, M. A. Maddaus, D. E. Niewoehner, G. A. Niehans, and J. Geradts. 1996. Rb and p16INK4a expression in resected non-small cell lung tumours. *Cancer Res* **56**:*3415*.

80. Huang, C. I., T. Taki, M. Higashiyama, N. Kohno, and M. Miyake. 2000. p16 protein expression is associated with a poor prognosis in squamous cell carcinoma of the lung. *Br J Cancer* **82**:*374*.

81. Goldstein, A. M., M. C. Fraser, J. P. Struewing, C. J. Hussussian, K. Ranade, D. P. Zametkin, L. S. Fontaine, S. M. Organic, N. C. Dracopoli, W. H. Clark, Jr. *et al.* 1995. Increased risk of pancreatic cancer in melanoma-prone kindreds with p16INK4 mutations. *N Engl J Med* **333**:*970*.

82. Hussussian, C. J., J. P. Struewing, A. M. Goldstein, P. A. Higgins, D. S. Ally, M. D. Sheahan, W. H. Clark, Jr., M. A. Tucker, and N. C. Dracopoli. 1994. Germline p16 mutations in familial melanoma. *Nat Genet* **8**:*15*.

83. Ruas, M., S. Brookes, N. Q. McDonald, and G. Peters. 1999. Functional evaluation of tumour-specific variants of p16INK4a/CDKN2A: correlation with protein structure information. *Oncogene* **18**:*5423*.

84. Ruas, M., and G. Peters. 1998. The p16INK4a/CDKN2A tumour suppressor and its relatives. *Biochim Biophys Acta* **1378**:*F115*.

85. Craig, C., M. Kim, E. Ohri, R. Wersto, D. Katayose, Z. Li, Y. H. Choi, B. Mudahar, S. Srivastava, P. Seth, and K. Cowan. 1998. Effects of adenovirus-mediated p16INK4A expression on cell cycle arrest are determined by endogenous p16 and Rb status in human cancer cells. *Oncogene* **16**:*265*.

86. Halvorsen, O. J., J. Hostmark, S. Haukaas, P. A. Hoisaeter, and L. A. Akslen. 2000. Prognostic significance of p16 and CDK4 proteins in localized prostate carcinoma. *Cancer* **88**:*416*.

87. Kanoe, H., T. Nakayama, H. Murakami, T. Hosaka, H. Yamamoto, Y. Nakashima, T. Tsuboyama, T. Nakamura, M. S. Sasaki, and J. Toguchida. 1998. Amplification of the CDK4 gene in sarcomas: tumour specificity and relationship with the RB gene mutation. *Anticancer Res* **18**:*2317*.

88. Wei, G., F. Lonardo, T. Ueda, T. Kim, A. G. Huvos, J. H. Healey, and M. Ladanyi. 1999. CDK4 gene amplification in osteosarcoma: reciprocal relationship with INK4A gene alterations and mapping of 12q13 amplicons. *Int J Cancer* **80***:199*.

89. Wölfel, T., M. Hauer, J. Schneider, M. Serrano, C. Wölfel, E. Klehmann-Hieb, E. De Plaen, T. Hankeln, K. H. Meyer zum Büschenfelde, and D. Beach. 1995. A p16INK4a-insensitive CDK4 mutant targeted by cytolytic T lymphocytes in a human melanoma. *Science* **269***:1281*.

90. Kataoka, M., S. Wiehle, F. Spitz, G. Schumacher, J. A. Roth, and R. J. Cristiano. 2000. Down-regulation of bcl-2 is associated with p16INK4-mediated apoptosis in non-small cell lung cancer cells. *Oncogene* **19***:1589*.

91. Kamijo, T., S. Bodner, E. van de Kamp, D. H. Randle, and C. J. Sherr. 1999. Tumour spectrum in ARF-deficient mice. *Cancer Res* **59***:2217*.

92. Kamijo, T., F. Zindy, M. F. Roussel, D. E. Quelle, J. R. Downing, R. A. Ashmun, G. Grosveld, and C. J. Sherr. 1997. Tumour suppression at the mouse INK4a locus mediated by the alternative reading frame product p19ARF. *Cell* **91***:649*.

93. Gomez-Manzano, C., J. Fueyo, F. Alameda, A. P. Kyritsis, and W. K. Yung. 1999. Gene therapy for gliomas: p53 and E2F-1 proteins and the target of apoptosis. *Int J Mol Med* **3***:81*.

94. Nevins, J. R. 1992. E2F: a link between the Rb tumour suppressor protein and viral oncoproteins. *Science* **258***:424*.

95. Leone, G., J. Degregori, L. Jakoi, J. G. Cook, and J. R. Nevins. 1999. Collaborative role of E2F transcriptional activity and G1 cyclin-dependent kinase activity in the induction of S phase. *Proc. Natl. Acad. Sci. USA* **96***:6626–6631*.

96. Barker, D. D., and A. J. Berk. 1987. Adenovirus proteins from both E1B reading frames are required for transformation of rodent cells by viral infection and DNA transfection. *Virology* **156***:107*.

97. Goodrum, F. D., and D. A. Ornelles. 1998. p53 status does not determine outcome of E1B 55-kDa mutant adenovirus lytic infection. *J Virol* **72***:9479*.

98. Harada, J. N., and A. J. Berk. 1999. p53-independent and -dependent requirements for E1B-55K in adenovirus type 5 replication. *J Virol* **73***:5333–5344*.

99. Rothmann, T., A. Hengstermann, N. J. Whitaker, M. Scheffner, and H. zur Hausen. 1998. Replication of ONYX-015, a potential anticancer adenovirus, is independent of p53 status in tumour cells. *J Virol* **72***:9470–9478*.

100. Turnell, A. S., R. J. Grand, and P. H. Gallimore. 1999. The replicative capacities of large E1B group A and group C adenoviruses are independent of host cell p53 status. *J Virol* **73***:2074*.

101. Ries, S. J., C. H. Brandts, A. S. Chung, C. H. Biederer, B. C. Hann, E. M. Lippner, F. McCormick, and W. M. Korn. 2000. Loss of p14ARF in tumour cells facilitates replication of the adenovirus mutant dl1520 (ONYX-015). *Nat Med* **6***:1128.*

102. Heise, C., T. Hermiston, L. Johnson, G. Brooks, A. Sampson-Johannes, A. Williams, L. Hawkins, and D. Kirn. 2000. An adenovirus E1A mutant that demonstrates potent and selective systemic anti-tumoural efficacy. *Nat Med* **6***:1134.*

103. Schuler, M., C. Rochlitz, J. A. Horowitz, J. Schlegel, A. P. Perruchoud, F. Kommoss, C. T. Bolliger, H. U. Kauczor, P. Dalquen, M. A. Fritz, S. Swanson, R. Herrmann, and C. Huber. 1998. A phase I study of adenovirus-mediated wild-type p53 gene transfer in patients with advanced non-small cell lung cancer. *Hum Gene Ther* **9***:2075.*

104. Swisher, S. G., and J. A. Roth. 1998. Gene therapy for human lung cancers. *Surg Oncol Clin N Am* **7***:603.*

105. Swisher, S. G., J. A. Roth, J. Nemunaitis, D. D. Lawrence, B. L. Kemp, C. H. Carrasco, D. G. Connors, A. K. El-Naggar, F. Fossella, B. S. Glisson, W. K. Hong, F. R. Khuri, J. M. Kurie, J. J. Lee, J. S. Lee, M. Mack, J. A. Merritt, D. M. Nguyen, J. C. Nesbitt, R. Perez-Soler, K. M. Pisters, J. B. Putnam, Jr., W. R. Richli, M. Savin, M. K. Waugh, and *et al.* 1999. Adenovirus-mediated p53 gene transfer in advanced non-small-cell lung cancer. *J Natl Cancer Inst* **91***:763.*

106. Pagliaro, L. C. 2000. Gene therapy for bladder cancer. *World J Urol* **18***:148.*

107. Habib, N. A., H. J. Hodgson, N. Lemoine, and M. Pignatelli. 1999. A phase I/II study of hepatic artery infusion with wtp53-CMV-Ad in metastatic malignant liver tumours. *Hum Gene Ther* **10***:2019.*

108. Herman, J. R., H. L. Adler, E. Aguilar-Cordova, A. Rojas-Martinez, S. Woo, T. L. Timme, T. M. Wheeler, T. C. Thompson, and P. T. Scardino. 1999. In situ gene therapy for adenocarcinoma of the prostate: a phase I clinical trial. *Hum Gene Ther* **10***:1239.*

109. Trask, T. W., R. P. Trask, E. Aguilar-Cordova, H. D. Shine, P. R. Wyde, J. C. Goodman, W. J. Hamilton, A. Rojas-Martinez, S. H. Chen, S. L. Woo, and R. G. Grossman. 2000. Phase I study of adenoviral delivery of the HSV-tk gene and ganciclovir administration in patients with current malignant brain tumours. *Mol Ther* **1***:195.*

110. Klatzmann, D., C. A. Valery, G. Bensimon, B. Marro, O. Boyer, K. Mokhtari, B. Diquet, J. L. Salzmann, and J. Philippon. 1998. A phase I/II study of herpes simplex virus type 1 thymidine kinase "suicide" gene therapy for recurrent glioblastoma. Study Group on Gene Therapy for Glioblastoma. *Hum Gene Ther* **9***:2595.*

111. Shand, N., F. Weber, L. Mariani, M. Bernstein, A. Gianella-Borradori, Z. Long, A. G. Sorensen, and N. Barbier. 1999. A phase 1-2 clinical trial of gene therapy for recurrent glioblastoma multiforme by tumour transduction with the herpes simplex thymidine kinase gene followed by ganciclovir. GLI328 European-Canadian Study Group. *Hum Gene Ther* **10***:2325.*

112. Klatzmann, D., P. Cherin, G. Bensimon, O. Boyer, A. Coutellier, F. Charlotte, C. Boccaccio, J. L. Salzmann, and S. Herson. 1998. A phase I/II dose-escalation study of herpes simplex virus type 1 thymidine kinase "suicide" gene therapy for metastatic melanoma. Study Group on Gene Therapy of Metastatic Melanoma. *Hum Gene Ther* **9**:*2585*.

113. Sterman, D. H., J. Treat, L. A. Litzky, K. M. Amin, L. Coonrod, K. Molnar-Kimber, A. Recio, L. Knox, J. M. Wilson, S. M. Albelda, and L. R. Kaiser. 1998. Adenovirus-mediated herpes simplex virus thymidine kinase/ganciclovir gene therapy in patients with localized malignancy: results of a phase I clinical trial in malignant mesothelioma. *Hum Gene Ther* **9**:*1083*.

114. Hasenburg, A., X. W. Tong, A. Rojas-Martinez, C. Nyberg-Hoffman, C. C. Kieback, A. L. Kaplan, R. H. Kaufman, I. Ramzy, E. Aguilar-Cordova, and D. G. Kieback. 1999. Thymidine kinase (TK) gene therapy of solid tumours: valacyclovir facilitates outpatient treatment. *Anticancer Res* **19**:*2163*.

115. Crystal, R. G., E. Hirschowitz, M. Lieberman, J. Daly, E. Kazam, C. Henschke, D. Yankelevitz, N. Kemeny, R. Silverstein, A. Ohwada, T. Russi, A. Mastrangeli, A. Sanders, J. Cooke, and B. G. Harvey. 1997. Phase I study of direct administration of a replication deficient adenovirus vector containing the E. coli cytosine deaminase gene to metastatic colon carcinoma of the liver in association with the oral administration of the pro-drug 5-fluorocytosine. *Hum Gene Ther* **8**:*985*.

116. Pandha, H. S., L. A. Martin, A. Rigg, H. C. Hurst, G. W. Stamp, K. Sikora, and N. R. Lemoine. 1999. Genetic prodrug activation therapy for breast cancer: A phase I clinical trial of erbB-2-directed suicide gene expression. *J Clin Oncol* **17**:*2180*.

117. Anonymous. 1999. Onyx plans phase III trial of ONYX-015 for head & neck cancer. *Oncologist* **4**:*432*.

118. Gilly, F. N., A. Beaujard, J. Bienvenu, V. Trillet Lenoir, O. Glehen, D. Thouvenot, C. Malcus, M. Favrot, C. Dumontet, C. Lombard-Bohas, F. Garbit, P. Y. Gueugniaud, J. Vignal, M. Aymard, F. Touraine Moulin, M. Roos, A. Pavirani, and M. Courtney. 1999. Gene therapy with Adv-IL-2 in unresectable digestive cancer: phase I-II study, intermediate report. *Hepatogastroenterology* **46** *Suppl 1:1268*.

119. Okada, H., I. F. Pollack, M. T. Lotze, L. D. Lunsford, D. Kondziolka, F. Lieberman, D. Schiff, J. Attanucci, H. Edington, W. Chambers, P. Robbins, J. Baar, D. Kinzler, T. Whiteside, and E. Elder. 2000. Gene therapy of malignant gliomas: a phase I study of IL-4-HSV-TK gene- modified autologous tumour to elicit an immune response. *Hum Gene Ther* **11**:*637*.

120. Stewart, A. K., N. J. Lassam, I. C. Quirt, D. J. Bailey, L. E. Rotstein, M. Krajden, S. Dessureault, S. Gallinger, D. Cappe, Y. Wan, C. L. Addison, R. C. Moen, J. Gauldie, and F. L. Graham. 1999. Adenovector-mediated gene delivery of interleukin-2 in metastatic breast cancer and melanoma: results of a phase 1 clinical trial. *Gene Ther* **6**:*350*.

121. Wollenberg, B., H. Kastenbauer, H. Mundl, J. Schaumberg, A. Mayer, M. Andratschke, S. Lang, C. Pauli, R. Zeidler, S. Ihrler, Lohrs, K. Naujoks, and R. Rollston. 1999. Gene

therapy--phase I trial for primary untreated head and neck squamous cell cancer (HNSCC) UICC stage II-IV with a single intratumoural injection of hIL-2 plasmids formulated in DOTMA/Chol. *Hum Gene Ther* **10**:*141*.

122. Rini, B. I., L. M. Selk, and N. J. Vogelzang. 1999. Phase I study of direct intralesional gene transfer of HLA-B7 into metastatic renal carcinoma lesions. *Clin Cancer Res* **5**:*2766*.

123. Daniel, P. T., A. Kroidl, S. Cayeux, R. Bargou, T. Blankenstein, and B. Dörken. 1997. Costimulatory signals through B7.1/CD28 prevent T cell apoptosis during target cell lysis. *J Immunol* **159**:*3808*.

124. Daniel, P. T., A. Kroidl, J. Kopp, I. Sturm, G. Moldenhauer, B. Dörken, and A. Pezzutto. 1998. Immunotherapy of B-cell lymphoma with CD3x19 bispecific antibodies: costimulation via CD28 prevents "veto" apoptosis of antibody-targeted cytotoxic T cells. *Blood* **92**:*4750*.

125. Daniel, P. T., C. Scholz, F. Essmann, J. Westermann, A. Pezzutto, and B. Dörken. 1999. CD95/Fas-triggered apoptosis of activated T lymphocytes is prevented by dendritic cells through a CD58-dependent mechanism. *Exp Hematol* **27**:*1402*.

126. Krammer, P. H. 1999. CD95(APO-1/Fas)-mediated apoptosis: live and let die. *Adv Immunol* **71**:*163*.

127. Sturm, I., H. Petrowsky, R. Volz, M. Lorenz, S. Radetzki, T. Hillebrand, G. Wolff, S. Hauptmann, B. Dörken, P.T. Daniel. 2001. Analysis of p53/BAX/p16$^{ink4a/CDKN2}$ in esophageal squamous cell carcinoma: High BAX and p16$^{ink4a/CDKN2}$ identifies patients with good prognosis. *J Clin Oncol: in press.*

128. Friedrich K, T. Wieder, C. von Haefen, S. Radetzki, K. Schulze-Osthoff, R. Jänicke, B. Dörken, P.T. Daniel. 2001. Overexpression of caspase-3 restores sensitivity for drug-induced apoptosis in breast cancer cells with acquired drug resistance. *Oncogene: in press.*

129. Nemunaitis, J., F. Khuri, I. Ganly, J. Arseneau, M. Posner, E. Vokes, J. Kuhn, T. McCarty, S. Landers, A. Blackburn, L. Romel, B. Randlev, S. Kaye, and D. Kirn. 2001. Phase II trial of intratumoural administration of ONYX-015, a replication-selective adenovirus, in patients with refractory head and neck cancer. *J Clin Oncol:***19**:*289*.

Protein Binding Matrices

Tools for phenol-free gene cloning and vector assembling

VLADIMIR I. EVTUSHENKO
Laboratory of Genetic Engineering, Research Institute of Roentgenology and Radiology, St-Petersburg 189646, Russia

1.   INTRODUCTION

Historically, the origin of gene therapy, the revolutionary new field of human medicine[1], was underlined by development of molecular genetic, recombinant DNA technology, and genetic engineering. Advantages is phage, E. coli, S. cerevisiae, D. melanogaster, M. musculus, and eventually H. sapience molecular genetics have created a rational background for practical human genetic engineering or gene therapy. Roughly speaking, human gene therapy, the correction or prevention of disease at the genetic level rather that at secondary metabolic one, has arisen as a branch of genetic engineering specifically applied to particular species, Homo sapience. Gene therapy is building up from laboratory, manufacturing, and clinical blocks that are very separate things in terms of specific goals, employed methods and applied regulations. The goal of laboratory research is the development of ideal vector: injectable, tissue-specific construct that will precisely target appropriate regions of the genome or persist as stable episome, will be regulated, cost-effective to manufacture, safe, and will cure disease[2]. Therapeutic vector design, which includes *in vitro* conversion of native genes into it beneficial cDNA derivative, assembling of expression cassette, followed by it fusion with delivery DNA backbone, is based upon conventional molecular biology and genetic engineering techniques. Regardless of desirable vector type, a large number of intermediate constructs will be necessary to assemble and evaluate in laboratory to design a desirable therapeutic construct for even few from about 300 human tissues.

As a result, for every successful construct that reaches at least phase I trial, many unsatisfactory constructs will have been rejected. Most of these rejections take place in laboratory very early on vector assembling and it subsequent evaluation. This dark side of laboratory life is not displayed on journal pages but it is inherited part of any project. Therefore, if substantial time and efforts are going to be placed on vector design, it is desirable to optimise protocol at least by replacement of multiple time-consuming purification steps with brief single-pass ones.

In parallel to continuous technical advances that facilitate genetic engineering, some key purification techniques remain the same as they were decades ago. Removal of enzymes from post-reaction mixture with modified nucleic acids is probably the most ubiquitous procedure, which until now relays mainly upon phenol-chloroform followed by ethanol precipitation. This method is relatively slow and labour-intensive to perform, nevertheless, in laboratory enzymes are typically removed by liquid organic extraction. It come to mind that in a recently published issue of Methods of Enzymology[3] focused on cDNA applications, fifteen out of totally twenty-nine chapters referred to the phenol-chloroform as enzyme removal tool. Web site search of protocols-on-line from leading companies (Ambion, CLONTECH Laboratories, Epicentre Technologies, Life Technologies, New England Biolabs, Promega, Stratagene) has reveal that phenolic extraction is widely employed in cDNA synthesis and gene cloning kits. Even highly sophisticated gene identification and cloning systems listed in Table 1 utilise phenol-chloroform deproteinisation as inherited part of protocol. Some of

Table 1. Advanced techniques that employ phenol-chloroform enzyme extraction

Technique	Reference
Site-directed mutagenesis	4
In vitro transposition	5, 6
Differential display	7
Rapid amplification of cDNA ends (RACE)	8
GeneRacer™ Kit for RNA ligase-mediated RACE	9
Selective amplification SABRE	10
DNA array	11
Genome-wide expression monitoring	12
Suppression subtractive hybridisation	13, 14
Topoisomerase I-mediated cloning	15
GATEWAY™ cloning technology	16
Full-length cDNA production employing:	
- template-switching oligonucleotide	17, 18
- mRNA 5'-end oligo-capping	19, 20
- RecA- mediated affinity capture	21
GeneTrapper® cDNA Positive Selection	22
Restriction landmark cDNA scanning (RLCS)	23

these techniques and related commercial kits apply at least three organic extractions in course of protocol[5,6,9,19,22]. Therefore, liquid organic extraction remains the commonly used tool for enzyme removal in both basic and highly advanced genetic engineering protocols.

Traditional phenol-chloroform deproteinisation is time- and labour-consuming procedure that requires multiple manipulations such as organic extraction, alcohol precipitation, DNA pellet washing steps and at least three centrifugations in between[24]. Solid-phase based purification systems and kits, that efficiently purify nucleic acids in microgram range, also requires several binding, washing and elution steps and are not quite consistent when quantitative recovery of short DNA fragments[25], A-T rich DNA sequences[26] or very small amount of DNA[27] is imperative. These factors stimulated us to develop fast, versatile, and universal phenol-free nucleic acids deproteinisation system that allow to remove unwanted enzymes and proteins from post-reaction mixture in a brief vortex-centrifugation or filtration step. In recent years, we have formulated and optimised to use four solid-phase protein binding matrices that highly selective bind proteins from a mixture with nucleic acids and are applicable for many genetic engineering protocols and high-throughput sample processing[28]. The PBM offer universal deproteinisation tool since it (i) quantitatively bind and remove all tested enzymes, totally over 100 species, (ii) compatible with any type of nucleic acid, and (iii) do not required for enzyme binding special buffer formulations. Extraction procedure is very fast, reproducible, and flexible, e.g. it can be performed manually or employing disposable devices and 96-well plate for increased throughput. Therefore, PBM provide convenient and reliable way of nucleic acids purification and thus will further facilitate developing of advanced gene cloning and vector assembling protocols.

2. DEVELOPMENT OF NUCLEIC ACID DEPROTEINISATION STRATEGIES

2.1 Main approaches

Development of techniques for nucleic acid isolation can be split into four main directions: i - liquid organic extraction; ii - selective precipitation; iii - buoyant-density centrifugation; iv - solid-phase extraction. All these approaches build upon, complement one another, and are in process of permanent development.

2.2   Liquid organic extraction

Isolation of highly purified undegraded DNA and RNA is the crucial starting point of many genetic engineering and gene therapy projects. Preparation high molecular weight nucleic acid samples was a challenge task till the mid-1950s when phenol-based liquid organic extraction has been employed. The ability of two-phase aqueous-phenol mixture extract undegraded protein-free polysaccharides and RNA from bacteria[29] was further optimised for isolation of cellular RNA[30] and DNA[31]. Successful combination of proteinase K digestion with phenol-chloroform deproteinisation has become for many years the most popular procedure for DNA isolation[32]. Introduction of guanidium thiocyanate[33,34] has significantly improved quality of isolated RNA from RNase-rich tissues and made extraction procedure more reliable. The most popular RNA extraction procedures have been re-evaluated in parallel [35,36,37,38] many times to select the best isolation method for particular needs. Yet, acid guanidine-phenol-chloroform (AGPC) extraction procedure introduced by Chomczynski and Sacchi[34] and its commercial derivatives (RNAzol, Trisolv, TRIzol, TRI Reagent, RNAgents, RNA Stat, and other reagents) thought is the best liquid organics-based method of total RNA isolation from a wide range sample sizes and tissue types[39,40]. In the continuing search and debate over the best RNA isolation method for microarray applications, the combination of AGPC extraction with solid-phase RNeasy kit (QIAGEN) purification seems to be the method of choice[41,42,43,44]. Therefore, liquid organic extraction has significantly advanced nucleic acids isolation and till now remains a powerful laboratory tool for DNA and RNA production.

Compare to cellular DNA and RNA isolation that requires harsh treatment, deproteinisation of *in vitro* modified nucleic acids can be achieved by mild phenol-chloroform extraction of reaction mixture without special buffer formulation. Nevertheless, due to corrosive nature of phenol and chloroform, employing of liquid organic solvents is associated with several well-known and unwanted drawbacks. These are organics toxicity, inhibition of downstream enzyme activity, partial losses of material in organic interphase, damage of nucleic acids by oxidised phenol derivatives, time and labour consuming of extraction procedure, waste disposal costs, and shark-skin fingers. Although improvement of organic extraction procedure can be achieved by employing a density barrier material, Phase Lock Gel™, that minimises loss of nucleic acids in organic interphase[45], nevertheless, the material does not eliminate utilising of volatile phenol in course of enzyme removal.

2.3   Selective precipitation

Apart of ubiquitous nucleic acids precipitation with alcohol (ethanol or isopropanol) and high concentration of salt (NaCl or LiCl), some compounds are able to precipitate DNA and RNA by compacting and condensation mechanism being added in relatively low concentration. These are trivalent cations[46,47], compacting agents[32,48] (spermine, spermidine, hexammine cobalt), polyethylene glycol[49], polyvinylpyrrolidone[50], and detergent cetavlon[51]. For selective RNA precipitation from biological liquids, cationic detergent Catrimox[52] probably is the most promising agent for diagnostic applications, even it use does present a number of limitations[53]. Selective precipitation is mainly used as intermediate step in plasmid DNA isolation on both laboratory [54] and large production[55] scale, thus being employed as integral part of complex purification protocol rather that stay-alone isolation method.

2.4   Density gradient centrifugation

Cesium chloride and cesium sulphate density gradient or cesium trifluoroacetate isopycnic ultracentrifugation are in use for a long time for efficient separation and isolation of nucleic acids[56,57]. CsCl buoyant-density separation till now remains a reliable way to produce highly purified plasmid DNA suitable for transfection experiments and the reference for achievable DNA purity[58,59]. The drawbacks of this method, apart of equipment cost, is employing of toxic components, time consuming, and low throughput. Introducing of coupled guanidium thiocyanate tissue homogenisation with subsequent CsCl ultracentrifugation[33] has made the next impact in efficiency and simplification of RNA isolation, but this method is also suffer from the same drawbacks as buoyant-density separation alone.

2.5   Nucleic acid binding matrices

Solid-phase extraction (SPE), based upon selective removal of substance of interest from biological liquid or chemical solution, has replaced liquid-liquid chromatography and thus revolutionised in recent years sample extraction and preparation for biochemical and pharmaceutical analysis[60,61]. In molecular biology, silica based matrixes that bind nucleic acid from aqueous solutions were successfully utilised as solid phase for DNA and RNA isolation. Extraction of DNA fragments from agarose gel[62], isolation of plasmid[63] and genomic[64] DNA was performed using silica, glass, or diatomaceous earth particles in the presence of chaotropic agents (NaI, NaClO$_4$, guanidium thiocyanate). Compare to solid-phase DNA isolation, development of similar system for RNA extraction was a more difficult task. Commercial silica-based RNaid® kit (BIO 101) was developed for purification of RNA from gels and solutions only. First reliable method that allows isolate undegraded RNA (mixture of RNA and DNA) from biological samples utilising silica particles and guanidium thiocyanate was published in 1990 by Boom *et al.*[65]. The next advanced system that allows isolate pure RNA by employed anion-exchange technology was introduced in 1994 by QIAGEN[66]. Since the late 1980th, nucleic acid isolation have become highly commercialised and a variety of kits have been designed by manufacturers[67,68,69] to produce high quality DNA and RNA. Some companies now offer nucleic acids isolation and purification systems in a convenient 96-well format. Finally, development of silica[70] and magnetic[71,72] based automated systems that minimise or even eliminate manual steps have converted nucleic acid isolation and manipulation into high throughput procedure, even these systems are rather expensive.

2.6   Protein binding matrices

Another promising approach to solid-phase nucleic acids deproteinisation is employing of matrices that bind and thereof remove enzymes and proteins from a mixture with DNA and RNA. Ion-exchange chromatography is one of the most suitable techniques for purifying of desirable macromolecules from crude fermentation feedstock. On the large scale, employing commercial anion-exchange adsorbents (Sepharose® and STREAMLINE®) is the reliable tool for chromatographic protein removal from preparations of purified pharmaceutical-grade therapeutic DNA[73]. Unfortunately, on laboratory scale liquid chromatography is impractical since it both low throughput and time consuming procedure that requires special buffer formulations and is associated with sample dilution and non-specific nucleic acid adsorption. In practice, phenol-chloroform extraction is widely employed in research laboratory for routine enzyme removal (Table 1), despite of it inherited limitations. Therefore, it is of high interest for researchers to avoid organic extraction and perform nucleic acid deproteinisation in a simple, preferable single-pass operation. In our knowledge, there are two commercial products that allow to remove enzymes from post-reaction mixture, leaving DNA in solution. These are StrataClean® Resin from Stratagene and Micropure™-EZ enzyme removal filter device from Amicon. Micropure-EZ device, which employs nitrocellulose membrane as protein-binding matrix[74] has two limitations: first, according to Amicon's bulletin it does not completely remove at least eleven enzymes[75]; second, according to information disclosed in patent[74], recovery of single-stranded DNA is low at pH under 11,0. StrataClean Resin efficiently extracts a variety of DNA modifying enzymes, but it does not intended by the vendor for RNA purification[76]. In practice, the resin does not employed for enzyme removal in the vendor own kits that utilise traditional phenol-chloroform extraction[77,78,79]. Therefore, both methods are not universal and are suitable for restricted application areas.

The necessity of "ideal" deproteinisation method has arisen from our research needs of fast enzyme removal from a mixture with any type of nucleic acid without employing of organic extraction accompanied with alcohol precipitation or using special high salt buffers for solid-phase purification. In a recent years, we have designed and optimised for use four solid-phase protein-binding matrices - BlueSorb, QuickClean, deENZYME, and EnzyLock - that selective capture enzymes from a mixture with nucleic acids and thus deproteinate modified DNA and RNA in a simple vortex-centrifugation or filtration step. Binding of protein efficiently occurs in water and in a variety of commonly used assay buffers, so that purification procedure does not require special salt additives. PBM deproteinise both concentrated and highly diluted nucleic acids without material losses, thus allow quantitative recover DNA and RNA from proteinaceous mixture. Figure 2 summarises the more important features of the PBM.

Table 2. List of Protein Binding Matrices Features

	BlueSorb	QuickClean	deENZYME	EnzyLock
Removal of bulk of protein	♦	♦	♦	♦
Quantitative enzyme extraction	♦	♦	♦	♦
Compatibility with DNA and RNA	♦	♦	♦	♦
Fast removal of aggressive enzymes		♦	♦	
96-wells plate format			♦	♦
High throughput liquid handling			♦	
Resistance to drying-rehydration cycling				♦

PBM-based enzyme extraction procedure is very simple: 0.1 volume of matrix suspension is added to reaction mixture to be deproteinised, the reaction tube is vortex mixed for 10-20 sec, followed by centrifugation at 14,000 rpm for 1 min to pellet PBM particles with bound enzyme. Otherwise, separation can be achieved by vacuum- or pressure-driven filtration. Supernatant, containing purified DNA or RNA, is transferred in a new tube and procedure is repeated one more time. In all employed assay procedures, enzyme removal was defined as the complete absence of detectable activity in PBM-extracted supernatant or filtrate.

3.   APPLICATIONS OF PBM

3.1   Restriction endonucleases extraction

Because of their unique sequence specificity, restriction endonucleases have become the fundamental tool for recombinant DNA technology and related biotechnology areas. Totally over 3000 restriction enzymes of type II specificity have been identified and more that 500 of it are commercially available[80]. In addition to commonly used methods like general mapping and cloning[24], a growing number of new applications that utilise restriction enzymes are appearing each year, with total number of over 5600 references[80]. Introducing of new type IIS rare cutting restriction enzymes[81] allows further advance genome manipulation[82], expression analysis[83], simplify vector assembling[84] and obtaining intronless genes[85].

Design of DNA vectors for therapeutic applications employs the same principles, techniques, and enzymes that have been accumulated by recombinant DNA technology. A variety of therapeutic DNA constructs are now under development and evaluation. Among these are viral vectors[86] (adenoviral, adeno-associated, lentiviral, retroviral, and some other), chimeric viral constructs[87], plasmid DNA[88], viral-plasmid recombinants[89], or even minimalistic non-viral DNA constructs that consist of transgene expression cassette flanked by two short hairpin oligos[90], so called MIDGE[91] vector (Mologen). Although most of individual DNA vehicle fragments are known and derived from existent recombinant constructs, the resulted vector might contains unique combination of regulatory elements that will improve cell targeting and transgene expression[92,93]. Construction of multicomponent vector is a time-consuming procedure that includes cloning and subcloning steps, excision of DNA fragments of interest with restriction endonucleases, it subsequent modification with specific enzymes, followed by re-ligation into desirable construct. Typical assembling procedure might involve up to ten different restriction enzymes[94,95,96] and even more[97,98]. Traditionally, in laboratory post-digestion inactivation of restriction enzymes includes heat treatment, if applicable, or/and phenol-chloroform extraction. Since many restriction endonucleases are resistant to heat denaturation or reveal only partial heat sensitivity, employing of conventional liquid organic extraction or DNA purification kits is more reliable way to remove enzyme from a

M   Pvu I   Mlu I   Hind III   Eco RI   Bgl II   Bam HI

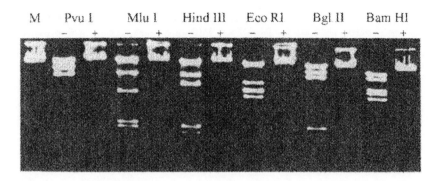

Figure 1. Extraction of restriction endonucleases. Unextracted (-) and BlueSorb extracted (+) restriction endonuclease assay mixtures were incubated with phage lambda DNA as a substrate to reveal recidual enzyme activity. M - untreated phage lambda DNA.

post-reaction mixture. In real life, many gene analysis and cloning techniques, which utilise restriction endonucleases for DNA cleavage, rely upon phenol-chloroform extraction as post-reaction purification tool (Table 1) even it is tedious procedure.

Historically, to evaluate the utility of PBM-based deproteinisation, first we try to extract the most commonly used enzymes - restriction endonucleases. Employed stringent assay conditions (50 units of enzyme per assay mixture and incubation of extracted mixture with lambda DNA for 16 hours) reveals the lack of remaining enzyme activity (Figure 1), thus demonstrating that the matrices quantitative remove restriction endonucleases. Extraction was performed in the presence of enzyme stabilisers and enhances (BSA, glycerol, Triton X-100, and Tween 20) that did not interfere it efficiency. It was also shown that PBM are tolerant to moderate level of common contaminants presenting in nucleic acids preparations (common salts, EDTA, SDS, urea), nevertheless, some of these compound are itself strong inhibitors of enzyme activity and its complete removal might be crucial for particular downstream application. Up to date, over 80 restriction enzymes that belong to several empirical groups have been successfully extracted and some examples are presented in Table 3.

Table 3. Examples of PBM extracted restriction endonucleases

Group	Enzyme
Cohesive end generating	Pvu I, Mlu I, Hind III, EcoR I, Bgl II, Bam HI
Blunt cutting	Alu I, Dra I, EcoR V, Hpa I, Rsa I, Sma I, Sca I, Ssp I
Rare cutting	Apa I, Kpn I, Nae I, Not I, Sac I, Sal I, Xba I, Xho I
Thermophiles	BssHII, Sfi I, Taq I, Tth1 III
"Aggressive" enzymes	Acc I, Ava II, Bgl II, Cla I, Hae III, Pal I, Sca I

Taken together, these results show that employing of the PBM provides a fast and reliable method for restriction endonuclease extraction that can be applied to a wide range of existing and newly developing methods where phenol-chloroform is used for enzyme removal from post-reaction mixture with digested nucleic acids.

3.2   Functional integrity of PBM-deproteinated nucleic acids

The efficiency of DNA subcloning procedure depends on integrity of both insert and vector ends that might be affected by residual enzyme activity and/or mechanical forces applied during purification procedure (extensive vortexing or shearing). Before subsequent ligation, excised DNA fragments must be released from active endonuclease by heat inactivation or DNA purification employing agarose gel electrophoresis, solid-phase kit, or liquid organic extraction. Post-ligation manipulations also include heat treatment to inactivate enzyme activity or phenol-chloroform treatment to physically remove T4 DNA ligase from post-reaction mixture[99].

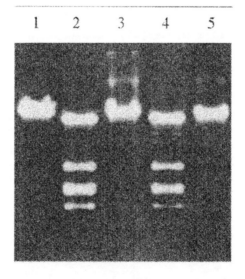

Figure 2. Functional integrity of PBM-extracted DNA. 1 - untreated lambda DNA; 2 - Eco RI digestion/QuickClean extraction; 3 - ligation/QuickClean extraction; 4 - Eco RI re-digestion/QuickClean extraction; 5 - re-ligation.

To simplify and accelerate subcloning procedure, next we thought to apply PBM extractions in a course of multi-step DNA modification protocol. To determine the functional ability of PBM-extracted DNA for subsequent manipulations, we have chose as feasibility test two rounds of consequent cleavage-ligation of phage lambda DNA since it is standard quality control assay. During the first cycle, Eco RI digested lambda DNA (Figure 2, line 2) was deproteinated with QuickClean and thereof purified DNA fragments were re-assembled with T4 DNA ligase (line 3), followed by ligase extraction with the matrix. The second cycle included re-digestion assembled lambda DNA fragments with Eco RI (line 4), followed by restriction enzyme removal with QuickClean and DNA fragments re-ligation (line 5). Electrophoretic analysis has shown both unaltered DNA banding pattern after second EcoR I digestion and successful final re-ligation of DNA fragments, thus ensure that sticky ends of original DNA fragments have not been damaged after applying of multiple PBM extractions. While we formulated universal adsorbents that might be applied to any type of nucleic acid, next we evaluate PBM compatibility with labile RNA molecules. In experiments presented in this chapter and reported earlier[101], it was demonstrated that the matrices are also friendly to RNA, e.g. it do not affect structural and functional integrity of RNA and its derivative - cDNA, which was successfully used for subsequent RT-PCR and cloning. Therefore, PBM-based extraction does not affect functional purity of both DNA and RNA or its termini integrity and, thus, can be conveniently employed for deproteination in a course of gene cloning and vector assembling protocols.

3.3   DNA thermopolymerases

The invention of PCR has lead to development of third generation of *in vitro* DNA manipulation - *de novo* gene synthesis[102] that enhances the ability to produce desired DNA molecules. PCR has become a general method of *in vitro* gene manipulation and extremely convenient technical tool with a variety of applications reviewed elsewhere[103, 104, 105]. In gene therapy, PCR is widely employed for gene cloning and vector construction, analysis of transgene expression in pharmacokinetics studies[106,107] and clinical protocol[108,109], and monitoring of host cell genomic DNA contamination in final recombinant therapeutic product[110].

On analytical scale, aliquots of post-PCR mixture are routinely directly applied on agarose gel without any purification. For subcloning or other downstream enzymatic manipulations, PCR products must be further purified to escape amplicon heterogeneity and/or residual polymerase activity. Since Taq DNA polymerase remains active at +37°C, traces of thermopolymerase in post-PCR mixture can interfere subsequent modification of amplified DNA and to avoid this problem, it is best to phenol extract DNA or at least inactivate enzyme[111]. Yet, in parallel to employing of commercial DNA purification kits, phenol-chloroform extraction of PCR products remains a routine procedure for many downstream applications: cohesive end generation[111,112], ends tailing[113], *in vitro* transcription[114], PCR mutagenesis[115,116], and cloning[117].

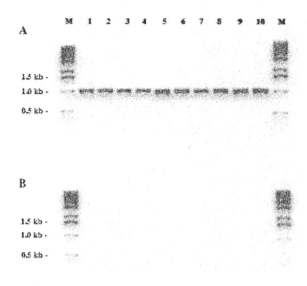

Figure 3. Extraction of DNA thermopolymerases. A - reaction mixtures without polymerase were extracted with deENZYME, followed by enzyme addition and PCR cycling. B - complete PCR mixtures were extracted with deENZYME, followed by PCR cycling. Lines: 1 - AmpliTaq (PE Applied Biosystems); 2 - Stoffel Fragment (Perkin Elmer); 3 - Taq (Boehringer); 4 - Expand High Fidelity (Boehringer); 5 - Taq2000 (Stratagene); 6 - TaqPlus (Stratagene); 7 - Pfu (Stratagene); 8 - PLATINUM™ Taq (Life Technologies); 9 - Taq (Promega); 10 - Tet-Z (Dailat). M: marker, 1 Kb DNA Ladder.

In our laboratory practice, the PBM have been employed for removal of different DNA thermopolymerases, both native and recombinant, totally over tested 15 species. Figure 3 illustrates efficiency of QuickClean extraction of ten different thermostable DNA polymerases that are in use for routine and specific PCR applications. All matrices work well in both low (10 mM ammonium sulfate or KCl) and high salt (60 mM KCl) PCR buffers in the presence of proteinaceous stabilisers, additives, or thermolabile polymerase inhibitor: BSA, Perfect Match™, Taq™ Extender, and antibody (PLATINUM™ Taq). PBM-extracted PCR products were successfully used for subsequent blunt ligation and restriction enzyme digestion to generate cohesive ends in a course of standard subcloning procedure.

3.4   Reverse transcription

The strategy of reverse transcription which is widely employed by viruses and dispersed genetic elements has become a key instrument for conversion of native genes into *in vitro* processed ones[118]. As the experimental tool, reverse transcription allows to convert labile mRNA molecule into stable cDNA copy and therefore generates *in vitro* intronless genes that are suitable for genetic engineering manipulations[119] and gene therapy applications[120,121,122]. In principle, cloning of beneficial genes for curing of diseases, including cancer, is based upon the strategy that has been evolutionary developed by retroviruses - acquiring of cDNA copies of essential cellular genes via reverse transcription for subsequent altering of targeted cell gene expression. Preparation of high quality cDNA is the crucial initial step for many subsequent cloning protocols. It consists of three consequent reactions: (i) first-strand cDNA synthesis, that employs MMLV or AMV reverse transcriptase[24]; (ii) second strand synthesis, that employs RNase H and DNA polymerase I or T7 DNA polymerase[123]; and (iii) blunting of resulted cDNA for efficient subcloning or adapter ligation that employs T4 DNA polymerase and Pfu, or generation of cohesive cDNA ends, employing restriction endonucleases[24,78]. In research laboratory, the common procedure to remove enzymes and proteins from cDNA post-reaction mixture is the phenol-chloroform extractions[124,125,126]. The real situation thought is reflected in above cited Methods of Enzymology[3] where fifteen chapters referred to phenol-chloroform enzyme extraction and only five referred to commercial purification kits.

In a course of our activity to replace liquid organic extraction with PBM-based procedure, next we evaluated the ability of the matrices to remove MMLV and AMV reverse transcriptases. 0.24-9.5 Kb RNA ladder (Life Technologies) containing a mixture of six polyadenylated RNA was used as a template. The extraction procedure was performed in traditional batch format or employing push-through mini-column that fit standard digital micro-pipette. Figure 4 illustrates efficacy of single-pass removal of MMLV and AMV reverse transcriptases using mini-column filled with 3 µl of swollen deENZYME. In a separate test it was shown that the matrix does not introduce inhibitors in reaction mixture, rather it bind and remove enzyme. The important feature of mini-column format it that it allows to process small volumes (10 µl) of reaction mixture without material loss and securely separate deproteinated filtrate from matrix particles. Since in the feasibility test both MMLV and AMV transcriptases were efficiently removed, we further applied multiple BlueSorb extractions for cDNA purification in a course of Lambda ZAPII[101] and ZAP Express (unpublished) cloning protocols (Stratagene). The titres of resulted cDNA libraries that utilise phenol-chloroform or PBM-based extraction were similar and high enough ($6-8 \times 10^8$ pfu/μg mRNA), nevertheless employing of PBM has significantly reduced library construction time and has made cloning procedure more comfortable.

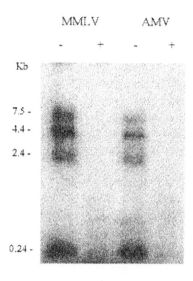

Figure 4. Extraction of MMLV and AMV reverse transcriptases. Untreated (-) and deENZYME purified (+) reverse transcription mixtures, containing 0.24-9.5 Kb RNA ladder as a template and ^{32}P-dCTP as a tracer.

Another extremely important area of cDNA applications is the DNA chip technology[127] as the general tool to inventory all expressed genes used to assemble a living creature[128]. The technology has evolved form so called Southern blot hybridisation method[129] and requires both high quality arrayed target DNA and hybridisation probe, e.g., radioactive- or fluorescent-labelled DNA, RNA, or oligonucleotide. Hybridisation probes are usually produced by direct labelling of random- or oligo(dT)-primed cDNA[24], reverse strand cDNA priming[130], labelling of cRNA transcribed on ds-cDNA template[11], and employing conventional[131] or asymmetric[132] PCR. Among a nebulous class of array troubles that is difficult to diagnose[133] is high fluorescent background, which is associated with poor probe preparation and purification. Post-labelling probe purification can be achieved by removal of impurities in Sephacryl® spin column[134], Microcon® concentrator[133], Qiagen nucleotide removal kit[135], or DNA precipitation with spermine[136]. While some modification enzymes and proteinaceous additives have nucleic acid binding properties (Perfect Match, PLATINUM Pfx, T4 gene 32 protein, RNase A and RNase inhibitor), it is looks reasonable to remove these components from post-reaction labelling mixture to avoid at least this source of background problems. Some purification protocols utilise deproteinisation of DNA template[11] and probe[132,137] with phenol-chloroform, but it was reported that liquid organic extraction results in substantial loss of fluorescent- and digoxigenin-labelled probes[137,138]. On the other hand, inclusion of phenol-chloroform extraction in Big-Dye™-labelled probe purification results in higher quality sequencing data[139]. Liquid organic extraction thought results in both probe purification and material loss. The losses are due to partial probe solubility in organic phase, that depends on the nature of the dye or haptene group, and/or non-specific probe capture by proteinaceous interphase, that is pour controlled. Since the PBM allow deproteinate nucleic acids in a nanogram range without material capture in organic interphase, it can be applied for probe cleaning. In a course of hybridisation probe purification, we have extracted $[^{32}P]$-, $[^{33}P]$- and fluorescent-labelled nucleic acids with PBM to eliminate enzymes and BSA from post-reaction mixture, followed by Microcon® centrifugation to cut-off small molecules. Combination of PBM deproteinisation with centrifugation-driven ultrafiltration efficiently removes both proteinaceous and low molecular weight contamination from post-labelling mixture without probe loss. Recently, CLONTECH Laboratories has included QuickClean resin as an alternative to liquid organic extraction into its new Atlas™ Glass Fluorescent Labelling Kit[140,141] that is intended for generation of fluorescently labelled cDNA probes for glass arrays.

3.5 DNase I extraction

Pancreatic DNase I has been purified fifty ears ago[142] and is widely used for molecular biology applications[24]. Yet, a very common DNase I application is RNA sample digestion before RT-PCR. In a course of PCR-based expression analysis and cDNA cloning, preparation of high quality RNA as the starting material is imperative. Even the majority of extraction kits generate highly deproteinated RNA with E_{260} 2.0 or even higher, most cellular RNA samples are contaminated with traces of genomic DNA that are undetectable with standard laboratory methods (spectrophotomety and fluorescence assay). Because of extremely high sensitivity of PCR, these DNA traces might generate false positive bands from RT-PCR and should be eliminated. As a result, DNase I digestion is routinely employed for removal of minute traces of genomic DNA from RNA preparations before RT-PCR or DD-PCR. On completion of RNA digestion, DNase I must be inactivated or removed prior to reverse transcription since even traces of residual activity will degrade newly synthesised cDNA. Heat treatment is a convenient way to kill enzyme activity, but it rather tricky procedure that balances between compete DNase I inactivation and RNA integrity[143]. Moreover, it was recently shown that in standard reaction buffers the enzyme is not completely inactivated even at +95°C and it could be efficiently heat-inactivated only when the pH is lower that 7 or higher that 9, e.g. pH should be adjusted before heating[144]. As a result, conventional phenol-chloroform extraction still be the most reliable procedure for DNase I removal from RNA samples[145].

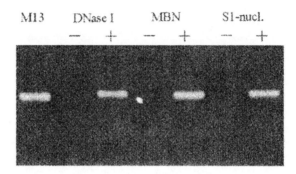

Figure 5. Extraction of DNase I, MBN, and S_1 nuclease. Unextracted (-) or QuickClean extracted (+) nuclease assay mixtures were incubated with single-stranded M13 DNA as substrate. M13 - untreated M13mp18ssDNA.

Since we are routinely eliminate DNA traces from RNA preparations, next we have evaluated feasibility of PBM extraction for DNase I removal. To monitor recidual nuclease activity in PBM-extracted supernatant, single-stranded M13 DNA was used as a substrate (Figure 5) to reveal even minute traces of DNase I nicking activity. This sensitive assay has demonstrated that DNase I can be completely extracted with deENZYME (non shown) or QuickClean[146] from restriction endonuclease, reverse transcription, and PCR buffers. Different lots of DNase I obtained from Ambion, Boehringer, CLONTECH, Life Technologies, Promega, and Stratagene were successfully removed with the PBM. In a course of RNA polishing, BlueSorb was also efficiently used for DNase I removal from digested RNA samples that were subsequently used for HIV-1 RNA splicing analysis, RT-PCR, and differential display[147].

Another routine application of DNase I is template DNA removal from *in vitro* transcribed RNA. Post-transcription RNA purification typically involves degradation of template DNA with DNase I, followed by nuclease extraction with phenol-chloroform[148]. Organic extraction of synthetic RNA is also employed for *in vitro* generation of biotinylated RNA-driver for construction of subtraction libraries[149] and internal standard[150] or competitor RNA template[151] for mRNA quantification by RT-PCR. Production of synthetic RNA for treatment of cancer cells, prevention of infectious diseases[152], and overexpression of physiologically important proteins[153] is achieved by *in vitro* transcription that utilises post-reaction DNA template removal with DNase I, followed by phenol-chloroform extraction[154,155,156]. In a course of developing "suicide genes" anti-cancer therapy via sensitisation of target cells to ganciclovir, HSVtk RNA for transfection was *in vitro* transcribed and purified by DNase I digestion and subsequent organic extraction[157]. Monitoring of anti-HIV-1 protective vector expression in target cells also utilises DNase I treatment and phenol-chloroform deproteinisation of RNA samples[94]. Non-viral gene therapy further employs DNase I for evaluation of liposomal protective effect on plasmid DNA[158,159] and for conformational analysis of peptide nucleic acid modified plasmid[160]. DNase I removal from DNA-liposome preparations was also achieved by conventional phenol-chloroform extraction[158,160] or employing Phase Lock Gel system[159] that used in conjunction with phenol-chloroform[45].

We have gradually replaced liquid organics with PBM extraction in a course of *in vitro* transcription protocol. First, BlueSorb was employed for removal T3-, T7-, and SP6 RNA polymerases from synthetic ^{32}P-RNA probe[161]. Next, PBM were applied for DNase I extraction from digested post-transcription samples. DNase I was removed from different *in vitro* transcription buffers (Ambion, Promega, Stratagene) containing proteinaceous compounds (RNA polymerase, BSA, and RNase inhibitor). Therefore, since the matrices are friendly to RNA and quantitative remove RNA polymerases and DNase I, it can be further conveniently employed for deproteinisation of *in vitro* transcribed RNA and purification of DNase I digested RNA samples suitable for a many downstream applications.

3.6 Exo- and single-strand nucleases removal

Various strand-specific nucleases, apart of restriction endonucleases and DNase, I have found extensive use in genetic engineering[162,163]. Among these are exonuclease III (Exo III), mung bean nuclease (MBN), and S1 nuclease, which provide a convenient way to remove unwanted nucleic acid molecules or single-stranded fragments. Exo III has been employed for production of single-stranded[164] and gradually deleted[165], and ligation independent cloning[167,168]. Continuous development of new techniques is based upon employing of both newly discovered and old-known enzymes. The new biotechnological application of Exo III is combinatorial protein engineering strategy, called iterative truncation for the creation of hybrid enzymes (ITCHY)[169]. In addition to "classical" MBN applications such as repairing

DNA fragment ends[170], generation new restriction sites[171], and PCR mutagenesis[172], the enzyme is also utilised for advanced cloning techniques SABRE[10] and linker-capture subtraction[173]. S1 nuclease is another enzyme that has been in use in nucleic acids research for a long time[174] and till now is used for hairpin removal from double-stranded cDNA[175], transcript mapping[176], and in nuclease sensitivity DNA assay in a course of non-viral gene therapy experiments[160]. Therefore, nucleases are widely used in a variety of DNA modification and gene cloning protocols.

Figure 6. Extraction of Exonuclease III. ΦX174/HaeIII DNA was added as substrate to unextracted (line 1) or QuickClean-extracted (line 2) Exo III assay mixture. 3 - untreated φX174 RF DNA/Hae III

Because even residual traces of the nucleases might affect downstream nucleic acid applications, post-digestion enzyme inactivation is the crucial step that relays upon phenol-chloroform extraction that has been employed in most above listed publications. We have evaluated utility of PBM extraction for removal of MBN, S1 nuclease (Figure 5), and Exo III (Figure 6) as an alternative to liquid organics. The integrity of single-stranded DNA, which was added as a substrate in MBN and S1 nuclease assay mixtures, demonstrates the absence of recidual nicking activity in the PBM-extracted supernatant.

Complete extraction of another important enzyme uracil-DNA glycosylase (UNG) can be achieved with two extractions of QuickClean[146] or single extraction with deENZYME (Figure 7). Additional experiments not shown here have indicated that the matrices are also efficient in removal of other enzymes that exhibit specific exonuclease activities (Klenow fragment, T4 DNA polymerase and Pfu thermopolymerase) and widely employed for cDNA and PCR products end-blunting for subsequent cloning. Efficient removal of nucleases is the challenging test for both traditional and newly developing deproteination methods and as it has been shown the PBM meet this stringent demand.

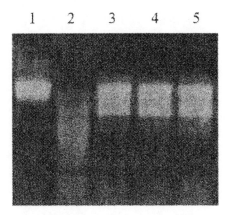

Figure 7. Extraction of uracil-DNA glycosylase. 1: UNG substrate - PCR generated 1 kb dU-DNA fragment; 2: dU-DNA digested with UNG, followed by DNA denaturation at +95°C; 3: dU-DNA denatured without UDG digestion; 4: dU-DNA incubated with double deENZYME-extracted UNG, followed by DNA denaturation; 5: dU-DNA incubated with single deENZYME-extracted UNG, followed by DNA denaturation.

3.7 High throughput

As recombinant DNA research and related areas move toward high throughput processing and automation[177,178,179], replacement of liquid organic extraction and other multi-step purification techniques with a single-pass procedure become very important. Yet, the 96-well format is the minimal requirement not only for drug discovery[180], but also for genomic libraries construction[181] and massive gene expression analysis[182,183]. In research laboratory, when 12 or more nucleic acid samples need to be deproteinated simultaneously, it becomes desirable to have an access to non-expensive medium- or high-throughput purification procedures. As an intermediate step, we first assembled push-through mini-columns and spin cartridges pre-loaded with wet or dried matrix. Employing of these medium-throughput devices is convenient, but is limited by multi-channel pipette or centrifuge rotor capacity of maximum 12-24 samples in an operation cycle. To meet oncoming needs for high throughput enzyme removal from post-reaction mixtures, we have further optimised PBM extraction procedure for 96-well plate format. Two variations of plate deproteinisation were developed: one using liquid slurry of deENZYME that was added to in-plate arrayed samples, and a second using dried EnzyLock pre-loaded in plate wells to which samples to be purified were added. We employ deENZYME for liquid manipulations, since being once mixed, the matrix forms a stable suspension for a period of time sufficient to fill several 96-well plates with arrayed samples. For dry format, EnzyLock is optimal as pre-filling matrix, since it intended to be dried and rehydrated without loosing of protein binding capacity. After applying of vacuum or centrifugation forces, deproteinated filtrates were collected in a fresh plate, being immediately ready for subsequent manipulations. To manufacture a stock of ready-to-use 96-well plates pre-loaded with dried EnzyLock, we fill plate wells with the matrix slurry, followed by it drying under vacuum or by heating at 50°C. Because environmental stability of dried matrix is a key requirement for convenient manufacturing, different conditions for EnzyLock storage were evaluated. Long-term storage experiments have revealed that dried matrix can be kept at ambient temperature ranging from 0°C to +50°C for a long period of time (at least one year) without detectable decrease of its protein binding capacity. To demonstrate the utility of PBM for high throughput deproteinisation, restriction enzyme assay was performed. Assay mixtures (32 different enzymes in triplicate, 50 units in 100 µl) were loaded using 12-

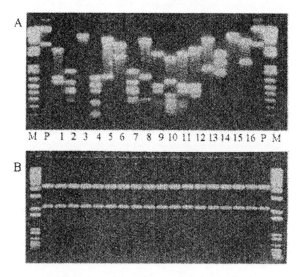

Figure 8. Restriction endonuclease extraction in 96-well plate format. Assay mixtures without enzyme (A) or with 50 units of enzyme (B) were filtered through plate wells pre-loaded with dried EnzyLock. Enzymes (1-16: Rsa I, Pal I, Sca I, Alu I, Dra I, Pvu II, Ssp I, Csp6 I, Msp I, Hpa II, Sau 3AI, Hinf I, Pvu I, Xho II, Nco I, Bgl II, resp.) were added to control filtrates (A) to ensure that the matrix did not introduce reaction inhibitors. M - marker, 1 Kb DNA ladder; P - untreated plasmid DNA.

channel pipette in Millipore multi-well plate fitted with membrane support and pre-loaded with dried EnzyLock, followed by vacuum-driven filtration. Electrophoretic analysis of filtrates reveals the absence of recidual cutting or nicking activity in both EnzyLock (Figure 7) and deENZYME (not shown) deproteinated assay mixtures. Therefore, the PBM can be easily adapted for a convenient 96-well format which provides one-plate sample digestion and processing, enabling considerably higher throughput without any compromise in the enzyme removal efficiently compare to traditional phenol-chloroform extraction. We further suggesting that deENZYME and EnzyLock might be also suitable for automated instrumentation.

3.8 DNA quantification assay

Safety is an important consideration in the development of gene therapy protocols and therapeutic biopharmaceuticals production. The increasing demands for the production of pharmaceutical-grade recombinant proteins and plasmid DNA[184] require employing of adequate stringent quantitative testing for impurities and contamination[185]. Traces of host cell genomic DNA in preparations of plasmid DNA and recombinant proteins are among common contaminants that have to be quantified precisely. DNA contamination in picogram range can be detected by labour consuming blot-hybridisation[186,187] and PCR[188], or, what is preferable, by direct measurement of DNA fluorescence in solution that is much more rapid and less expensive. Commercially available homogeneous fluorescence based systems allow to measure double-stranded DNA with detection limit between 98 pg/ml[189] and 25 pg/ml[190] in buffers containing no protein. However, in practice high concentration of protein in sample affects fluorescent assay sensitivity, so that DNA detection limit has decreased from 98 pg/ml in buffer alone to 1 ng/ml in solutions containing 1 mg/ml of BSA[189]. Therefore, current detection demand of 10 pg/ml DNA and lower cannot be achieved in preparations with high protein load without preliminary sample deproteinisation. In dot-hybridisation DNA quantitation assay, samples to be measured were first extracted with phenol-chloroform, followed by DNA ethanol precipitated before blotting on filter[186]. Employing of liquid organic extraction for protein removal before DNA quantitation is undesirable since it labour-consume low throughput procedure with a risk of DNA lost in organic interphase or/and during subsequent alcohol precipitation. Sample processing with solid-phase DNA isolation/purification kits is not reliable while quantitative DNA recovery is not consistent enough in sub-nanogram range. In an attempt to develop fast and versatile sample deproteination method, which is fully compatible with solution-based fluorescent DNA quantification, the feasibility of QuickClean and BlueSorb extraction was evaluated in conjunction with PicoGreen® dsDNA Quantification kit (Molecular Probes). The following BlueSorb-PicoGreen DNA assay was developed: proteinaceous samples containing 330 μg/ml of purified E. coli heat-labile toxin were spiked with gradually diluted concentrations of phage lambda DNA, followed by BlueSorb extraction and DNA quantification. Here we present preliminary data of DNA fluorescent quantitation experiments that are now in progress. Both high-range (non-shown) and low-range (Figure 8) standard curves look quite good in the range of DNA concentrations tested. Since PBM allow to deproteinise virtually any type of nucleic acid, we believe that it might be employed as an initial cleaning step in conjunction with already existing quantification systems intended for method validation of host cell nucleic acids removal during recombinant protein manufacturing.

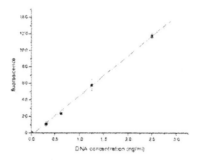

Figure 9. Low-range standard curve of DNA concentration. Curve and bars represent the linear fit and standard deviations, respectively, corresponding to six independent experiments. Data courtesy of Dr. D. Favre (Swiss Serum and Vaccine Institute, Bern, Switzerland)

3.9   Other applications

Electrophoretic analysis of modified nucleic acids without purification is a very common analytical procedure, which sometimes is associated with unwanted artefacts. Some well known (RNase A) and new (PLATINUM Pfx DNA polymerase) enzymes remain bound to DNA on completion of reaction and thus affect DNA migration pattern[191,192]. When non-purified restriction digested DNA is directly loaded on polyacrylamide gel, enzymes (Msp I, Hae III, Sau 3A, Taq I) co-migrate with DNA fragments on gel and are stained by silver, thereof masking DNA bands[193]. Applying of a single BlueSorb extraction to post-restriction mixture before electrophoresis removes restriction enzymes and significantly improves DNA band pattern[194]. Therefore, since PBM allow extraction of the bulk of proteins in a single operation step both in manual and multi-well plate format, it can be conveniently employed for deproteinisation of nucleic acid samples before electrophoretic and chromatographic analysis.

Figure 10. Binding and re-sorption of total cellular proteins. Aliquot of total cellular lysate of HEK-293 cell line (line 1) was extracted with deENZYME (line 2), followed by protein re-sorption in Laemmli SDS-buffer (line 3) and protein band pattern analysis in 10% PAAG. Amido black staining. Data is the courtesy of Dr. J. Cheburkin (Max Planck Institute for Biochemistry, Munich, Germany).

There is another, not so obvious, application of the PBM. It was observed that PBM quantitative bind and re-sorb cellular proteins (Figure 10) in a molecular weight range of 30-200 kDa, thereof allows to concentrate diluted proteins and enzymes from aqueous solution. In preliminary experiments it was shown that PBM-captured enzymes and proteins exhibit, at least in part, its functional and immunological activity (unpublished) and thus might be further employed for some applications in the area of molecular biology.

3.10   Limitations of the PBM

(i) It should be noted, however, that although PBM efficiently remove *in vitro* modification enzymes from post-reaction mixture, the matrices do not intended for direct DNA and RNA isolation from living cells and tissues. Dissociation of tightly bonded cellular nucleoprotein complexes requires harsh treatment with chaotropic agents or liquid organics.

(ii) Compared to nucleic acid purification kits, PBM do not remove nucleotides and oligonucleotides from post-reaction mixture.

(iii) Strong detergents (SDS) and chaotropic agents (guanidium salts) can interfere the PBM protein binding capacity and therefore should be avoided.

(iv) We have evaluated efficacy of the PBM extraction for a small scale only that does not exceed 1 ml of assay mixture.

(v) We did not investigate PBM regeneration for it re-use, instead, we discard the matrices after each run of extraction.

4.   CONCLUSIONS

In conclusion, we summarise advantages of PBM that make it attractive to use in genetic engineering and laboratory gene therapy.
- Universality: the PBM extract all tested enzymes and are compatible with any type of nucleic acids.
- High yield and reproducibility: no nucleic acid loss occurs in organic interphase or during accompanied alcohol precipitation.
- Fast and accuracy: few minute protocol leaves less time for occasional cross-contamination or nucleic acids damage.
- Flexibility: PBM allows extraction of enzymes in manual format or employing disposable devices and multi-well plates.
- High throughput: up to 96 nucleic acids samples can be simultaneously processed in a single operation step.
- Stable: at least one year shelf-life.
- Safety: the PBM are toxicologically and environmentally safe.
- Economy: employing of environmental-friendly matrices do not require waste disposal care.
- Growing potential: PBM are undergoing further optimisation and extending of application area.
- Beauty: the matrixes are pretty blue coloured.

ACKNOWLEDGEMENTS

I would like to thank my family and the Editor for tolerance.
This work was supported by the private funds of the author (V.I.E.)

REFERENCES

1.   Friedmann, T., 1995, Rigor in gene therapy studies. *Gene Therapy* 2:355-366

2. Anderson, W.F., 1998, Human gene therapy. *Nature* **392**:25-30

3. Weissman, S.M., 1999, Methods of Enzymology. cDNA Preparation and Characterization, v. 303. Academic Press

4. Nadin-Davis, S.A. and Chang, S.C., 1998, Site-directed mutagenesis of large plasmids. *BioTechniques* **25**:1015-1019

5. Goryshin, I.Y. and Reznikoff, W.S., 1998, Tn5 *in vitro* transposition. *J. Biol. Chem.* **273**:7367-7374

6. York, D., Welch, K., Goryshin, I.Y., and Reznikoff, W.S., 1998, Simple and efficient generation *in vitro* of nested deletions and inversions: Tn5 intramolecular transposition. *Nucleic Acids Res.* **26**:1927-1933

7. Liang, P. and Pardee, A.B., 1997, Differential display methods and protocols. Methods in Molecular Biology, Vol. 85. Humana Press, Totowa, New Jersey

8. Chen, Z., 1996, Simple modifications to increase specificity of the 5' RACE procedure. *Trends Genet.* **12**:87-88

9. Invitrogen, 2000, GeneRacer™ Kit for lull-length, RNA ligase-mediated rapid amplification of 5' and 3' cDNA ends (RLM-RACE)

10. Lavery, D.J., Lopes-Molina, L., Fleury-Olela, F., and Schibler, U., 1997, Selective amplification via biotin- and restriction-mediated enrichment (SABRE), a novel selective amplification procedure for detection of differentially expressed mRNAs. *Proc. Natl. Acad. Sci.* **94**:6831-6836

11. Lockhart, D.J., Dong, H., Byrne, M.C., Follettie, M.T., Gallo, M.V., Chee, M.S., Mittmann, M., Wang, C., Kobayashi, M., Horton, H., and Brown, E.L., 1996, Expression monitoring by hybridization to high-density oligonucleotide arrays. *Nature Biotech.* **14**:1675-1680

12. Wodicka, L., Dong, H., Mittmann, M., Ho, M.H., and Lockhart, D.L., 1997, Genome-wide expression monitoring in Saccharomyces cerevisiae. *Nature Biotech.* **15**:1359-1367

13. Diatchenko, L., Lau, Y.-F.C., Campbell, A.P., Chenchik, A., Moqadam F., Huang, B., Lukyanov, S., Lukyanov, K., Gurskaya, N., Sverdlov, E.D., and Siebert, P.D., 1996, Suppression subtractive hybridization: A method for generating differentially regulated or tissue-specific cDNA probes and libraries. *Proc. Natl. Acad. Sci. USA* **93**:6025-6030

14. Diatchenko, L., Lukyanov, S., Lau, Y.-F.C., and Siebert P.D., 1999, Suppression subtractive hybridization: A versatile method for identifying differentially expressed genes. *Methods Enzymol.* **303**:349-380

15. Heyman, J.A., Cornthwaite, J., Foncerrada, L., Gilmore, J.R., Gontang, E., Hartman, K.J., Hernandez, C.L., Hood, R., Hull, H.M., Lee, W.-Y., et al., 1999, Genome-scale cloning and expression of individual open reading frames using topoisomerase I-mediated ligation. *Genome Res.* **9**:383-392

16. Life Technologies, 2000, GATEWAY™ cloning technology. Instruction manual, Version 1

17. Chenchik, A., Zhu, Y., Diatchenko, L., and Siebert, P., inventors; Clontech Laboratories, assignee. 1999, Methods and compositions for full-length cDNA cloning using a template-switching oligonucleotide. US Patent 5,962,272

18. Clontech Laboratories, 1999, SMART™ PCR cDNA Synthesis Kit. User Manual, Protocol #PT3041-1

19. Maruyama, K. and Sugano, S., 1994, Oligo-capping: a simple method to replace the cap structure of eukaryotic mRNA with oligoribonucleotides. *Gene* **138**:171-174

20. Volloch, V., Schweitzer, B., and Ritz, S., 1994, Ligation-mediated amplification of RNA from murine erythroid cells reveals a novel class of β globin mRNA with an extended 5'-untranslated region. *Nucleic Acids Res.* **22**:2507-2511

21. Hakvoort, T.B.M., Vermeulen, J.L.M., and Lamers, W.H., 1998, Enriched full-length cDNA expression library by RecA-mediated affinity capture. In *Gene Cloning and Analysis by RT-PCR* (P.D. Siebert and J.W. Larrick, eds.), BioTechniques Molecular Laboratory Methods Series, pp.259-269

22. Life Technologies, 1999, GeneTrapper® cDNA Positive Selection System. Instruction Manual

23. Suzuki, H., Yaoi, T., Kawai, J., Hara, A., Kuwajima, G., and Watanabe, S., 1996, Restriction landmark cDNA scanning (RLCS): A novel cDNA display system using two-dimensional gel electrophoresis. *Nucleic Acids Res.* **24**:289-294

24. Sambrook, J., Fritch, E.F., and Maniatis T., 1989, Molecular cloning: A laboratory manual. 2nd edition. Cold Spring Harbor Laboratory Press, Cold Spring Harbor, New York

25. Smith, L.S., Lewis, T.L., and Matsui, S.M., 1995, Increased yield of small fragments purified by silica binding. *BioTechniques* **18**:970-975

26. Kaur, R., Kumar, R., and Bachhawat, A.K., 1995, Selective recovery of DNA fragments from silica particles: effect of A-T content and elution conditions. *Nucleic Acids Res.* **23**:4932-4933

27. Dr. Octavian Henegariu, Yale University School of Medicine, New Haven, USA., personal communication

28. Evtushenko, V.I., 2000, Protein-binding matrices for phenol-free extraction of DNA and RNA modifying enzymes: applications for genetic engineering and high throughput. *BioTech International* **12**:10

29. Westphal, O., Luderitz ,O., und Bistern, F., 1952, Uber die Extraktion von Bakterien mit Phenol/Wasser. *Z. Naturforschg.* **7b**:148-155

30. Kirby, K.S., 1956, A new method for isolation of ribonucleic acids from mammalian tissues. *Biochem. J.* **64**:405-408

31. Kirby, K.S., 1957, A new method for the isolation of deoxyribonucleic acids: evidence on the nature of bonds between deoxyribonucleic acid and protein. *Biochem. J.* **66**:495-504

32. Maniatis, E.F., T., Fritsch, E.F. and Sambrook, J., 1982, Molecular Cloning. A Laboratory Manual. Cold Spring Harbor Laboratory, Cold Spring Harbor, New York

33. Chirgwin, J.M., Przybyla, A.E., MacDonald, R.J., and Rutter, W.J., 1979, Isolation of biologically active ribonucleic acid from sources enriched in ribonuclease. *Biochemistry* **18**:5294-5299

34. Chomczynski, P. and Sacchi, N., 1987, Single-step method of RNA isolation by acid guanidium thiocyanate-phenol-chloroform extraction. *Anal. Biochem.* **162**:156-159

35. Puissant, C. and Houdebine, L.-M., 1990, An improvement of the single-step method of RNA isolation by acid guanidium thiocyanate-phenol-chloroform extraction. *BioTechniques* **8**:148-149

36. Huang, N. and Durica, D.S., 1993, Differential recovery of mRNA transcripts using acid organic extraction techniques: Quantification of abl mRNA abundance in the blowfly Calliphora erythrocephala. *SAAS Bull. Biochem Biotech.* 6:21-30

37. Gruffat, D., Piot, C., Durand, D., and Bauchart, D., 1996, Comparison of four methods for isolating large mRNA: Apolipoprotein B mRNA in bovine and rat liver. *Anal. Biochem.* 242:77-83

38. Gruffat, D., 1998, Isolation of RNA from mammalian cells: Applications to large mRNA. In *Gene Cloning and Analysis by RT-PCR* (P.D. Siebert and J.W. Larrick, eds.), BioTechniques Molecular Laboratory Methods Series, pp.35-55

39. Monstein, H.-J., Nylander, A.-G., Chen, D., 1995, RNA extraction from gastrointestinal tract and pancreas by a modified Chomczynski and Sacchi method. *BioTechniques* 19:240-244

40. Clark, M.D., Panopoulou, G.D., Cahill, D.J., Bussow, K., and Lehrach, H., 1999, Construction and analysis of arrayed cDNA libraries. *Methods Enzymol.* 303:205-233

41. Loftus, S.K., Chen, Y., Gooden, G., Birznieks, G., Hilliard, M., Baxevanis, A.D., Bittner, M., Meltzer, P., Trent, J., and Pavan, W., 1999, Informatic selection of a neutral crest-melanocyte cDNA set for microarray analysis. *Proc. Natl. Acad. Sci. USA* 96:9277-9280

42. Mahadevappa, M. and Warrington, J.A., 1999, A high-density probe array sample preparation method using 10- to 100-fold fewer cells. *Nature Biotech.* 17:1134-1136

43. Carlsson, B., Jernas, M., Lindell, K., and Carlsson, L.M.S., 2000, Total RNA and array-based expression monitoring. *Nature Biotech.* 18:579

44. Mahadevappa, M. and Warrington, J.A. respond, 2000. *Nature Biotech.* 18:579

45. Murphy, N.R. and Hellwig, R.J., 1996, Improved nucleic acid organic extraction through use of a unique gel barrier material. *BioTechniques* 21:934-939

46. Arscott, P.G., Li, A.-Z., and Bloomfield V.A., 1990, Condensation of DNA by trivalent cations. 1. Effects of DNA length and topology on the size and shape of condensed particles. *Biopolymers* 30:619-630

47. Kejnovsky, E. and Kypr, J., 1998, Millimolar concentrations of zinc and other metal cations cause sedimentation of DNA. *Nucleic Acids Res.* 26:5295-5299

48. Altermann, E., Klein, J.R., Henrich, B., 1999, Synthesis and automated detection of fluorescently labelled primer extension product. *BioTechniques* 26:96-101

49. Lis, J.T., 1980, Fractionation of DNA fragments by polyethylene glycol induced precipitation. *Methods Enzymol.* 65:347-353

50. Kim, C.S., Lee, C.H., Shin, J.S., Chung, Y.S., and Nyung, N.I., 1997, A simple method for isolation of high quality genomic DNA from fruit trees and conifers using PVP. *Nucleic Acids Res.* 25:1085-1086

51. Doctor, B.P., 1967, Fractionation of RNA's by countercurrent distribution. *Methods Enzymol.* 12:644-657

52. Dahle C.E., Macfarlane D.E., 1993, Isolation of RNA from cells in culture using Catrimox-14™ cationic surfactant. *BioTechnology* 15:1102-1105

53. Ali S.A., Kubik B., Gulle H., Eibl M.M., Steinkasserer A., 1998, Rapid isolation of HCV RNA from Catrimox-lysed whole blood using QIAamp® spin columns. *BioTechniques* **25**:975-978

54. Murphy, J.C., Wibbenmeyer, J.A., Fox, G.E., and Willson, R.C., 1999, Purification of plasmid DNA using selective precipitation by compacting agents. *Nature Biotech.* **17**:822-823

55. Marquet, M., Horn, N., Meek, J., Budahazi, G, inventors; Vical Incorporated, assignee; 1996, Production of pharmaceutical-grade plasmid DNA. US patent #5561064

56. Freifelder, D., 1971, Isolation of extrachromosomal DNA from Bacteria. *Methods Enzymol.* **21**:153-163

57. Perbal, B., 1984, *A Practical Guide to Molecular Biology.* John Wiley & Sons, New York, pp.150-165

58. Frauke, E., 1995, Effect of plasmid quality on transfection experiments. *J. NIH Res.* **7**:64

59. Davis, H.L., Schleef, M., Moritz, P., Mancini, M., Schorr, J., and Whalen, R.G., 1996, Comparison of plasmid DNA preparation methods for direct gene transfer and genetic immunization. *BioTechniques* **21**:92-99

60. Simpson, N., 1992, Solid-phase extraction: Disposable chromatography. *Intl. Chromat. Lab.* **11**:7-14

61. Majors, R.E. and Raynie, D.E., 1997, Sample preparation and solid phase extraction. *LG-GC* **15**:1106-1117

62. Vogelstein, B. and Gillespie, D., 1979, Preparative and analytical purification of DNA from agarose. *Proc. Natl. Acad. USA* **76**:615-619

63. Marko, M.A., Chipperfield, R., and Birnboim, H.C., 1982, A procedure for the large scale isolation of highly purified plasmid DNA using alkaline extraction and binding to glass powder. *Anal. Biochem.* **121**:382-387

64. Thompson, J.D., Cuddy, K.K., Haines D.S., and Gillespie, 1990, Extraction of cellular DNA from crude cell lysate with glass. *Nucleic Acids Res.* **18**:1074

65. Boom, R., Sol, C.J.A., Salimans, M.M.M., Jansen, C.L., Wertheim-van Dillen P.M.E., and van der Noordaa J., 1990, Rapid and simple method for purification of nucleic acids. *J. Clin. Microbiol.* **28**:495-503

66. QIAGEN, 1995, QIAGEN-tips for RNA Midi- and Maxipreps. *QIAGEN News for Biochemistry and Molecular Biology* **3**:1-2.

67. Planner, B., 1994, Purification and labelling kits. *J. NIH Res.* **6**:78-82

68. Planner, B., 1994, DNA purification kits. *J. NIH Res.* **6**:48-53,

69. Burnick, L., Martinez, C.Y., Sosa, J.L., Mishkin, D.H., Brush, M.D., and Zavala, M.E., 1996, Poly-A-tales. *BIO Consumer Review* **3**:68-87

70. Itoh, M., Kitsunai, T., Akiyama, J., Shibata, K., Izawa, M., Kawai, J., Tomaru, Y., Carninci, P., Shibata, Y., Ozawa, Y., Muramatsu, M., Okazaki, Y., and Hayashizaki, Y., 1999, Automated filtration-based high-throughput plasmid preparation system. *Genomic Res.* **9**:463-470

71. Dynal, 1995. *Biomagnetic Techniques in Molecular Biology. Technical Handbook.* Second ed.

72. Hawkins, T.L., McKernan, K.J., Jakotot, L.B., MacKenzie, J.B., Richardson, P.M., and Lander, E.S., 1997, A magnetic attraction to high-throughput genomics. *Science* **276**:1887-1889

73. Levy, M.S., O'Kennedy, R.D., Ayazi-Shamlou, P., and Dunnill, P., 2000, Biochemical engineering approaches to the challenges of producing pure plasmid DNA. *Trends Biotech.* **18**:296-305

74. Kung, V.T., inventor; Molecular Devices Corporation, assignee; 1991, Nitrocellulose filtration to remove proteins from polynucleotides. US patent #5004806

75. Leonard, J.T., Levasseur, V.E. and Lurantos, M.H.A., Micropure-EZ™: An enzyme removal device for molecular biology. *Amicon bioSolutions* **4**:1-3

76. Stratagene, 1999, StrataClean® Resin. Instruction Manual, Revision 059001, p.1

77. Weiner, M.P., 1993, Directional cloning of blunt-ended PCR products. *BioTechniques* **15**:502-505

78. Stratagene, 1999, ZAP Express® cDNA synthesis kit and ZAP Express® cDNA Gigapack® III Gold cloning kit. Instruction Manual, Revision 119011/200403

79. Stratagene, 2000, AdEasy™ Adenoviral Vector System. Instruction Manual, Revision 060002

80. Roberts, R.J. and Macelis, D., 2000, REBASE - restriction enzymes and methylases. *Nucleic Acids Res.* **28**:306-307

81. Stewart, F.J., 1999, Homing endonucleases: Applications for these rare cutters. The NEB Transcript 10:5

82. Jasin, M., 1996, Genetic manipulation of genomes with rare-cutting endonucleases. *Trends Genet.* **12**:224-228

83. Kato, K., 1995, Description of the entire mRNA population by a 3' end cDNA fragment generation by class IIS restriction enzymes. *Nucleic Acids Res.* **23**:3685-3690

84. Souza D.W. and Armentano D., 1999, New cloning method for recombinant adenovirus construction in Escherichia coli. *BioTechniques* **26**:502-508]

85. Berlin, Y.A., 1998, DNA splicing by directed ligation (SDL). In *Genetic Engineering with PCR* (Tait, R.C. and Horton, R.M., eds.), Vol. 5: Current Innovations in Molecular Biology series, Horizon Scientific Press, pp.71-82

86. Friedmann, T., ed., 1999, *The development of human gene therapy.* Cold Spring Harbor Laboratory Press, Cold Spring Harbor, New York

87. Reynolds, P.N., Feng, M., and Curiel, D.T., 1999, Chimeric viral vectors - the best of both worlds? *Molec. Med. Today* **5**:25-31

88. Mahato, R.I., Smith, L.C., and Rolland, A., 1999, Pharmaceutical perspectives of nonviral gene therapy. *Adv. Genetics* **41**:95-156

89. Souza, D.W. and Armentano, D., 1999, Novel cloning method for recombinant adenovirus construct in Escherichia coli. *BioTechniques* **26**:502-508

90. Zanta, M.A., Belguise-Valladier, P., Behr, J.-P., 1999, Gene delivery: A single nuclear localization signal peptide is sufficient to carry DNA to the cell nucleus. *Proc. Natl. Acad. Sci. USA* **96**:91-96

91. Gorschluter, M., Schakowski, F., Risse, F., Junghans, C., Schroff, M., Kleinschmidt, R., Sauerbruch, T., Wittig, B., Schmidt-Wolf, I, A novel type of minimal size non-viral vector with improved safety properties for clinical trials and enhances transgene expression. http://www.mologen.com/english/mologen.html

92. Nabel, G.J., 1999, Development of optimized vector for gene therapy. *Proc. Natl. Acad. Sci. USA* **96**:324-326

93. Scherman, D., Bessodes, M., Cameron, B., Herscovici, J., Hofland, H., Pitard, B., Soubrier, F., Wils, P., and Crouzet J., 1998, Application of lipids and plasmid design for gene delivery to mammalian cells. *Curr. Opin. Biotechnol.* **9**:480-485

94. Cara, A., Rybak, S.M., Newton, D.L., Crowley, R., Rottschafer, S.E., Reitz, M.S., and Gusella, G.L., 1998, Inhibition of HIV-1 replication by combined expression of gag dominant negative mutant and a human ribonuclease in a tightly controlled HIV-1 inducible vector. *Gene Therapy* **5**:65-75

95. Vacik, J., Dean, B.S., Zimmer W.E., and Dean D.A., 1999, Cell-specific nuclear import of plasmid DNA. *Gene Ther.* **6**:1006-1014

96. Hobart, P.M., Margalith, M., Parker, S.E., and Khatihi S., inventors; Vical Incorporated, assignee; 1997, Plasmids suitable for IL-2 expression. US patent #5641665

97. Okada, H., Miyamura, K., Itoh, T., Hagiwara, M., Wakabayashi, T., Mizuno, M., Colosi, P., Kurtzman, G., and Yoshida, J., 1996, Gene therapy against an experimental glioma using adeno-associated virus vectors. *Gene Ther.* **3**:957-964

98. Nabel, G.J., Nabel, E.G., Lew, D., and Marquet M., inventors; Vical Incorporated, assignee. 1999, Plasmids suitable for gene therapy. US patent #5910488

99. Simmons, A.D. and Lovett, M., 1999, Direct cDNA selection using large genomic DNA template. *Methods Enzymol.* **303**:111-126

100. Marsh, E., 1995, Stringent quality control of Stratagene restriction endonucleases, *Strategies in Molecular Biology* **8**:32-33

101. Evtushenko, V., 1994, Cloning in λ ZAPII/pBluescript. I. Construction of cDNA libraries. *Molecular Biology* **28**:510-516

102. Tait, R.C. and Horton, R.M., 1998, An introduction to genetic engineering with PCR. In *Genetic Engineering with PCR* (Tait, R.C. and Horton, R.M., eds.), Vol. 5: Current Innovations in Molecular Biology series, Horizon Scientific Press, pp.13-23

103. White, B.A., ed., 1993, *Methods in Molecular Biology, Vol. 15: PCR Protocols: Current Methods and Applications*, Humana Press, Totowa

104. McPherson, M.J., Hames, B.D., and Taylor, G.R., eds., 1995, *PCR 2: A Practical Approach*, IRL Press, Oxford, New York, Tokyo

105. Tait, R.C. and Horton, R.M., eds., 1998, *Genetic Engineering with PCR, Vol. 5*: Current Innovations in Molecular Biology series, Horizon Scientific Press

106. Parker, S.E., Borellini, F., Wenk, M.L., Hobart, P., Hoffman, S.L., Hedstom, R., Le, T., and Norman, J.A., 1999, Plasmid DNA malaria vaccine: Tissue distribution and safety studies in mice and rabbit. *Hum. Gene Ther.* **10**:741-758

107. Meyer, K.B., Thompson, M.M., Levy, M.J., Barron, L.G., and Szoka, F.C., 1995, Intratracheal gene delivery to the mouse airway: characterization of plasmid DNA expression and pharmacokinetics. *Gene Ther.* **2**:450-460

108. Long, Z., Lu, P., Grooms, T., Mychkovsky, I., Westley, T., Fitzgerald, T., Sharma-Chibber, S., Shand, N., McGarrity, G., and Otto, E., 1999, Molecular evaluation of biopsy and autopsy specimens from patients receiving *in vivo* retroviral gene therapy. *Hum. Gene Ther.* **10**:733-740

109. Welsh, M.J. and Zabner, J., 1999, Clinical Protocol. Cationic lipid mediated gene transfer of CFTR: Safety of a single administration to the nasal epithelia. *Hum. Gene Ther.* **10**:1559-1572

110. Lahijani, R., Duhon, M., Lusby E., Betita, H., and Marquet, M., 1998, Quantitation of host cell DNA contaminate in pharmaceutical-grade plasmid DNA using competitive polymerase chain reaction and enzyme-linked immunosorbent assay. *Hum. Gene Ther.* **9**:1173-1180].

111. Life Technologies, 2000, TECH-ONLINE. FAQs for Restriction Endonucleases. http://www.lifetech.com

112. Kaufman, D.L. and Evans, G.A., 1990, Restriction endonuclease cleavage at the termini of PCR products. *BioTechniques* **9**:304-306

113. Rudi, K., Fossheim, T., and Jakobsen, K., 1999, Restriction cutting independent method for cloning genomic DNA segments outside the boundaries of known sequences. *BioTechniques* **27**:1770-1777

114. Ilyin, S.E. and Planta-Salaman, C.R., 1999, Probe generation by PCR coupled with ligation. *Nature Biotechnol.* **17**:608-609

115. Seraphin, B. and Kandels-Lewis, S., 1996, An efficient PCR mutagenesis strategy without gel purification step that is amenable for automation. *Nucleic Acids Res.* **24**:3276-3277

116. Nadin-Davis, S.A. and Chang, S.C., 1998, Site-directed mutagenesis of large plasmids. *BioTechniques* **25**:1015-1019

117. Jiang, S.-W., Trujillo, M.A., and Eberhardt N.L., 1996, An efficient method for generation and subcloning of tandemly repeated DNA sequences with defined length, orientation and spacing. *Nucleic Acids Res.* **24**:3278-3279

118. Skalka, A.M. and Goff, S.P., 1993. *Reverse Transcriptase.* Cold Spring Harbor Laboratory Press, Cold Spring Harbor, New York

119. Gubler, U. and Hoffman, B.J., 1983, A simple and very efficient method for generating cDNA libraries. *Gene* **25**:263-269

120. Blau, H.M. and Springer, M.L., 1995, Gene therapy - a novel form of drug delivery. *New Eng. J. Med.* **333**:1204-1207

121. Spooner, R.A., Deonarain, M.P., and Epenetos, A.A., 1995, DNA vaccination for cancer treatment. *Gene Ther.* **2**:173-180

122. Pardoll, D. and Nabel, G.J., 1999, Cancer Immunotherapy. In *The development of human gene therapy* (T. Friedmann, ed.), Cold Spring Harbor Laboratory Press, Cold Spring Harbor, New York, pp.427-457

123. Bodescot, M. and Brison, O., 1994, Efficient second-strand synthesis using T7 DNA polymerase. *DNA Cell Biol.* **9**:977-985

124. Chenchik, A., Diatchenko, L., Moqadam, F., Tarabykin, V., Lukyanov, S., and Siebert, P.D., 1996, Full-length cDNA cloning and determination of mRNA 5' and 3' ends by amplification of adaptor-ligated cDNA. *BioTechniques* **21**:526-534

125. Bodescot, M. and Brison, O., 1997, Analysis of the size distribution of first-strand cDNA molecules. *BioTechniques* **22**:1119-1126

126. Prashar Y. and Weissman S.M. READS: A method for display of 3'-end fragments of restriction enzyme-digested cDNAs for analysis of differential gene expression, 1999, *Methods Enzymol.* **303**:258-272

127. Marshall, A. and Hodgson, J., 1998, DNA chips: An array of possibilities. *Nature Biotechnol.* **16**:27-32

128. Lander, E.S., 1999, Array of hope. *Nature Genet. Suppl.* **21**:3-4.

129. Southern, E.M., 1975, Detection of specific sequences among DNA fragments separated by gel electrophoresis. *J. Mol. Biol.* **98**:503-517

130. Mikulits, W., Dolznig, H., Hofbauer, R., and Mullner, E.W., 1999, Reverse strand priming: A versatile cDNA radiolabeling method for differential hybridization on nucleic acid arrays. *BioTechniques* **26**:846-850

131. Battaglia, C., Salani, G., Consolandi, C., Bernardi, L.R., De Bellis, G., Analysis of DNA microarrays by non-destructive fluorescent staining using SYBR® Green II. *BioTechniques* **29**:78-81

132. Ferguson, J.A., Boles, T.C., Adams, C.P., and Walt, D.R., 1996, A fiber-optic DNA biosensor microarray for the analysis of gene expression. *Nature Biotechnol.* **14**:1681-1684

133. Eisen, M.B. and Brown, P.O., 1999, DNA arrays for analysis of gene expression. *Methods Enzymol.* **303**:179-205

134. Battaglia, C., Salani, G., Consolandi, C., Bernardi, L.R., De Bellis, G., Eisen, M.B. and Brown, P.O., 1999, DNA Analysis, Microcon® concentrator[133]

135. Trenkle, T., Mathieu-Daude, F., Welsh, J., and McClelland, M., 1999, Reduced complexity probes for DNA arrays. *Methods Enzymol.* **303**:380-392

136. Altermann, E., Klein, J.R., and Henrich, B., 1999, Synthesis and automated detection of fluorescently labelled primer extension products. *BioTechniques* **26**:96-101

137. Henegariu, O., Bray-Ward, P., and Ward, D.C., 2000, Custom fluorescent-nucleotide synthesis as an alternative method for nucleic acid labelling. *Nature Biotechnol.* **18**:345-348

138. Fartmann, B., MWG-Biotech, cited in Altermann, E., Klein, J.R., and Henrich, B., 1999, Synthesis and automated detection of fluorescently labelled primer extension products. *BioTechniques* **26**:96-101

139. Tillett, D. and Neilan, B.A. 1999, n-Butanol purification of dye terminator sequencing reaction. *BioTechniques* **26**:606-610

140. CLONTECH Laboratories, 2000, Atlas™ Glass Fluorescent Labelling Kit. CLONTECHniques XV (2):3

141. CLONTECH Laboratories, 2000, Atlas™ Glass Fluorescent Labelling Kit. User Manual PT3452-1

142. Kunitz, M., 1950, Crystalline deoxyribonuclease I: Isolation and general properties - Spectrophotometric method for the measurement of deoxyribonuclease activity. *J. Gen. Physiol.* **33**:349-362

143. Huang, Z., Fasco, M.J., and Kaminsky, L.S., 1996, Optimization of DNase I removal of contaminating DNA from RNA for use in quantitative RNA-PCR. *BioTechniques* **20**:1012-1020

144. Hanaki, K., Nakatake, H., Yamamoto, K., Odawara, T., and Yoshikura H., 2000, DNase I activity retained after heat inactivation in standard buffers. *BioTechniques* **29**:38-42

145. Birren, B., Green, E.D., Klapholz, S., Myers, R.M., and Roskams J., eds., 1997, *Genome Analysis. A Laboratory Manual.* Vol. 1. Analyzing DNA. Cold Spring Harbor Laboratory Press, pp.486-431

146. Evtushenko, V.I., 1999, Rapid removal of nucleases using QuickClean™ Resin. CLONTECHniques, XIV(2):30]

147. Dr. Oksana Barabitskaya, Institute of Human Virology, Baltimore, USA

148. Krieg, P.A. and Johnson, A.D., 1996, *In vitro* synthesis of mRNA. In *A Laboratory Guide to RNA. Isolation, Analysis, and Synthesis* (Ed. By P.A. Krieg), Wiley-Liss, Inc., New York, pp.141-153

149. Ghosh S., 1996, A novel ligation mediated-PCR based strategy for construction of subtraction libraries from limiting amounts of mRNA. *Nucleic Acids Res.* **24**:795-796

150. Dostal D.E., Kempinski A.M., Motel T.J., Baker K.M., 1998, Absolute quantification of messenger RNA multiplex RT-PCR. In *Gene Cloning and Analysis by RT-PCR* (P.D. Siebert and J.W. Larrick, eds.), BioTechniques Molecular Laboratory Methods Series, pp.71-90

151. Tsai S.-J., Wiltbank M.C., 1998, Standard curve quantitative competitive (SC-QC-RT-PCR): A simple method to quantify absolute concentration of mRNA from limited amounts of sample. In *Gene Cloning and Analysis by RT-PCR* (P.D. Siebert and J.W. Larrick, eds.), BioTechniques Molecular Laboratory Methods Series, pp.91-101

152. Leitner W.W., Ying H., Restifo N.P., 1999, DNA and RNA-based vaccines: principles, progress and prospects. *Vaccine* **18**:765-777

153. Kariko K., Kuo A., Barnathan E.S., 1999, Overexpression of urokinase receptor in mammalian cells following administration of the in vitro transcribed encoding mRNA. *Gene Therapy* **6**:1092-1100

154. Boczkowski D., Nair S.K., Snyder D., Gilboa E., 1996, Dendritic cells pulsed with RNA are potent antigen-presenting cells in vitro and in vivo. *J. Exp. Med.* **184**:465-472

155. Mandl C.W., Aberle J.H., Aberle S.W., Holzmann H., Allison S., Heinz F.X., 1998, In vitro-synthesized infectious RNA as an attenuated live vaccine in a flavivirus model. *Nature Med.* **4**:1438-1440

156. Dr. Christian Mandl, Institute of Virology, Vienna, Austria, Personal communication

157. Yarovoi S.V., Mouawad R., Colbere-Garapin F., Khayat D., Rixe D., 1996, In vitro sensitization of the B16 murine melanoma cells to ganciclovir by different RNA and plasmid DNA constructions encoding HSVtk. *Gene Therapy* **3**:913-918

158. Thierry A.R., Rabinovitch P., Peng B., Mahan L.C., Bryan J.L., Gallo R.C., 1997, Characterization of liposome-mediated gene delivery: expression, stability and pharmacokinetics of plasmid DNA. *Gene Therapy* **4**:226-237

159. Ferrari M.E., Nguyen C.M., Zelphati O., Tsai Y., Felgner P.L., 1998, Analytical methods for the characterization of cationic lipid-nucleic acid complexes. *Hum. Gene Ther.* **9**:341-351

160. Zelphati O., Liang X., Hobart P., Felgner P.L., 1999, Gene chemistry: Functionally and conformationally intact fluorescent plasmid DNA. *Hum. Gene Ther.* **10**:15-24

161. Yakubovitch E., Chavchich M., Yarmolovitch M., Volkov K., Evtushenko V., 1994, Applications of solid phase matrix "BlueSorb" for fast phenol-free extraction of genetic engineering enzymes [abstracts]. International Conference "Biotechnology St. Petersburg' 94. St. Petersburg, Russia, September 21-23, 1994, pp.84-85

162. Chirikjian J.G., Papas T.S., eds., 1981, *Gene Amplification and Analysis.* Vol. 2. *Structural Analysis of Nucleic Acids*, Elsevier/North-Holland, New York, Amsterdam, Oxford

163. Linn S.M., Lloyd R.S., Roberts R.J., eds., 1993, *Nucleases, Second Edition.* Cold Spring Harbor Laboratory Press. New York

164. Straus N.A., Zagursky R.J., 1991, In vitro production of large single-stranded templates for DNA sequencing. *BioTechniques* **10**:376-384

165. Henikoff, S., 1987, Unidirectional digestion with exonuclease III in DNA sequence analysis. *Methods Enzymol.* **155**:156-165

166. Deng W.P. and Henikoloff J.A., 1992, Site-directed mutagenesis of virtually any plasmid by eliminating a unique site. *Anal. Biochem.* **200**:81-88

167. Hsiao, K.-C., 1993, Exonuclease III induced ligase-free directional subcloning of PCR products. *Nucleic Acids Res.* **21**:5528-5529

168. Li C. and Evans R.M., 1997, Ligation independent cloning irrespective of restriction site compatibility. *Nucleic Acids Res.* **25**:4165-4166

169. Ostermeier M., Shim J.H., Benkovic S.J., 1999, A combinatorial approach to hybrid enzymes independent of DNA homology. *Nature Biotech.* **17**:1205-1209

170. Birren B., Green E.D., Klapholz S., Myers R.M., Roskams J., eds., 1997, *Genome Analysis. A Laboratory Manual.* Vol. 1. Analyzing DNA. Cold Spring Harbor Laboratory Press, pp.428-431

171. New England Biolabs, 1995, Mung Bean Nuclease, Technical Bulletin #250

172. Chattopadhyay D., Raha T., Chattopadhyay D., 1997, PCR mutagenesis: treatment of the megaprimer with mung bean nuclease improves yield. *BioTechniques* **22**:1054-1056

173. Yang M., Sytkowski A.J., 1998, Linker-capture subtraction. In *Gene Cloning and Analysis by RT-PCR* (P.D. Siebert and J.W. Larrick, eds.), BioTechniques Molecular Laboratory Methods Series, pp.203-211

174. Rushizky G.W., 1981, S_1 nuclease of Aspergillus oryzae. In *Gene Amplification and Analysis. Vol. 2. Structural Analysis of Nucleic Acids* (Chirikjian J.G., Papas T.S., eds.), Elsevier/North-Holland, New York, Amsterdam, Oxford, pp.205-215

175. Kacharmina J.E., Crino P.B., Eberwine J., 1999, Preparation of cDNA from single cells and subcellular regions. *Methods Enzymol.* **303**:3-18

176. Flouriot G., P. Nestor, M.-R. Kenealy, C. Pope, Gannon F., 1996, An S1 nuclease mapping method for detection of low abundance transcripts. *Anal. Biochem.* **237**:159-161

177. Garner, H.R., Armstrong, B., and Kramarsky, D.A., 1992, High-throughput DNA preparation system. *GATA* **9**:134-139

178. Marziali, A., Willis, T.D., Federspiel, N.A., and Davis, R. W. 1999, An automated sample preparation system for large-scale DNA sequencing. *Genome Res.* **9**:457-462

179. Jacotot, L., Madarapu, S., and Pytlik, M., 2000, Automating the normalization of DNA samples. *Gen. Eng. News* **20**:32

180. Broach, J.R., and Thorner, J. High-throughput screening for drug discovery. *Nature* **384** (Suppl.):14-16

181. Wattler S., Kelly M., and Nehls M., 1999, Construction of gene targeting vectors from λKOS genomic libraries. *BioTechniques* **26**:1150-1160

182. Wan, J.S., Sharp, S.J., Poirier, G.M.-C., Wagaman, P.C., Chambers, J., Pyati, J., Horn, Y.-L., Galindo, J.E., Huvar, A., Peterson, P.A., Jackson, M.R., and Erlander, M.G., Cloning differentially expressed mRNAs. *Nature Biotech* **14**:1685-1691

183. Ross, D.T., Scherf, U., Eisen, M.B., Perou, C.M., Rees, C., Spellman, P., Iyer, V., Jeffrey S.S., Van de Rijn, M., Waltham, M., et al., 2000, Systematic variation in gene expression pattern in human cancer cell lines. *Nature Genet.* **24**:227-235

184. Horn, N.A., Meek, J.A., Budahazi, G., and Marquet, M., 1995, Cancer gene therapy using plasmid DNA: Purification of DNA for human clinical trials. *Hum. Gene Ther.* **6**:565-573

185. Briggs, J. and Panfili, P.R., 1991, Quantification of DNA and protein impurities in biopharmaceuticals. *Anal. Chem.* **63**:850-859

186. Kuroda, S., Itoh, Y., Miyazaki, T., and Fujisawa, Y., 1988, A supersensitive dot-hybridization method: rapid and quantitative detection of host-derived DNA in recombinant products at the one picogram level. *Biochem. Biophys. Res. Commun.* **152**:9-14

187. Pepin R.A., Lukas D.J., Lang R.B., Lee N., Liao M.-J., Testa D., 1990, Detection of picogram amounts of nucleic acid by blot hybridization. *BioTechniques* **8**:628-632

188. Lahijani, R., Duhon, M., Lusby, E., Betita, H., and Marquet, M., 1998, Quantification of host cell DNA contaminate in pharmaceutical-grade plasmid DNA using competitive polymerase chain reaction and enzyme-linked immunosorbent assay. *Hum. Gene Ther.* **9**:1173-1180]

189. Bolger, R., Lenoch, F., Allen, E., Meiklejohn, B., and Burke, T., 1997, Fluorescent dye assay for detection of DNA in recombinant protein products. *BioTechniques* **23**:532-537

190. Haugland, R. P., 1996. *Handbook of fluorescent probes and research chemicals.* 6th ed. Molecular probes, Inc. Eugene, OR, USA

191. Benore-Parsons, M. and Ayoub, M.A., 1977, Presence of RNase A causes aberrant DNA band shifts. *BioTechniques* **23**:128-131

192. Bogetto, P., Wainder, L., and Anderson, H., 2000, Helpful tips for PCR. *FOCUS* **2**:12

193. Dr. Eugene Sokolov, Institute of Zoology, St. Petersburg, Russia, personal communication

194. Sokolov, E.P., 1998, Analysis of genomic repetitive DNA sequences in closely related bird species. *Molecular Genetics* **34**: 871-874

Gene Transfer into Eukaryotic Cells

MARTIN WEBER
QIAGEN GmbH, Hilden, Germany

1.   INTRODUCTION

1.1   Gene transfer *in vitro* and *in vivo*

Gene transfer into eukaryotic cells has become a powerful tool in the study and control of gene expression, for example in biochemical characterisation, mutational analysis, or investigation of the effects of regulatory elements or cell growth behaviour. Gene transfer can be performed either into cultured cells *in vitro*, or *in vivo*, directly into cells living in their natural environment in the body.

In *in vitro* cell culture systems, the most common aim of a gene transfer, "transfection", is to study the effects of the newly introduced gene on the cell metabolism, signalling pathways, or other biological features within the particular cell type. The methods used to perform such *in vitro* transfections aim to introduce the DNA of interest in a highly efficient manner into a large percentage of the target cells, with high reproducibility. Furthermore, it is desirable that these methods only alter cell behaviour minimally, and that changes in the cell metabolism, signalling or cell viability are not caused by the transfection method itself. Another important point in the design of gene transfer vectors for *in vitro* systems, is their ease of use. This point is especially relevant for high throughput systems, which are becoming increasingly important for the study of phenotypes of large banks of sequenced genes and their mutants. In high throughput systems, each

135

additional step which has to be performed during the course of transfection results in a large increase in workload.

In comparison to *in vitro* gene transfer, *in vivo* gene transfer involves introducing genes into specific cells and tissues in the body so that they provide a complementary action. For example, to complement for missing genes in individuals with inherited disorders, or to code for proteins which can act as antigens in tissues, as is the case in genetic vaccination. Therefore, in addition to the above mentioned features for *in vitro* transfection reagents, the shuttles for *in vivo* gene transfer also have to fulfil criteria, such as bioavailability, targeting features and pharmacological qualities.

In contrast to *in vitro* transfection, vectors for *in vivo* gene transfer have to be designed in a specific way, so that the genetic message reaches the target area within the body, and does not end up in an undesired tissue. Additionally, vectors should be designed to prevent the genetic message being masked during transport through the body, as transfection potency would be lost if this occurred. Irrespective of whether vectors are used in *in vitro* or *in vivo* systems, the carried genes of interest must be transcription competent when they reach the nucleus within the cell, as this is the site where DNA is transcribed to RNA.

During its way through the cell, the DNA has to cross certain barriers, such as the cell membrane, the endosomal membrane and the nuclear envelope. If, as with most methods, the DNA is intermediately present in the endosomal compartment, it also has to withstand the acidic and nuclease containing environment within this compartment to remain viable for transcription. In designing vectors for gene transfer, all of the general points have to be taken into consideration together with the specific points mentioned above.

This chapter describes the basic principles of the most commonly used gene transfer methods for *in vitro* and *in vivo* applications, and summarises the advantages and disadvantages of each of these technologies as well as the main areas of applications.

In general, two different approaches for the design of vectors for gene transfer currently exist. One of these approaches is based on naturally occurring viruses. Using this approach, gene transfer vectors are designed on the basis of a particular virus where the unneeded and dangerous genetic information of the virus has been deleted, and the genetic information which is to be introduced into the cell is incorporated into the virus structure to form the vector. Such vectors are mainly used in *in vivo* systems, and not very frequently in *in vitro* systems. Viral vectors will only be mentioned briefly in this chapter, as other chapters of this book discuss these systems in more detail.

The second approach for designing gene transfer vectors is based on nonviral systems, and this approach is used equally as often in *in vivo* as in *in vitro* systems. The basic principles of a nonviral vector approach will be the focus of this chapter. Physical as well as chemical nonviral gene transfer vectors will be reviewed side by side, and classified as either "classic" or "advanced" methods based on there actual scientific relevance.

1.2 The efficiency of gene transfer and its side effects are strongly influenced by DNA quality

Gene transfer efficiency as well as unwanted effects on the cell or tissue of interest, are not only dependent on the transfection method applied but also on the purity of the basic reagents, which are combined at the beginning of a transfection experiment to form the transfection complex. Therefore, it is agreed that transfection reagents and buffers used for gene transfer experiments must pass stringent quality controls at the place of manufacture. Impurities in these reagents are widely accepted as being responsible for poor transfection results and nonreproducibility in an experimental set-up.

On the contrary, substances present in the plasmid DNA preparation have only recently come into focus as a source of contamination for transfection, despite plasmid DNA being one of the main components in a transfection reaction away from the starting point. Reports on the influence of contaminating lipopolysaccharides on both cell viability[1,2] and transfection efficiency[3] have now brought broader attention to this point.

1.3 The efficiency of gene transfer *in vitro* is highly dependent on the quality of cell culture

Besides the quality of the DNA and the transfection method used, a third main factor of successful *in vitro* transfection is the quality of the cell culture. In particular, a higher passage number of established cell lines, in particular, can lead to altered behaviour of the cells under investigation, and subsequently, to a changed transfection behaviour.

Another main reason for altered cell behaviour is the contamination of cell cultures *in vitro*. Contaminating substances which are not easily visible to the eye are especially dangerous. For example, contamination of cell cultures with mycoplasmas can lead to undetected and therefore "chronic" changes in transfectability, as well as a modified behavioural response to the trigger DNA by the cells under investigation.

2.   CLASSICAL NONVIRAL TRANSFECTION TECHNOLOGIES

In the following section the "classical transfection technologies", DEAE dextran transfection, calcium phosphate based transfection and electroporation, are summarised. These classical methods are still widely used for transfections *in vitro* in established, robust, cell lines.

2.1   DEAE dextran mediated transfection

In vitro gene transfer into cultured eukaryotic cells with the help of diethylaminoethyl (DEAE)-dextran was first described in 1965[4]. It is one of the oldest methods available for the transfection of eukaryotic cells.

In this method, positively charged DEAE-dextran molecules interact electrostatically with the negatively charged phosphate backbone of the nucleic acid. As the positively charged DEAE-dextran is used in excess of the negatively charged DNA, the complexes retain an overall positive charge. They thereby attach to the negatively charged surfaces of eukaryotic cells and are subsequently taken up by unspecific endocytosis.

The advantage of this technique is its relative simplicity and the fact that in a large number of laboratories it is very well established. Disadvantages of this method include the strong cytotoxic effects of the DEAE-dextran and also that serum in the culture medium must be temporarily reduced during the transfection procedure. Additionally, the DEAE-dextran method is only preferable for use with transient transfections, and is not recommended for stable transfections[5].

The basic steps of DEAE-dextran mediated gene transfer are summarised in figure 1. In some cases an additional protocol step, a glycerol shock[6] or DMSO shock[6], is performed during DEAE-dextran transfections. This exposure to either glycerol or DMSO for a few minutes increases transfection efficiencies in some cell lines *in vitro,* but on the other hand also further increases the high toxicity of this method.

2.2   Calcium phosphate mediated transfection

The second classical method, calcium phosphate transfection, was first used in 1973 to introduce adenovirus DNA into mammalian cells[7]. The basic principle of this technique is to mix DNA in a phosphate buffer with calcium chloride. The resulting calcium phosphate-DNA complexes then adhere to the cell membrane of cells, and enter the cytoplasm by endocytosis. Advantages of the calcium phosphate-based transfection are its high grade of establishment throughout most laboratories, and a much

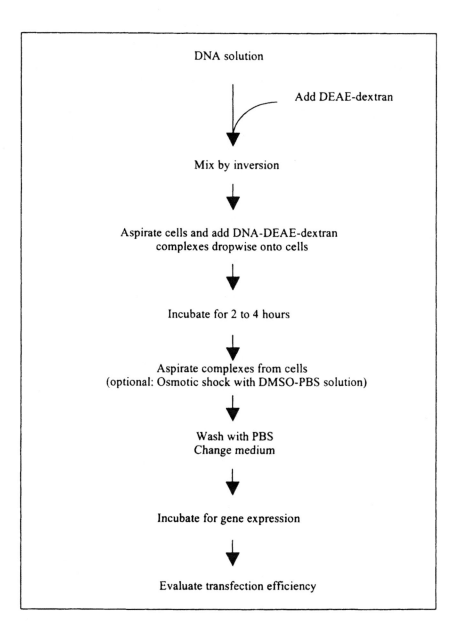

Figure 1. DEAE dextran method

higher suitability for stable transfections than the DEAE-dextran method. A common disadvantage of this technique is low reproducibility, which is mainly due to variation in transfection-complex size and shape (caused by inconsistent handling and also dependent on the pH-values of the different buffers). A further drawback of the calcium-phosphate method is that some cell types, including primary cells, resist this form of DNA transfer and a generally high toxicity level results. The basic steps of calcium phosphate mediated transfection are shown in figure 2.

In addition to *in vitro* gene transfer by calcium phosphate, a similar technique based on strontium phosphate has also been described[8].

2.3   Electroporation

The use of high-voltage pulses to introduce DNA into cultured cells was first described in 1982[9,10]. Cells in a suitable cuvette are subjected to a short, high-voltage pulse that causes the membrane potential of the cells to break down. As a result, transient pores are formed through which macromolecules such as DNA can enter. The main advantage of electroporation is its suitability for transient and stable transfection of most cell types. One disadvantage is the fact that approximately 5-fold greater quantities of DNA and cells are needed than in either DEAE-dextran or calcium-phosphate methods. A major drawback of electroporation is the high cell mortality, that can result in up to 50-70% cell death. Moreover, the optimal settings of voltage, capacitance, pulse length, and gap width are cell-type dependent, and it is therefore necessary to perform a high number of optimisation experiments before the optimal settings for a specific cell line are found. The basic steps of gene transfer by electroporation are summarised in figure 3. Interestingly, some recent reports also describe the use of electroporation in *in vivo* gene transfer systems[11-13].

3.   VIRUS-BASED GENE TRANSFER

Viruses are natural carriers of nucleic acids and can transport their genetic materials into eukaryotic cells for replication. Therefore, the cloning of desired nucleic acids into viral vectors, the packaging of these vectors into viral envelopes, and the transport into eukaryotic cells via these vectors seems to be an optimal system.

The advantage of such virus-based gene transfer technologies are clearly their efficiency and the high percentage of the target cells which are reached by such vectors.

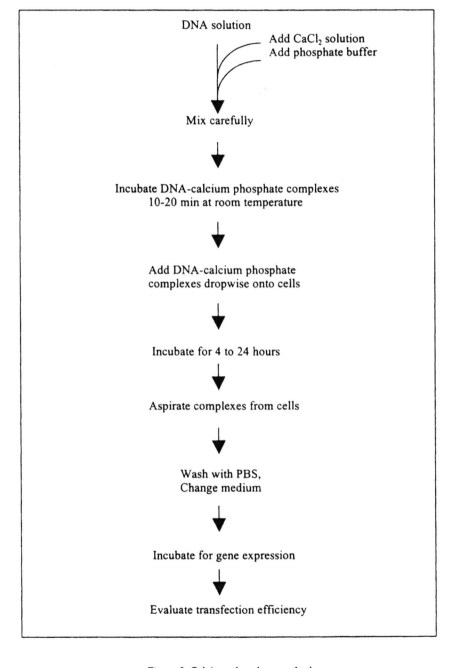

Figure 2. Calcium phosphate method

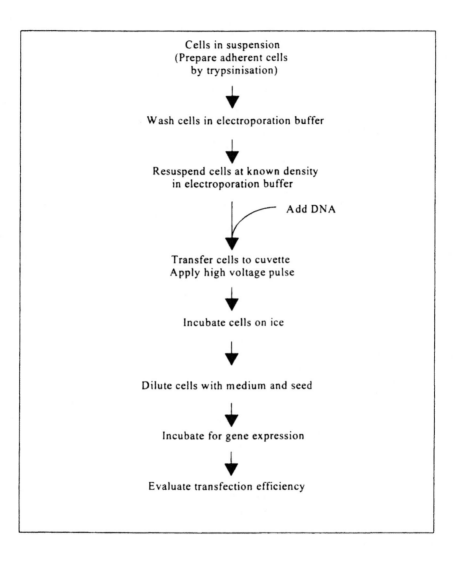

Figure 3. Electroporation

Disadvantages in *in vitro* systems include the complicated handling and the potentially biohazardous procedure. Therefore, virus-based systems are not commonly used for *in vitro* cell culture applications, however relatively frequently for *in vivo* applications.

Disadvantages in *in vivo* systems are the general risk of recombination and complementation with cell coded sequences and subsequent generation of hazardous viral material[14]. Furthermore, many viral systems are limited with respect to the length of the nucleic acid that can be introduced into their genome. Immunogenicity and toxicity of such systems have also been reported.

3.1 Retrovirally mediated gene transfer

Retroviruses can be used as vectors to transfer a nonviral gene into mitotic cells[15]. Retroviruses can precisely integrate their own genes into specific sites in the host genome, an ability which can be exploited to modify cells. Retroviruses are only used in *in vivo* applications, and carry the inherent danger of recombination and complementation with cell coded pseudoretroviral sequences, thereby creating infectious species. Moreover, gene transfer by retroviral vectors is limited by the fact that these vectors attach only to, and subsequently are taken up by, cells carrying the appropriate cell surface receptors. Retroviral vectors are described in detail in other chapters of this volume.

3.2 Adenovirus and AAV mediated gene transfer

Adenoviral as well as AAV (Adeno-Associated Virus) based vectors are used for *in vivo* gene transfer applications only.

In general adenoviral vectors show the same kind of limitations as retroviral vectors. Again specific cell surface receptors on the target cells are necessary for attachment of the virus and subsequent uptake of the nucleic acids. Again considerations with respect to biohazardous effects have to be taken into account. In addition the possibility exists, that after repeated *in vivo* treatment with adenoviral vectors immunological reactions occur in the body, which can lead to massive immune reactions of the body[16].

Adenoviral vectors show, on the other hand, a high gene transfer efficiency in permissive cells.

Gene transfer with adenoviruses is described in detail in chapter 3 of this book.

Gene transfer vectors based on AAV have demonstrated prolonged gene expression in the absence of toxicity or immunological clearance. The efficiency of AAV-mediated gene delivery, however, can vary dramatically dependent of the target cell type and in some instances target cells are not permissive to vector infection at all.

4.   ADVANCED NONVIRAL TRANSFECTION TECHNOLOGIES

Advanced nonviral transfection technologies are used for both *in vitro* and *in vivo* applications.

Compared to virus-based gene transfer technologies *in vivo*, they show the disadvantage of generally lower gene transfer efficiency. Advantages are, on the other hand, no risk with respect to the generation of infectious material by recombination events, as well as a lower risk for massive immune reactions of the body. In addition there is no limitation of the length of the DNA used for the gene transfer.

Compared to classical transfection technologies *in vitro* many advanced nonviral transfection technologies have the advantages of generally higher transfection efficiencies, a more easy and reproducible handling, a much higher reproducibility as well as lower toxicity within cultured cells.

4.1   Biolistic particle delivery

This technique involves precipitating DNA onto heavy metal (gold or tungsten) microprojectiles, which normally have a diameter of 1 to 5 μm and projecting the coated particles into cells using a ballistic device. The high velocity reached is sufficient to also penetrate cell walls e.g. in plant-cells. Once inside the cell the DNA is slowly released from the particles and the genes subsequently expressed. Biolistic bombardment can be used in *in vivo* (gene gun)[17] or *in vitro* applications like in human epithelial, fibroblast, and lymphocyte cell lines, in plant cells, as well as in primary cells[18-20].

One advantage of the method is, that DNA can be introduced into all cell types, as the method only relies on physical parameters and not on biological procedures for uptake. Disadvantages of the method are, that it is relatively labour intensive, and shows a relatively high toxicity.

A similar method is transfection by cell wounding, which was first described in 1987[21]. Here uncoated glass particles with a diameter of 75 to 300 μm are used to cause transient wounding of the target cells, which had been covered before by a thin film of DNA solution. The DNA can be taken up by the cells through this transient wounds.

One advantage of the method again is its applicability to a broad range of cells. On the other hand the method is relatively unreproducible, as the glass beads are sprinkled onto the cells by hand and again shows relatively high toxicity.

4.2 Microinjection

Gene delivery by direct microinjection of naked DNA into cell nuclei leads to expression and occasional chromosomal integration, resulting in recombinant genes. This method is the most direct approach of gene delivery, as the gene of interest is directly delivered to the place where it acts, namely the nucleus of the cell. The biggest disadvantage of the method is, that the number of cells that can be transfected by microinjection is limited. Therefore the method cannot be used for experimental questions that require a certain expression level of the DNA to get a readable signal. The method, therefore is limited to experimental question on a single cell level or is used to introduce genes into embryos to engineer modified or transgenic animals[22].

4.3 First generation cationic lipid transfection technologies

Cationic liposomes for transfection of eukaryotic cells were first introduced by Felgner and co-workers in 1987[23].

Basically cationic liposomes are formed by sonication, heating or membrane extrusion of a mixture of cationic lipids and neutral helper lipids such as DOPE. DOPE increases transfection efficiency of many cationic lipids by destabilisation of the endosomal membrane[24,25]. Among cationic lipids currently in use are e.g. DOTMA[23], DOGS[26], DOTAP[27], DC-CHOL[28]. During the transfection procedure preformed liposomes are combined with DNA. Based on electrostatic interactions between the positively charged liposome and the negatively charged phosphate backbone of the DNA the transfection complexes[29] forms. The uptake of the liposome-DNA complexes into the cells takes place by endocytosis after the overall positively charged complex has adhered to negatively charged structures on the cell surface. The basic steps of cationic liposome mediated transfection are summarised in figure 4.

Compared to the classical transfection methods, transfection by cationic liposomes often offers higher transfection efficiency and better reproducibility in *in vitro* transfections. In addition, cationic liposome based transfection sometimes shows lower toxicity than the classical transfection technologies and therefore can be used in cell types, that cannot be transfected by the DEAE dextran or calcium-phosphate method[30]. Cationic lipid based transfection can be used both for transient and stable gene

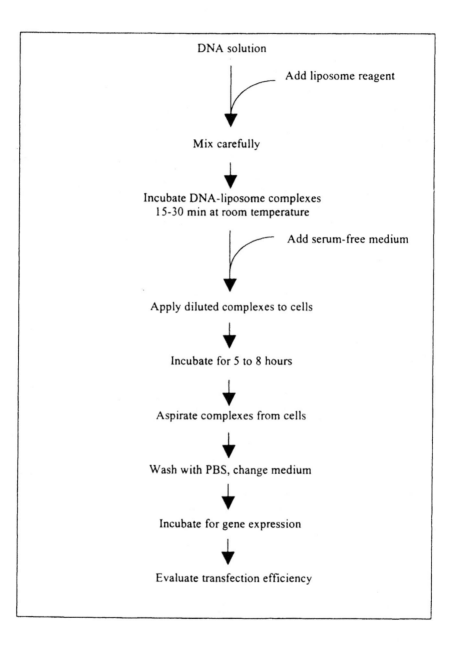

Figure 4. Transfection with cationic liposomes

transfer. A major drawback is the fact, that most of these first generation cationic lipid based methods show decreased transfection efficiencies in the presence of serum. Therefore serum has to be removed during transfection, which increases the toxicity of the method especially with sensitive cells. Another drawback is the fact, that transfection efficiencies with a given lipid vary dramatically from cell line to cell line. Therefore it is often necessary to use different liposomes for different cell lines.

Cationic liposomes also have been applied in a number of *in vivo* studies[31-33]. Here the big advantages in comparison with viral methods is the fact, that all biosafety issues are better fulfilled by liposomes than by viral vectors.

Drawbacks in *in vivo* applications are the low transfection efficiency compared to viral vectors as well as the rapid opsonisation of the complexes within the body fluids by proteins, which render the overall complex untransfectable. Moreover such complexes show an untargeted behaviour within the body, meaning, that gene transfer takes place not only in the target cells but also in undesired tissue. It has been shown, that toxicity is associated to cationic lipid / DNA complexes, when these complexes are delivered systemically[34].

4.4 Second generation cationic lipid transfection technologies

The use of second generation cationic lipid transfection technologies[35] for gene transfer *in vitro* eliminates some of the disadvantages that first generation cationic lipids have.

This new technologies are commonly based on one sort of a cationic lipid, it is not necessary to add a helper lipid like DOPE any more. Furthermore it is not necessary to prepare preformed liposomes out of this lipids. In a watery environment the lipids spontaneously form transfection competent membranous aggregates. As a consequence potential variables that are known to have influences on transfection efficiencies like e.g. DOPE content and the size of the preformed liposomes built during sonication, are eliminated from the production process, which leads to a much higher batch to batch reproducibility.

In addition second generation cationic lipids are not sensitive to the addition of serum any more. Transfections of eukaryotic cells *in vitro* can be performed in the presence of serum, which in some cases dramatically reduces the toxicity of the overall method and makes the method applicable to sensitive cells e.g. several primary cells.

The reduced toxicity of second generation cationic lipids allows *in vitro* transfections without medium change after addition of the DNA-liposome complexes, which simplifies the overall procedure, being especially useful in high throughput transfection applications.

Moreover second generation cationic lipid transfection technologies use basic principles copied from viruses[36] to increase transfection efficiencies. This are e.g. the compaction of DNA to dense complexes[35,37] before the lipids are added, the inclusion of endosomolytic peptides[38] to overcome the endosomal barrier, the inclusion of targeting peptides to the cell nucleus to overcome the nuclear membrane barrier[39], or a combination thereof. The inclusion of such basic viral principles into a synthetic systems leads to much higher transfection efficiencies compared to first generation cationic lipid transfection technologies in *in vitro* cell culture systems.

The described features also make these systems to promising candidates for future *in vivo* applications.

4.5   Cationic polymer based transfection technologies

In addition to transfection by cationic lipid based vectors, efficient *in vitro* transfection with synthetic cationic polymer based vectors has been described.

The most commonly used vectors of this class are polyethylenimmin (PEI)[40], polyamidoamine (PAMAM) dendrimers[41] and fractured PAMAM dendrimers[42,43]. Whereas PEI is either a branched or a linear polymer without defined structure, dendrimers are highly structured synthetic polymers. Partly degraded or fractured dendrimers show a transfection efficiency which is 2 to 3 orders of magnitude higher than the transfection efficiency of intact dendrimers. They still have the size and shape of intact dendrimers, differ however from them by their more flexible structure[43]. The negatively charged DNA is attracted and condensed by these cationic polymers in the first step of transfection. The resulting complexes either build aggregates of 2500 ± 2000nm diameter in the case of intact dendrimers or do not build aggregates and show diameters of 130 ± 30nm in the case of fractured dendrimers[44], which may explain the different transfection behaviour.

Complexes consisting of fractured PAMAM dendrimers or PEI and DNA are then taken up by the cells through endocytosis. Within the endosome the cationic polymers serves as a buffer, thus rendering the pH of the endosome to more neutral conditions thereby inhibiting endosomal nucleases and protecting the DNA from degradation. As release mechanism of the DNA from the endosome to the cytosol an osmotic swelling mechanism is discussed[43]. The basic steps of fractured dendrimer mediated transfection are summarised in figure 5.

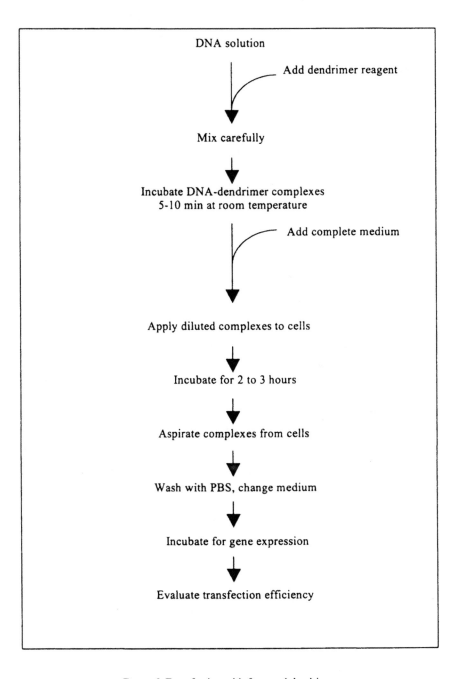

Figure 5. Transfection with fractured dendrimers

Synthetic cationic polymers and especially fractured dendrimers are versatile vectors for *in vitro* and *in vivo* applications. Advantages for *in vitro* applications are high transfection efficiency, low toxicity, ease of use and no inhibition by serum present during transfection.

Fractured PAMAM dendrimers have been used for *ex vivo* and *in vivo* gene transfer studies already[45,46] and maybe of future interest for this applications, too.

In addition there have been some recent reports on the usage of fractured dendrimers in combination with adenoviral vectors, leading to extremely high gene transfer rates *in vitro* and *in vivo*[47,48].

REFERENCES

1. Cotten, M., Baker, A., Saltik, M., Wagner, E., and Buschle, M., 1994, Lipopolysaccharide is a frequent contaminant of plasmid DNA preparations and can be toxic to primary human cells in the presence of adenovirus. *Gene Ther.* **1**:239-246

2. Cotten, M. and Saltik, M., 1997, Intracellular delivery of lipopolysaccharide during DNA transfection activates a lipid A-dependent cell death response that can be prevented by polymyxin B. *Hum Gene Ther.* **8**:555-561

3. Weber, M., Möller, K., Welzeck, M., and Schorr, J., 1995, Effect of lipopolysaccharide on transfection efficiency in eukaryotic cells. *BioTechniques* **19**:930-939

4. Vaheri, A. and Pagano, J.S., 1965, Infectious poliovirus RNA: A sensitive method of assay. *Virology* **27**:434-436

5. Gluzman, Y., 1981, SV40-transformed simian cells support the replication of early SV40 mutants. *Cell* **23**:175-182

6. Lopata, M.A., Cleveland, D.W., and Sollner-Webb, B., 1984, High level transient expression of a chloramphenicol acetyltransferase gene by DEAE-dextran mediated DNA transfection coupled with a dimethylsulfoxid or glycerol shock treatment. *Nucl. Acids Res.* **12**:5707-5717

7. Graham, F.L. and Van der Eb, A.J., 1973, A new technique for the assay of infectivity of human adenovirus 5 DNA. *Virology* **52**:456-467

8. Brash, D.E., Reddel, R.R., Quanrud, M., Yang, K., Farrell, M.P., and Harris, C.C., 1987, Strontium phosphate transfection of human cells in primary culture: stable expression of the Simian Virus 40 Large-T-Antigen gene in primary human bronchial epithelial cells. *Mol. Cell Biol.* **7**:2031-2034

9. Wong, T.K. and Neumann, E., 1982, Electric field mediated gene transfer. *Biochem. Biophys. Res. Commun.* **107**:584-587

10. Neumann, E., Schaefer-Ridder, M., Wang, Y., and Hofschneider, P.H., 1982, Gene transfer into mouse glyoma cells by electroporation in high electric fields. *EMBO J.* 1:841-845

11. Banga, A.K., and Prausnitz, M.R., 1998, Assessing the potential of skin electroporation for the delivery of protein- and gene-based drugs. *Trends Biotechnol.* 16:408-412

12. Oshima, Y., Sakamoto, T., Yamanaka, I., Nishi, T., Ishibashi, T., and Inomata, H., 1998, Targeted gene transfer to corneal endothelium in vivo by electric pulse. *Gene Ther.* 5:1347-1354

13. Aihara, H., and Miyazaki, J., 1998, Gene transfer into muscle by electroporation in vivo. *Nature Biotechnol.* 16:867-870.

14. Ali, M., Lemoine, N.R., and Ring, C.J.A., 1994, The use of DNA virus as vectors for gene therapy. Gene Ther. 1:367-384

15. Grandgenett, D.P. and Mumm, S.R., 1990, Unravelling retrovirus integration. *Cell* 60:3-4

16. Herz, J. and Gerard, R.D., 1993, Adenovirus-mediated low-density accelerated cholesterol clearence in mice. *Proc. Natl. Acad. Sci. USA* 90:2812-2816

17. Qiu, P., Ziegelhoffer, P., Sun, J., and Yang, N.S., 1996, Gene gun delivery of mRNA in situ results in efficient transgene expression and genetic immunisation. *Gene Ther.* 3:262-268

18. Yang, N.S., Burkholder, J., Roberts, B., Martinell, B., and McCabe, D., 1990, In vivo and in vitro gene transfer to mammalian somatic cells by particle bombardment. *Proc. Natl. Acad. Sci. USA* 87:9568-9572

19. Burkholder, J.K., Decker, J., and Yang, N.S., 1993, Rapid transgene expression in lymphocyte and macrophage primary cultures after particle bombardment-mediated gene transfer. *J. Immunol. Methods* 165:149-156

20. Ye, G.N., Daniell, H. and Sanford, J.C., Optimisation of delivery of foreign DNA into higher-plant chloroplasts. *Plant Mol. Biol.* 15:809-819

21. McNeill, P.L. and Warder, E., 1987, Glass beads load macromolecules into living cells. *J. Cell Science* 88:669-678

22. Capecchi, M.R., 1989, The new mouse genetics: altering the genome by gene targeting. *Trends Genet.* 5:70-76

23. Felgner, P.L., Gadek, T.R., Holm, M., Roman, R., Chan, H.W., Wenz, M., Northrop, J.P., Ringold, G.M., and Danielsen, M., 1987, Lipofection: A highly efficient, lipid-mediated DNA-transfection procedure. *Proc. Natl. Acad. Sci. USA* 84: 7413-7417

24. Litzinger, D.C., and Huang. L., 1992, Phosphatidethanolamine liposomes: drug delivery, gene transfer and immunodiagnostic applications. *Biochim. Biophys. Acta* 1113:201-227

25. Farhood, H., Servina, N.S., and Huang, L., 1995, The role of dioleoylphosphatidyl-ethanolamine in cationic liposome mediated gene transfer. *Biochim. Biophys. Acta* **1235**:289-295

26. Behr, J.P., Demeneix, B., Loeffler, J.P., and Mutul, J.P., 1989, Efficient gene transfer into mammalian primary endocrine cells with lipopolyamine-coated DNA. *Proc. Natl. Acad. Sci. USA* **86**:6982-6986

27. Leventis, R., and Silvius, J.R., 1990, Interactions of mammalian cells with lipid dispersions containing novel metabolizable cationic amphiphiles. *Biochem. Biophys. Acta* **1023**:124-132

28. Gao, X., and Huang. L., 1991, A novel cationic liposome reagent for efficient transfection of mammalian cells. *Biophys. Res. Comm.* **179**:280-285

29. Sternberg, B., Sorgi, F.L., and Huang, L., 1994, New structures in complex formation between DNA and cationic liposomes visualized by freeze-fracture electron microscopy. *FEBS Letters* **356**:362-366

30. Loeffler, J.P., Batthel, F., Feltz, P., Behr. J.P., Sassone-Corsi, P., and Feltz, A., 1990, Lipopolyamin-mediated transfection allows gene expression studies in primary neuronal cells. *J. Neurochem.* **54**:1812-1815

31. Caplen, N.J., Alton, E.W.F.W., Middleton, P.G., Dorin, J.R., Stevenson, B.J., Gao, X., Durham, S.R., Jeffery, P.K., Hodson, M.E., Coutelle, C., Huang, L., Porteous, D.J., Williamson, R., and Geddes, D.M., 1995, Liposome-mediated CFTR gene transfer to the nasal epithelium of patients with cystic fibrosis. *Nature Medicine* **1**:39-46

32. Barron, L.G., Gagné, L., and Szoka, F.C., 1999, Lipoplex-mediated gene delivery to the lung occurs within 60 minutes of intravenous administration. *Hum Gene Ther.* **10**:1683-1694

33. Smith, J. Zhang. Y.L., and Niven, R., 1997, Toward development of a non-viral gene-therapeutic. *Adv. Drug Del. Rev.* **26**:135-150

34. Filion, M.C., and Phillips, N.C., 1998, Major limitations in the use of cationic liposomes for DNA delivery. *Int. J. Pjarmaceut.* **162**:159-170

35. The QIAGEN transfection resource book, 1999, QIAGEN, 12-18

36. Weber, M., 2000, Neue Techniken zum Gentransfer in Eukaryontenzellen. *Nachrichten aus der Chemie* **48**:18-23

37. Schwartz, B., Ivanov, M.A., Pitard, B., Escriou, V., Rangara, R., Byk, G., Wils, P., Crouzet, J., and Scherman, D., 1999, Synthetic DNA-compacting peptides derived from human sequence enhance cationic lipid-mediated gene transfer in vitro and in vivo. *Gene Ther.* **6**:282-292

38. Plank, C., Oberhauser, B., Mechtler, K., Koch C., and Wagner, E., 1994, The influence of endosome-disruptive peptides on gene transfer using synthetic virus-like gene transfer systems. *J. Biol. Chem* **269**:12918-12924

39. Subramaniam, A., Ranganathan, P, and Diamond, S.L., 1999, Nuclear targeting peptide scaffolds for lipofection of nondividing mammalian cells. *Nat. Biotechnol.* **17**:873-877

40. Boussif, O., Lezoualc`h, F., Zanta, M.A., Mergny, M.D., Scherman, D., Demeneix, B., and Behr, J.P, 1995, A versatile vector for gene and oligonucleatide transfer into cells in culture and in vivo: Polyethylenimine. *Proc. Natl. Acad. Sci. USA* **92**:7297-7301

41. Kukowska-Latallo, J.F., Bielinska, A.U., Johnson, J., Spindler, R., Tomalia, D.A., and Baker, J.R, 1996, Efficient transfer of genetic material into mammalian cells using Starburst polyamidoamine dendrimers. *Proc. Natl. Acad. Sci. USA* **93**:4897-4902

42. Haensler, J., and Szoka, F.C., 1993, Polyamidoamine cascade polymers mediate efficient transfection of cells in culture. *Bioconjug. Chem.* **4**:372-379

43. Tang, M.X., Redemann, C.T., and Szoka, F.C., 1996, In vitro gene delivery by degraded polyamidoamine dendrimers. *Bioconjug. Chem.* **7**:703-714

44. Tang, M.X., and Szoka, F.C., 1997, The influence of polymer structure on the interactions of cationic polymers with DNA and morphology of the resulting complex. *Gene Ther.* **4**:823-832

45. Hudde, T., Rayner, S.A., Comer, R.M., Weber, M., Isaacs, J.D., Waldmann, H., Larkin, D.F.P., and George, A.J.T, 1999, Activated polyamidoamine dendrimers, a non-viral vector for gene transfer to the corneal endothelium. *Gene Ther.* **6**:939-943

46. Turunen, M.P., Hiltunen, M.O., Ruponen, M., Vikamäki, L., Szoka, F.C., Urtti, A., and Ylä-Herttuala, S., 1999, Efficient adventitial gene delivery to rabbit carotid artery with cationic polymer-plasmid complexes. *Gene Ther.* **6**:6-11

47. Howard, D.S., Rizzierri, D.A., Grimes, B., Upchurch, D., Phillips, G.L., Stewart, A.K., Yannelli, J.R., Jordan, C.T., 1999, Genetic manipulation of primitive leukemic and normal hematopoietic cells using a neovel method of adenovirus-mediated gene transfer. *Leukemia* **13**:1608-1616

48. Dunphy, E.J., Redman, R.A., Herweijer, H., Cripe, T.P., 1999, Reciprocal Enhancement of Gene Transfer by Combinatorial Adenovirus Transduction and Plasmid DNA Transfection in Vitro and in Vivo. *Hum. Gene Ther.* **10**:2407-2417

Plasmid DNA Manufacturing

MARTIN SCHLEEF[1], TORSTEN SCHMIDT[1], KARL FRIEHS[2], and ERWIN FLASCHEL[1,2]
[1]PlasmidFactory GmbH & Co. KG, Meisenstrasse 96, D-33607 Bielefeld, Germany (www.PlasmidFactory.com); [2]University of Bielefeld, Postfach 10 01 31, D-33501 Bielefeld, Germany

1. INTRODUCTION

Substantial progress in medicine and biopharmaceutical research may once enable to cure diseases at the level of the specific gene defects rather than at the conventional phenotype level[1,2]. In addition, the preventive or curative vaccination against pathogenes like bacteria or viruses[3,4,5,6,7]; (review: Ref. 8) turns out to be at least working in animal studies. Recent examples for the application of plasmid DNA show that the process of regeneration after surgery can be supported with gene therapeutic techniques[9,10].

The state of the art of this genetic medical technology with respect to science and application has been reviewed[11,12], and guidelines for clinical DNA-vaccine trials are available[13,14,15,16] (for review see Ref.[17]). The economic perspectives of this fast developing field underlined the need for the development of industrial scale processes for the production of plasmid DNA in an adequate quality. Such a process has to fulfil current GMP guidelines and be acceptable to the regulatory agencies. Genetic vaccination with other than DNA molecules is also investigated for mRNA-liposome complexes. One advantage of this nucleic acid is that there is no risk of genome integration (as known from certain viruses and which is not completely excluded for plasmids).

Traditional methods of purifying plasmids usually require sophisticated methodology, if the DNA is to be separated from contaminating organic components. Plasmids have been produced as a drug substance for some years now[18]. However, the development of such processes is still at the beginning. Compared with proteins these products require less time for development and validation of different individual plasmids. Process design can make use of a technology that is generic for at least plasmids of up to 10 kbp capacity. Actually, a requirement for producing plasmids with higher capacities is observed. This is due to larger sequences within the vectors, which need further modifications within the manufacturing process.

Any process development has to be accompanied by a powerful in-process-control (IPC) system to generate data on the characteristics of the plasmid molecules. In particular, methods are required for obtaining supercoiled covalently closed circular (ccc) plasmid DNA in pure form. Commonly, other plasmid topologies appear as well, which have to be separated from the desired product.

The innovative technology of capillary gel electrophoresis[19,20] (CGE) is a necessary tool to be added to the actual short list of applicable quality control assays for clinical grade plasmid DNA because it is superior to simple agarose gel electrophoresis (AGE) assays. We will summarise the large scale technologies disclosed so far and describe the general steps required in plasmid manufacturing processes. The focus will be on those aspects that turned out to be of significant importance for quality assurance.

2.   PLASMID-DNA PRODUCTION

2.1   Plasmid-DNA vectors

Plasmid DNA has become quite important as a pharmaceutical substance since it was shown that naked DNA injected into muscle tissue was expressed *in vivo*. Thus, the introduction of immunogenic sequences may result in vaccination against an encoded peptide in an animal model[21,22,23].

Plasmids are circular duplex molecules, which may be stably maintained as episomal genetic information within bacteria[24,25,26]. Their size ranges from 1.5 to appr. 120 kbp, and the plasmid copy number per bacterial cell may vary considerably[27]. In the case of small plasmids, copy numbers as high as 1000 (copies per cell) have been reported. The replication (amplification) does not depend on any plasmid-encoded protein, and is not synchronised with the replication of the bacterial host chromosome[27].

The plasmid dimension depends on its form. A linear plasmid of 3 kbp has a molecular mass of 2×10^6 Da and a length of 1 μm[27]. The exact form of a plasmid molecule depends on its integrity. While the ccc-form is in a supercoiled state, the "open circular" (oc) form is in a relaxed or nicked state. In addition, monomeric and multimeric forms are distinguished[19,28,29]. A potential influence of the plasmid form on the efficacy of DNA vaccines is under investigation (M.Schleef, unpublished).

Plasmid DNA may be used as a novel vaccine. The major difference of classical vaccines and DNA vaccines is the opportunity of saving time in the development of different vaccines. Classical approaches need years each time - about 10 years for the development of a drug[29,30], a substantial part of which is spent on process development. No generic process is suitable for manufacturing different protein vaccines, since they are usually different and, hence, require different processing conditions. By contrast, the characteristics of plasmids are not much depending on the particular target sequence. Therefore, a generic process may be developed for the manufacturing of different plasmid-DNA vaccines.

While the DNA-vaccine approach starts to be transformed into applicable pharmaceuticals, only little is known about the process technology for generating the material needed for clinical trials in a fast, economic and consistent way. The common use of animal-derived and complex raw materials like bovine RNase or fermentation media based on meat peptones may represent severe problems (see also sections 3.2 and 3.3 of this chapter). The formulation and stabilisation of plasmid DNA for long term applications has to be assessed, and improvements of the vector system for preventive applications are required in order to avoid potential side effects due to the presence of non-relevant sequences of the vector system. In addition, the modification of the vector sequence may result in major improvements, especially with respect to CpG motifs or the codon usage.

One further very important advantage of plasmid-DNA vaccines is the opportunity of combining vaccine-encoding sequences within one vaccine formulation. This would not be possible in the majority of cases with classical vaccines based on proteins or pathogen-inactivated antigens due to e.g. different storage- or formulation-buffer requirements[31]. Especially in veterinary DNA-vaccine applications, the costs for the application per animal are substantial and add up quickly because of the number of different vaccinations required. It would represent a great advantage if all vaccines could be applied in one single injection. No difficulty of this type is expected in mixing different plasmids within one cocktail. However, such an approach may be result of individual studies performed with the different vectors in individual series of tests. With respect to manufacturing and developmental costs it makes sense to design vectors containing more than one epitope for vaccination[32].

In human DNA vaccination certain clinical trials have already been performed[17,33] and are actually heading to phase III applications. Recent literature even points out the performance of DNA vaccines taken up via hair follicles[34].

2.2   Plasmid manufacturing step-by-step

The manufacturing of plasmid DNA is divided into two major phases. The first phase (Fig. 1a) starts with the transformation of the fully characterised vector plasmid into appropriate and characterised host cells. The resulting "genetically modified organism" (GMO) is to be checked carefully for the expected characteristics. Subsequently, it will be transferred into the GMP environment for GMP-conformal processing. This includes the generation of a master (MCB) and working cell bank (WCB), which are required for reproducible large scale cultivation of the bacterial biomass. This biomass is subject to QC tests for product content and absence of contamination and will be processed, if released for manufacturing. Plasmid molecules are usually released from cells by alkaline lysis. The resulting lysate is separated from insoluble matter and cell debris. This "cleared lysate" is sterilised by filtration for further downstream processing (DSP).

This DSP comprises the second phase of plasmid manufacturing (Fig. 1b). Here the plasmid molecules should be separated from soluble biomolecules (e.g. host chromosomal DNA, RNA, nucleotides, lipids, residual proteins, amino acids, saccharides), salts and buffer components. A classical approach is chromatography. The design of the chromatographic process is depending on the required purity of the plasmid product. Actually, only a few techniques are used in pharmaceutical grade processing of plasmids. Anion exchange chromatographic techniques were described for certain processes[29,35,36,37,38,39]. Alternative chromatographic technologies make use of size exclusion chromatography for small scales[40] or reverse phase chromatography[41].

The second phase of plasmid manufacturing ends with bulk purified plasmid DNA being resuspended within the appropriate buffer or solution for further processing or storage and application.

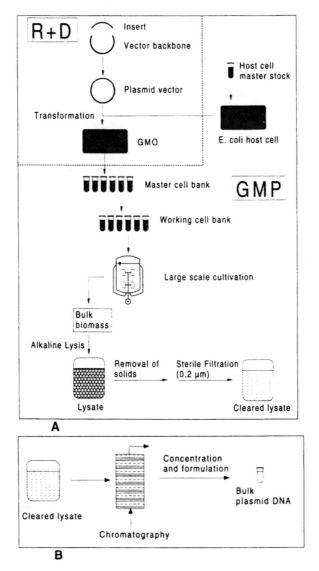

Figure 1. Plasmid manufacturing steps. (a) Construction and transformation of the plasmid vector is performed under research and development conditions (R&D). The host cells used for transformation are from a host cell master stock and all further work is performed under GMP. (b) Chromatography of DNA is performed within GMP clean room working conditions.

2.3   Improvements in plasmid-DNA quality control and quality assurance

Plasmids of identical nucleotide sequence isolated from *Escherichia coli* may exist in different shapes and forms. Previous literature[25,42,43,44,45,46,47,48,49] used a nomenclature which was significantly consolidated by Schmidt and co-workers[19]. The ccc isoform exhibits the most compact structure. If one strand is broken (nicked), the oc form results and coiling is lost. Linear forms are generated when both strands are cleaved once at approximately the same position. In addition, plasmids may appear as oligomeric forms, e.g. concatemers[50], which may in turn exist as different isoforms. All forms may be isolated in plasmid DNA manufacturing processes and should be considered as contaminations as long as their mode of action is unknown.

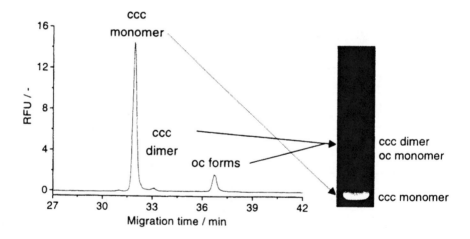

Figure 2. Comparison of CGE and AGE technology. The electropherogram (left) shows the individual peaks of the monomeric supercoiled plasmid vector (pUK21-CMVß, 7,6 kbp), the supercoil dimer and the open circular form. The agarose gel electrophoresis is not able to distinguish between the ccc dimer and the oc form.

Furthermore, the appearance of such forms requires ensuring that this happens reproducibly if it cannot be excluded completely. The reason for this is that the plasmid product should be as homogeneous as possible since its use as a drug is intended. Further forms are catenanes, which are oligomeric plasmids connected like chain links. However, these forms appear mainly in kinetoplasts and rarely in bacterial plasmid preparations[48].

The usual method applied to determine the plasmid form distribution is by agarose gel electrophoresis (AGE). Different bands in AGE from an in-process control sample or a plasmid product sample may be assigned to different plasmid isoforms. The assignment of bands to forms, however, is not easy since the electrophoretic mobility of plasmids of different shape changes with the electrophoretic operating conditions[43,49,51,52]. In addition, the quantification of isoforms based on the signal intensity of stained bands in AGE may not be reliable due to no linear responses. Adequate equipment is required in order to obtain reproducible results. In cases, when only some restriction enzyme digestions are performed to evaluate the fragment sizes of a plasmid, the isoforms cannot be distinguished. We recommend an AGE assay with undigested plasmid samples at least.

It is well known that typically only one band, the ccc form, is observed when only a small amount of a plasmid sample is applied to an agarose gel. Standard AGE usually reveals two prominent bands: the ccc form and another slower migrating form, commonly thought to be the oc form (see Fig. 2). Schmidt and co-workers[19] showed that this is not always the case. Earlier publications reported the separation of monomeric forms only[53,54,55]. CGE enables to identify the other forms mentioned as well as to quantify them individually[19,56] (see also www.CGEservice.com).

Another aspect that makes CGE outstanding is that the characterisation of a plasmid sample can be performed in less than 35 minutes and requires only 10 to 50 ng of plasmid sample.

3. FUTURE PERSPECTIVES

Approximately 25% of all gene-therapy protocols performed so far in clinical studies were based on plasmid DNA vectors[33]. The market share of plasmid-DNA vectors with respect to all vaccines is expected to rise to about 60%[57]. However, diverse aspects of producing plasmid DNA need still to be addressed in order to develop processes for the safe and economic production of large amounts of plasmid DNA.

3.1 Subcontracting plasmid manufacturing

Outsourcing research, development and biopharmaceutical manu-facturing for pre-clinical and the first steps in clinical research has become an attractive way around the risk of time consuming and expensive construction of manufacturing plants. Especially small start-up biotech and pharmaceutical companies with a strong scientific background but limited financial resources benefit from this development within their first years in developing innovative biotech-based therapeutics.

Limiting the economic risk in drug development is a necessary aspect for larger pharmaceutical companies. Hence, also these large companies benefit from the availability of subcontractors for the manufacturing of their products for clinical phases I through II, maybe even III. If successful until this point, commercial manufacturing may require dedicated facilities. With respect to data summarised by Edwards[58] the field of contract manufacturing with a focus on biotechnology products is rapidly growing.

3.2   Critical aspects: animal derived growth media

Culture media for the growth of microorganisms were developed in the 19th and 20th century. Liebig extracted beef and concentrated the essence for storage and later use in media preparation for the cultivation of microorganisms. The history of development, manufacture and control of microbiological culture media was summarised by Bridson[59].

One major improvement for such media was the addition of peptones and salts, which led to the increased supplementation with amino acids and an enhanced osmolarity. These peptones were generated by enzymatic digest of meat. Other protein sources were also used to be digested in order to generate peptones.

Today's technology for the generation of complex bacterial growth media uses soy bean peptones to avoid animal-derived protein sources. If it is indicated not to have any potential source of animal derived component within a plasmid manufacturing process, complex media should be avoided at all. However, only little data are available on cultivation processes with the intention to manufacture plasmid DNA by using synthetic growth media (see www.PlasmidFactory.com) and being able to ensure at least comparable yields of product. High cell density cultivation processes for plasmid production (Schmidt *et al.*, in preparation) commonly use at least one complex component.

If vegetable peptones shall be used it should be considered that no enzymes of animal origin are applied (e.g. pepsin, pancreatin or trypsin). Hydrolysis with acids avoids the use of such enzymes but leads to higher salt content.

A further alternative is the use of yeast extract. If no enzymatic digestion was performed in production, this may be a complex component of choice. However, even this yeast may have been grown on culture medium containing components of animal origin.

3.3 Critical aspects: animal derived processing aids

Ribonucleases (RNases) are able to hydrolyse phosphodiester bonds within RNA molecules. These enzymes are needed in cells to modify and process RNA molecules. In plasmid production processes one common major contaminant is RNA from *E.coli*. In general, RNA has a short half life, but is a substantial contamination in plasmid preparations. This RNA may block the binding capacity in e.g. anion exchange chromatography. Therefore, an enzymatic digestion of RNA prior to the chromatographic step is usually applied[35,37,38]. If this is carried out in pharmaceutical manufacturing processes, the question must be raised, if the quality of such processing aids, which normally are derived from animal sources, is sufficient. Since RNase A is typically prepared from bovine pancreas the origin of these glands has to be assessed and certified.

Some plasmid manufacturing processes completely avoid the use of RNase. In this cases, the RNA is removed by specific precipitation techniques[41,60,61]. However, this is only applicable in a laboratory scale today.

3.4 Critical aspects: chromosomal DNA

A major contaminant of plasmid preparations commonly is the host chromosomal or genomic DNA (gDNA). It has to be completely removed by the process steps applied. The most critical step is the alkaline lysis, were the gDNA as well as the plasmid DNA are denatured by a pH-shift to alkaline buffer conditions. A subsequent neutralisation step allows the reannealing of plasmid DNA within short time. The gDNA however does not reanneal completely to a DNA double strand again. Therefore, the major part of gDNA will be a component of the flaky material which is generated by the alkaline lysis. It mainly consists of potassium dodecyl sulphate, insoluble proteins, cell debris, lipopolysaccharides (LPS) and DNA. Chromosomal DNA is extremely shear sensitive, what may easily result in DNA fragmentation. A removal of DNA by subsequently applied process steps is difficult. Some gDNA fragments are large enough to migrate in one distinct band in AGE. Smaller fragments can be detected as an undefined smear by overloading the AGE gel. Below a certain size these fragments leave the gel without being detected. More sensitive assays like Southern blot hybridisation[62] can demonstrate this - depending on the hybridisation and washing conditions applied. A most sensitive assay was recently presented by Smith et al.[63], for which a kinetic PCR method using a TaqMan-probe was used to quantify gDNA contaminations. This technique makes use of the detection of the 7-copy-23S-rDNA-gene in *E.coli*.

Levy *et al.*[64] claim to be able to remove chromosomal *E. coli* DNA from plasmid containing process liquids to a residual content of less than 1%. These authors describe, how cleared lysate, resuspended PEG precipitates or anion exchange chromatography eluates were filtered through nitrocellulose membranes to selectively remove chromosomal DNA.

ACKNOWLEDGEMENTS

We thank Carsten Voß, Kirsten Lanfermann and Thomas Schäffer, University of Bielefeld, Faculty for Technology for their support with the research projects on DNA manufacturing, the whole manufacturing team of PlasmidFactory, Bielefeld, Germany for their discussion.

REFERENCES

1. Alton, E.W.F.W., Middleton, P.G., Caplen, N.J., Smith, S.N., Steel, D.M., Munkonge, F.M., Jeffery, P.K., Geddes, D.M., Hart, S.L., Williamson, R., Fasold, K.I., Miller, A.D., Dickinsons, P., Stevenson, B.J., McLachlan, G., Dorins, J.R., and Porteous, D.J. (1993): Non-invasive liposome-mediated gene delivery can correct the ion transport defect in cystic fibrosis mutant mice. *Nature Genetics* **5**:135-142.

2. Caplen, N.J., Gao, X., Hayers, P., Elaswarapu, R., Fisher, G. *et al.* (1994): Gene therapy for cystic fibrosis in humans by liposome-mediated DNA transfer: U.K. regulatory process and production of resources. *Gene Ther.* **1**:139-147.

3. Michel, M.-L., Davis, H.L., Schleef, M., Mancini, M., Tiollais, P., and Whalen, R.G. (1995): DNA-mediated immunization to the hepatitis B surface antigen in mice: Aspects of the humoral response mimic hepatitis B viral infection in humans. *Proc. Natl. Acad. Sci. USA* **92**:5307-5311.

4. Major, M.E., Vitvitski, L., Mink, M.A., Schleef, M., Whalen, R.G., Trépo, C., and Inchauspé, G. (1995): DNA based immunisation using chimeric vectors for the induction of immune responses against the hepatitis C virus nucleocapsid. *J. Virology* **69**:5798-5805.

5. Le Borgne, S., Mancini, M., Le Grand, R., Schleef, M., Dormont, D., Tiollais, P., Rivière, Y., and Michel, M.-L. (1998): In vivo induction of specific cytotoxic T lymphocytes in mice and rhesus macaques immunized with DNA vector encoding HIV epitope fused with hepatitis B surface antigen. *Virology* **240**:304-315.

6. Gregoriadis, G. (1998): Genetic vaccines: strategies for optimization. *Pharm. Res.* **15**:661-670.

7. Schirmbeck, R., van Kampen, J., Metzger, K., Wild, J., Grüner, B., Schleef, M., Hauser, H., and Reimann, J. (1999): DNA-based vaccination with polycistronic expression

plasmids. *In:* Lowrie, D.B., and Whalen, R.G. (Eds.) "DNA Vaccines: Methods and Protocols", pp. 313-322, *Humana Press*, Totowa, NJ.

8. Liljeqvist, S., and Stahl, S. (1999): Production of recombinant subunit vaccines: protein immunogens, live delivery systems and nucleic acid vaccines. *J. Biotechnol.* 73:1-33.

9. Bonadio, J., Smiley, E., Patil, P., and Goldstein, S. (1999): Localized, direct plasmid gene delivery in vivo: prolonged therapy results in reproducible tissue regeneration. *Nature Medicine* 5:753-759.

10. Shea, L.D., Smiley, E., Bonadio, J., and Mooney, D.J. (1999): DNA delivery from polymer matrices for tissue engineering. *Nature Biotechnology* 17:551-554.

11. Nikol, S., and Höfling, B. (1996): Aktueller Stand der Gentherapie. *Dt. Ärztebl.* 93:A-2620-A2628.

12. Gottschalk, U., and Chan, S. (1998): Somatic gene therapy, present situation and future perspective. *Arzneim.-Forsch./Drug Res.* 48:1111-1120.

13. CBER (1996): Points to consider on plasmid DNA vaccines for preventive infectious disease indications. (HFM-630), Center for Biologics Evaluation and Research, FDA, Rockville, MD.

14. CBER (1998): Guidance for industry: guidance for human somatic cell therapy and gene therapy. *Center for Biologics Evaluation and Research, FDA,* Rockville, MD.

15. Meager, A., Robertson, J.S. (1998): Regulatory and standardization issues for DNA and vectored vaccines. *Curr. Res. Mol. Ther.* 1:262-265.

16. Robertson, J., and Griffiths, E. (1998): WHO guidelines for assuring the quality of DNA vaccines. *Biologicals* 26:205-212.

17. Schleef, M. (2001), *"Plasmids for therapy and vaccination"*, *Wiley-VCH*, Weinheim

18. Schleef, M. (1999): Issues of large-scale plasmid manufacturing. *In:* Rehm, H.-J., Reed, G., Pühler, A., and Stadler, P. (Eds.) *Biotechnology Vol. 5a: Recombinant proteins, monoclonal antibodies and therapeutic genes* (Volume Eds.: A. Mountain, U. Ney, and D. Schomburg), pp. 443-470, *Wiley-VCH*, Weinheim.

19. Schmidt, T., Friehs, K., Schleef, M., Voss, K., and Flaschel, E. (1999): Quantitative analysis of plasmid forms by agarose and capillary gel electrophoresis. *Analyt. Biochem.* 274:235-240.

20. Schmidt, T., Friehs, K., and Flaschel, E. (2001): Structures of plasmid DNA, in: M. Schleef (Ed.): *Plasmids for therapy and vaccination*, pp.29-43, Wiley-VCH, Weinheim.

21. Wolff, J.A., Williams, P., Acsadi, G., Jiao, S., Jani, A., and Chong, W. (1991): Conditions affecting direct gene transfer into redent muscle in vivo. *Biotechniques* 11:474-485.

22. Vogel, F.R., Sarver, H. (1995): Nucleic acid vaccines. *Clin. Microbiol. Rev.* 8:406-410.

23. Donnelly, J.J., Ulmer, J.B., Liu, M. (1997): DNA Vaccines. *Life Sci.* **60:**163-172.

24. Helinski, D. (1979): Bacterial plasmids: autonomous replication and vehicles for gene cloning. *CRC Crit. Rev. Biochem.* 7:83-101.

25. Summers, D.K. (1996): *The Biology of Plasmids.* Oxford: Blackwell Science.

26. Schumann, W. (2001): The biology of plasmids, in: M. Schleef (Ed.): *Plasmids for therapy and vaccination,* pp. 1-43, Wiley-VCH, Weinheim.

27. Davis, B.D., Dulbecco, R., Eisen, H.H. (1980): *Microbiology.* 3rd. Edn. Philadelphia, PA: Harper and Row.

28. Devlin, T.M. (1997): *Textbook of biochemistry with clinical correlations.* 4th Edition, Wiley-Liss, New York.

29. Schleef, M., Schmidt, T., and Flaschel, E. (2000): Plasmid DNA for pharmaceutical applications, in: *Development and Clinical Progress of DNA Vaccines,* Brown, F., Cichutek, K., and Robertson, J. (Eds.), Dev. Biol., vol. 104, 2000, pp.25-31, Karger, Basel,

30. Bio World (1999): Entstehungsgeschichte eines neuen Medikamentes. *Bio World* **2:**31-34.

31. Braun, R., Babiuk, L.A., and van Drunen Littel-van den Hurk, S. (1998): Compatibility of plasmids expressing different antigens in a single DNA vaccine formulation. *J. Gen. Virol.* **79:**2965-2970.

32. Schneider, J., Gilbert, S.C., and Hill, A. (2001): pSG.MEPfTRAP – a first generation malaria DNA vaccine vector, in: M. Schleef (Ed.): *Plasmids for therapy and vaccination,* pp. 103-117, Wiley-VCH, Weinheim.

33. Clinical Trial Database (1999): Clinical trials - charts and statistics. www.wiley.co.uk/wileychi/genmed

34. Fan, H., Lin, Q, Morrissey, G.R., and Khavari, P.A. (1999): Immunization via hair follicles by topical application of naked DNA to normal skin. *Nature Biotechnology* **17:**870-872.

35. Colpan, M., Schorr, J., and Moritz, P. (1995): Process for producing endotoxin-free or endotoxin-poor nucleic acids and/or oligonucleotides for gene therapy. *WO 95/21177.*

36. Thatcher, D.R., Hitchcock, A.G., Hanak, J.A., and Varley, D.L. (1997): Method of plasmid DNA production and purification. *WO 97/29190.*

37. Bussey, L., Adamson, R., and Atchley, A. (1998): Methods for purifying nucleic acids. *WO 98/05673.*

38. Schorr, J., Moritz, P., and Schleef, M. (1999): Production of plasmid DNA in industrial quantities according to cGMP guidelines. *In:* Lowrie, D.B., and Whalen, R.G. (Eds.) *DNA Vaccines: Methods and Protocols,* pp. 11-21, Humana Press, Totowa, NJ.

39. Ferreira,, G.N.M., Prazeres, D.M.F., Cabral, J.M.S., and Schleef, M. (2001): Plasmid manufacturing – an overview; in: M. Schleef (Ed.): *Plasmids for therapy and vaccination*, pp. 193-236, Wiley-VCH, Weinheim.

40. Horn, N., Budahazi, G., Marquet, M. (1998) Purification of plasmid DNA during column chromatography. U.S. Patent 5 707 812

41. Green, A.P. (1999): Purification of supercoiled plasmid. *In:* Lowrie, D.B., and Whalen, R.G. (Eds.) *DNA Vaccines: Methods and Protocols*, pp. 1-9, Humana Press, Totowa, NJ.

42. DeLeys, R.J., and Jackson, D.A. (1975): Dye titrations of covalently closed supercoiled DNA analysed by agarose gel electrophoresis. *Biochem. Biophys. Res. Commun.* **69**:446-454.

43. Johnson, P.H., and Grossmann, L.I. (1977): Electrophoresis of DNA in agarose gels. Optimizing separations of conformational isomers of double- and single-stranded DNAs. *Biochemistry* **16**:4217-4225.

44. Johnson, P.H., and Grossman, L.I. (1977), Electrophoresis of DNA in agarose gels. Optimizing separations of conformational isomers of double- and single-stranded DNAs. *Biochemistry* **16**:4217-4225

45. Meyers, J.A., Sanchez, D., Elwell, L.P., and Falkow, S. (1976): Simple agarose gel electrophoretic method for the identification and characterization of plasmid deoxyribonucleic acid. *J. Bacteriol.* **127**:1529-1537.

46. Pulleyblank, D.E., and Morgan, A.R. (1975): The sense of naturally occuring superhelices and the unwinding angle of intercalated ethidium. *J. Mol. Biol.* **91**:1-13.

47. Tse, Y.-C., and Wang, J.C. (1980): *E. coli* and *M. luteus* DNA topoisomerase I can catalyze catenation or decatenation of double-stranded rings. *Cell* **22**:269-276.

48. Martin, R. (1996): *Gel Electrophoresis: Nucleic Acids*, Bios Scientific, London, UK.

49. Sinden, R.R. (1994): *DNA structure and function*, Academic Press, San Diego, CA.

50. Oliver, S.G., and Ward, J.M. (1985): *A dictionary of genetic engineering*, Cambridge Univ. Press, Cambridge, UK.

51. Serwer, P., and Allen, J.A. (1984): : Conformation of double-stranded DNA during agarose gel electrophoresis: Fractionation of linear and circular molecules with molecular weights between $3*10^6$ and $25*10^6$. *Biochemistry* **23**:922-927.

52. Garner, M.M., and Chrambach, A. (1992): Resolution of circular, nicked circular and linear DNA, 4 kb in length, by electrophoresis in polyacrylamide solutions. *Electrophoresis* **13**:176-178.

53. Courtney, B.C., Williams, K.C., Bing, Q.A., and Schlager, J. (1995): Capillary gel electrophoresis as a method to determine ligation efficency. *Anal. Biochem.* **228**:281-286.

54. Nackerdien, Z., Morris, S., Choquette, S., Ramos, B., and Atha, D. (1996): Analysis of laser-induced plasmid DNA photolysis by capillary electrophoresis. *J. Chromatogr. B* **683**:91-96.

55. Hammond, R.W., Oana, H., Schwinefus, J.J., Bonadio, J., Levy, R.J., and MOrris, M.D. (1997). Capillary electrophoresis of supercoiled and linear DNA in dilute hydroxyethyl cellulose solution. *Anal. Chem.* **69**:1192-1196.

56. Schmidt, T., Friehs, K., Flaschel, E. (1996). Rapid determination of plasmid copy number. *J. Biotechnol.* **49**:219-229.

57. Scrip Report (1995): *Vectors for gene therapy: Current status and future prospects.* PJB Publications Ltd.

58. Edwards, P.M. (1999): Contract manufacturing - the way ahead for biopharma?. *Helix* 05/1999 11-13.

59. Bridson, E. (1994): *The development, manufacture and control of microbiological culture media.* Unipath Ltd., Basingstoke, UK.

60. Chen, Z., and Ruffner, D. (1998): Compositions and methods for rapid isolation of plasmid DNA. *WO 98/16653.*

61. Murphy, J.C., Wibbenmeyer, J.A., Fax, G.E., and Willson, R.C. (1999): Purification of plasmid DNA using selective precipitation by compaction agents. *Nature Biotechnol.* **17**:822-823.

62. Southern, E.M. (1975): Detection of specific sequences among DNA fragments separated by gel electrophoresis. *J. Mol. Biol.* **98**:503-517.

63. Smith III, G.J., Helf, M., Nesbet, C., Betita, H.A., Mek, J., and Ferre, F. (1999): Fast and accurate method for quantitating E.coli host-cell DNA contamination in plasmid DNA preparations *BioTechniques* **26**:518-526.

64. Levy, M.S., Collins, I.J., Tsai, J.T., Shamlou, P.A., Ward, J.M., and Dunnill, P. (2000): Removal of contaminant nucleic acids by nitrocellulose filtration during pharmaceutical-grade plasmid DNA processing. *J. Biotech.* **76**:197-205.

Quality Assurance and Quality Control for Viral Therapeutics

STEVEN S. KUWAHARA
Kuwahara Consulting, PMB #506, 1669-2 Hollenbeck Avenue, Sunnyvale, CA 94087-5042, USA

1.   INTRODUCTION

It is the job of the Quality Control (QC) function to provide the data that a company uses to assure the public and regulators that its products are safe and efficacious. A second function is to provide good data from which good management decisions may be made. The Quality Assurance (QA) function is charged with providing the review and monitoring (not supervisory) activities that are needed to provide management and customers with a high level of confidence that the company can provide a product that is safe and efficacious. In the case of gene therapy, Quality (QA/QC) activities are affected by the type of gene therapy that a company is engaged in developing.

Gene therapy has been defined as a medical intervention based on the modification of the genetic material of living cells[1]. The two primary types of gene therapy are characterised by the location of the cells when they are modified. *Ex-vivo* gene therapy involves the modification of living cells while outside of the patient with their subsequent return to the patients' body. In this case the process is also regarded as being a type of somatic cell therapy. (Note that the modification of germ cells is currently beyond the scope of regulation and will not be covered here.) The second type, known as *in vivo* gene therapy involves the modification of the cellular genome while the cells are located within the patient's body, usually in their normal location. These distinctions create a difference in the approach that is taken with respect to the quality assurance and quality control (QA/QC) of these products.

In the case of *ex-vivo* therapy, the modified cells are returned to the patient and the finished product is considered to be this preparation. Therefore, when the cells are taken from and returned to the same patient (homologous cells) each preparation is considered to be a separate production lot and the viral or other transforming agent is used as an intermediate in the process. If the cells are a single preparation derived from a single individual and destined for administration into a particular individual in an allogeneic process, each patient may receive cells from the same lot, but a lot is defined as arising from a single set of cells that may have been derived from a single or multiple donors. In contrast, an *in vivo* therapeutic treatment involves the direct administration of the preparation of viral or transforming agent into the body of the patient. In this situation, the final product may be composed of many units and be administered to many different patients. Thus the product used for *in vivo* procedures is more like a traditional pharmaceutical lot. Viral agents have been employed as the transforming agent for both types of therapy. While the virus preparations are considered to be intermediate products in one case and final products in the other, the actual QA/QC process is not that different. The reason is that the preparations used in *ex-vivo* processes may be looked upon as material that is incorporated into the final product, and, as such, must be given the same level of scrutiny as if they were the final product.

The purpose of this chapter is to discuss the QA/QC procedures that are relevant to the viral product that will be used in gene therapy. Consequently, subjects such as cell line characterisations and process validation that are not lot specific or manufacturing run specific have been left for other chapters and authors.

1.1   Virus preparations

With viral preparations there are three primary QA/QC concerns:

First is the transmission of disease or the production of a disease state as a result of the action of the virus vector. Second is the prevention of contamination of the product while maintaining its ability to function in the manner intended. Third is the assurance of clinical safety and efficacy.

These concerns are not necessarily confined to viral preparations used in gene therapy. They and the following discussion are generally applicable to any agent used for the modification of the cellular genome.

2.   RAW MATERIAL TESTING

One of the major functions of the QC laboratory is to perform raw material testing. This testing is normally conducted on material that will be used in the manufacturing process to ensure that it possesses the properties that are needed to make it useful in manufacturing. Generally, this involves testing on process water and the reagents and substances that are components used for manufacturing. The discussions that follow on cell banks and other biological raw material are of that type. For this part of this chapter the discussion will be limited to the testing of non-cellular or viral material.

One of the most critical components is the culture media that will be used to grow the producer cell line that produces the vector. In addition, with *ex vivo* methods, there will be a point where the patient's cells will be incubated with the vector, and later, the cells will be formulated for re-infusion into the patient. In both of these situations, it is common to suspend the cells in media of the type that are used in tissue culture, and these media need to undergo testing that will at least establish their identity before the media are used in processing. In the case of media that are used for transduction or cell storage, if these media become a part of the material that is infused into the patient, the media then become excepients and must be considered to be in-process or final product material. This, in turn, creates a need for a more extensive examination of the properties of these components.

3.   CELLULAR TARGETS

Before proceeding to a discussion of the virus preparations, it is necessary to consider the cell preparations used in *ex-vivo* therapy and also the condition of the patient's cells during *in vivo* therapy. The properties of these preparations can have a major effect on the safety and efficacy of the therapeutic procedure. For example, the state of the cells within the cell division cycle can affect the ability of the viral vector to infect the cells and incorporate the therapeutic genes into the cellular genome. The condition of the cells or their ability to pass through the required state is usually established in a process validation study and does not become a matter for QA/QC release testing.

On the other hand, it is important to have some understanding of the possibility of viral infections in the cells that will be used in manufacturing or targeted for therapy. Normally, the patient is tested before therapy is initiated. These test results should be on hand and reviewed as a part of the lot release review in situations where individual lots are administered to individual patients. A very important question that must be answered deals with the potential for a recombination event between the viral vector and the genome of the target cells. This is due to the fact that viral vectors are normally engineered to be replication defective. In other words, the virus can infect the target cells but cannot replicate within those cells. This is normally accomplished by engineering the viral genome so that it will contain a deletion or a defective gene that results in the inability to produce a protein required for viral replication. If, however, a wild-type virus is already present in the patient's cells either as a free virus or as a part of the cellular genome, it is possible that a recombination event will create a replication competent virus that can multiply while replicating the therapeutic gene.

The primary concern in the event that a replication competent virus should be generated has to do with the lack of control over the dosage of the therapeutic gene and over the effects of a generalised viremia. In most therapeutic settings, the level of concern is not high since the dosage of the vector is controlled. However when a possibility for a recombination is present, there is a possibility that what was once a quiescent infection with the virus resting more or less inactive in the host cell, could now become an active infection with high levels of viruses through out the body. A special concern here arises from the possibility that some of these viruses may carry genes that have an oncogenic potential or may, themselves be toxic to cells.

To minimise the potential for recombination events, it may be necessary to test the patient with serologic or PCR-based methods to detect the current or past presence of wild-type viruses. In addition, it may be useful to detect the presence of latent viruses that could be activated to produce viremias as a result of the regimen imposed on the patient during therapy.

In the case of procedures that involve allogeneic cell therapy, where cells taken from one patient are modified and then administered to a second patient, the donor of the cells should be tested for the presence of infectious diseases that may be transmitted to the second patient. In addition to the usual viruses such as HAV, HIV, and HTLV and the wild types related to the vector, it may be necessary to test for the presence of other viruses depending upon the status of the recipient. For instance, an immunosuppressed recipient would need to receive cells from a more heavily screened donor than a recipient with a normal immune system. At a minimum, the cell donor should receive screening equivalent to that for a blood donor[1].

In most cases the virus testing required here would not be a part of the company's QC program but would be performed at the clinical site in their laboratories. In these situations the company's QA department needs to verify the clinical laboratory's ability to perform the tests, and also obtain a copy of the data from each patient. This would involve the collection of test method procedures, their corresponding validation studies, data related to their participation in proficiency testing, and an audit of the procedures used for verifying the testing and clearing the patient before initiating therapy. In many situations there will be some confusion since the clinical laboratory will not be operating under GMP or GLP rules. However, it should be remembered that in all developed countries, clinical laboratories are subject to regulation and oversight. In the United States there are local licensing regulations as well as a need for compliance with the provisions of the Clinical Laboratory Improvement Act (CLIA). In those situations where the compliance with these standards and regulations are considered to be sufficient for meet the needs of Good Clinical Practices or the other regulations, given the stage of product development being considered, it may not be necessary to pursue full GMP compliance. However the Quality Assurance function should verify the existence of the screening data and have some evidence that the data were reviewed and the patient cleared for treatment before the cells or vectors were released for use with that individual.

4.   CELL BANKS

The cells that are used for growing the viruses that become vector preparations need to be characterised with respect to their ability to produce viruses containing the desired gene, their ability to allow replication of the virus, and for the possibility that the cells may contain unwanted sequences that may be introduced into the virus. This is especially important if the cells have been genetically engineered to contain a desired gene. These properties are normally established during the development process and are not a part of the routine QA/QC of the products. However certain information should be made available to the QA/QC function before a decision is made to release a particular viral vector preparation for use.

There are three sets of cells that are normally employed in the production of viral vector preparations. These are the Master Cell Bank (MCB), Working Cell Bank (WCB) and the production cells. Production cells are normally tested either at the limit of *in vitro* cell life or at some time point beyond the end of production culturing. This is done to ensure the maximum opportunity for the expression or growth of contaminating viruses or other adventitious agents such as mycoplasmas.

The QC testing of cell banks is normally done as a matter of qualifying a raw material and involves tests for cellular identity, contaminating microorganisms, and even a characterisation of the growth medium. These raw material qualification tests are often done by the QA/QC laboratory, and the results should be reviewed before the material is released for further use. The results also need to be incorporated into the records so that they will be available when other aliquots are withdrawn from the cell bank.

With respect to the eventual virus vector preparations, it is important that the cell banks be tested for the presence of adventitious viruses before the cultivation of the desired virus is undertaken. The origin and history of the cells is important in determining the testing that is required. Also, since the eventual vector preparation will contain a live virus at high titre, testing of the uninfected cells may offer the only opportunity to detect the presence of certain types of contaminating viruses.

4.1   Master cell bank (MCB)

The MCB has the function of providing the manufacturer with a set of uniform cells with reproducible characteristics that may be used for manufacturing over the lifetime of the product. The cells should be screened for the presence of endogenous and nonendogenous viruses. The testing should be governed by the types of viruses that may be expected as contaminants. For instance, porcine cells should be tested for the presence of porcine adventitious viruses as well as for specific viruses, such as porcine B9 parvovirus, that are of specific concern. If the cells are of human or primate origin, the testing should be more extensive since the viruses that may contaminate such preparations are more likely to be infectious to human patients.

The QC scientist needs to have data covering the history of the cell line as this will affect the nature of the raw material testing that will be needed on the MCB. For instance, human tumour cells that were once grown in a medium containing equine plasma will require testing for equine adventitious viruses as well as for possible oncogenic viruses. It is possible that the cells may have been exposed to a virus to immortalise the line, and it would be important to establish the latency of this virus.

In addition to data on the history of the cell line, it is necessary to collect data and information on the history of any modifications that may have been made to the cells. This is especially important for cells that have been engineered to contain a specific gene that will be incorporated into viruses along with the modifications that will result in the production of a virus that will not replicate in the patient. Such a packaging cell line is the result of a modification of the original cell line, and it is important to show that the

MCB contains cells with the required properties. The presence of unmodified cells or partially modified cells should also be a concern as the growth of these cells may affect the overall properties of the producer cell line.

The QA/QC worker may face some difficulties here as the early history of the cell line may be buried in notebooks and the memory of early workers in the Development Group. As soon as a cell line becomes important in the planning for a product, quality assurance auditors should review the documents and data. The goal here is to ensure a future ability to account for the original cell line and the events that converted it into the packaging cell line that will be used for manufacturing. An early audit of this information is critical, as it should occur before the urgency of a regulatory submission arises. The early audit is also important given the high turnover rate in the biotech industry. If the audit occurs too late, the personnel who generated the original data may be gone, and the ability to translate the contents of a poorly written laboratory notebook may be dependent on the memories of individuals who were not directly involved with the work.

In general, testing of the MCB should also cover the possible types of contamination that might have occurred before the cells came under the control of the manufacturer. The testing should also confirm the modifications made by the manufacturer and the stability of these modifications. The idea here is to provide assurances that the cells are suitable for their intended use and will continue to be suitable over the lifetime of the product.

4.2   Working cell bank (WCB)

The WCB is usually an expanded cell preparation generated from the MCB and used to initiate the actual manufacturing process. In some cases, where the demand for inoculum is small, the MCB and WCB may be the same. On the other hand there are situations where cells from the MCB undergo modification to make a specific WCB. In these cases, it may be necessary of prepare a secondary MCB. In many cases, however, the WCB is prepared by adapting MCB cells to the medium that will be employed in production. If the change in the medium is significant, such as going to a protein-free medium, it may be necessary to reclone the cells and characterise the WCB. On the other hand an adaptation to a minor change in the medium will usually not require additional characterisation of the cells in the WCB. Differences exist between U.S. and ICH guidelines with respect to the need for re-cloning and re-characterising adapted cell lines. The worker should consult the regulators most directly involved with the company.

In many cases, the testing conducted on the MCB and end-of-production cells may be sufficient to allow the elimination of testing on the WCB. For example, virus antibody production tests done on the MCB do not need to be repeated on the WCB, and other tests have similar considerations. On the other hand, cells in the WCB will have undergone greater expansion than cells in the MCB, and, therefore, testing may be done on the cells from the WCB rather than the MCB. The decision here will be based on two factors. First, if the potential contamination were such that one would not want to expand the cell culture, it would be wise to test the cells as soon as possible at the MCB stage. Second, if the test requires a large amount of sample or cells that have undergone more divisions, it may be better to wait until the culture has been expanded to produce the WCB, and derive samples from that preparation. In addition, a portion of the cells produced while preparing a WCB could be grown to the limit of their *in vitro* age to obtain information before initiating production, although this procedure will not really replicate a full production preparation.

The idea here is that tests for the presence of endogenous viruses are more sensitive if conducted on end-of-production cells that have undergone extended culture beyond the time usually required for a production run. Consequently, tests performed on these cells (presumably at the limit of their *in vitro* cell age) will be on a system that has had a maximum opportunity to express any contaminating viruses.

4.3   End-of-production cells

While not a proper cell bank, cells taken at the end of a production run are usually available in great quantity and represent an opportunity for testing the cells under extreme conditions. The consideration here is that if there are contaminants whose concentration should increase during the cultivation process, one should expect them to be at a maximum in the cells at the end of the production process. In fact, it is normal to extend the cultivation time of the cells to a time beyond that normally used for the production process so that it may be argued that an additional opportunity was provided for the amplification of a contaminant. While not a test of the cell banks, a test for replication competent viruses is normally performed on the extended culture cell supernatants to allow the greatest opportunity for the amplification of the replication competent viruses.

As a practical matter, and to derive the greatest sensitivity, it is normal to have the end-of-production cells subjected to extended culture times to reach the expected limit of *in vitro* cell age. However, it is possible that the limit may be far beyond the normal length of cultivation or the cell line may be immortal. It is also possible that the cultivation process is designed to end when the cells have attained stationary growth. In these cases the culture should be extended only for a "reasonable" length of time before the cells are tested. It should be remembered that conditions change greatly when the culture has entered a stationary phase or when nutrients have been depleted. Consequently, there may be situations where the use of material from extended cultures may not be appropriate.

4.4   Testing on cell banks

The general goal, with respect to the eventual virus vector preparations, for testing of the cell banks is to establish the absence of contaminants that may affect the safety of the final preparations. This goal should be remembered, as the following list is not comprehensive enough to cover all possible tests that may be required for any particular virus preparation. Also, developments in biotechnology are such that testing methods will change and the procedures described below may be superseded in a short time.

4.4.1   Polymerase chain reaction (PCR)

PCR testing is especially useful for detecting the presence of low levels of contaminating viruses. The usual procedure is to use lysed cells or cell supernatants to insure the exposure of all of the nucleic acids to the test system. The method sensitivity is based on the ability to detect a specific number of copies of the nucleic acid sequence of interest. Most workers employ detection limits of 50 to 100 copies of the sequence of interest, but many workers have extended their sensitivity studies to the level of 10 copies. Specificity is usually established by a digestion of any PCR products with restriction endonucleases to generate specific fragments that are detected by electrophoresis. The specificity of the test is a combination of the priming ability of a specific nucleotide sequence and the finding of an expected polynucleotide fragment after digestion.

A major problem for the PCR procedure arises from its specificity. The analyst must have a specific primer available and know what fragment to expect after digestion. Thus the procedure only works with expected viruses for which primer sequences are available.

4.4.2   *In vitro* adventitious virus tests

These tests are designed to detect the presence of cultivable viruses by exposing whole or lysed cells or culture supernatant to indicator cell lines that are known to be susceptible to infections by a broad range of viruses. Since culture supernatants often contain the contents of lysed cells, it is normal to have one sample represent both sample types. When the exposure is through whole cells, these are sometimes known as co-cultivation tests. The specificity of these tests is often poor, but this is not a negative factor as the analyst should be looking for the broadest possible range of potentially infectious viruses for a particular cell line. The sensitivity is normally assumed to be one infectious particle, but this is a theoretical consideration and reality is often quite different.

The detection of a virus is usually based upon the finding of a cytopathic effect (e.g. plaque) in a cell monolayer, but the analyst should also consider the possibility that the virus may not be cytopathic and look for haemagglutination as well. The problem here is similar to that encountered in general sterility tests where the organism of interest may not grow well or at all in the culture system used. Consequently, it is usually necessary to use culture supernatant containing lysed cells and whole cells with several cell lines to cover all of the possibilities. If it is possible that a specific type of virus may be present (e.g. bovine viruses), a corresponding cell line (bovine testicular cells) should be employed to provide confidence that such viruses, if present, would be detected.

4.4.3   In vivo tests

This test involves the administration of the culture supernatant to a test animal. The animal is then observed for evidence of illness. Because of the variation in the sensitivity of individual animals to a particular disease, it is desirable to inoculate more than one (usually at least three) test animals. Also, to ensure the broadest possible range of virus detection, it is usual to inoculate several species of animals, and embryonated eggs along with suckling animals. The resulting sample requirements may be quite large and create a need to limit the extent of the test.

In addition to checking the animals for overt signs of illness (weight loss, fever, changes in behaviour) it is usually valuable to perform a necropsy on the animals at the end of the test period. The gross and microscopic examination of the tissues may reveal the presence of disease processes when there were no overt symptoms.

4.4.4   Antibody production tests

In this test, the sample (usually lysed cells in the culture supernatant) is administered to an animal and the animal is subsequently tested for the presence of a specific antibody. A large number of different viruses with a potential for infecting humans may be detected by this method. It may be possible to combine this test with the *in vivo* test described above. The laboratory that is seeking cost reductions should consider this possibility.

The specificity of this test is high, but it also creates problems, as the analyst must possess the appropriate antigen to use in detecting the antibody. Also it is possible that a virus may be present at levels that do not induce an immune response in the animal.

4.4.5 Transmission electron microscopy (TEM)

TEM is often employed to detect retroviruses or retrovirus-like particles in cells. The sample could be a preparation of intact or lysed cells that is used to create a pellet that is sectioned and negatively stained. In actuality just about any virus can be detected by this method. If a structure resembling a virus is observed, it must be a cause for concern and comment. There have been situations where potential problems were missed because workers ignored structures that did not look like the structures that they had been trained to seek. Because of this potential, it is important that workers who review electron micrographs should be trained to recognise an eclectic selection of viruses.

There are two limitations to this method. First, the small field of view makes it possible that the microscopist may not see viruses that are present at a low level. This problem arises from sampling effects and places a severe limitation on the ability of the method to detect low levels of contamination. Also, any attempt to quantitate the virus using TEM is also subject to a high degree of uncertainty. The second problem is that the simple observation of the presence of a virus does not ensure the viability of that virus. Therefore it is possible to see arrays of noninfectious or nonviable particles. This has happened repeatedly with murine hybridoma, Chinese hamster ovary (CHO) and baby hamster kidney (BHK) cells. Extensive studies have been performed on these particles, and, at present, their finding in a TEM is not considered to be a major problem for manufacturing processes. However if particles with an unusual morphology or from a cell line of concern (human or nonhuman primate, for instance) were to be observed, that should result in a need for extensive studies of the properties of such particles.

4.4.6 Tests for retroviruses

The cells should be screened for the presence of retroviruses through an array of sensitive tests. In addition to the general testing described previously, specific tests such as co-cultivation with mink or feline cells should be performed along with the use of specific PCR and antibody tests for retroviruses that are of concern. The reverse transcriptase assay should also be performed to detect noninfectious retroviruses.

One of the reasons for special concern here has to do with the fact that many gene therapy vectors are derived from retroviruses, and the presence of a wild-type virus in the production cells can create a potential for recombination or activation and a general viremia. Consequently, when a different type of vector such as an adenovirus or lentivirus is being employed as a vector, tests for corresponding wild-type viruses should be added to the screening of the production cells.

5.   TESTING THE PARENT VIRUS

In addition to the cell line, it is important to characterise and develop a history on the parental virus line and the modifications that were made to it to produce the desired viral vector. This may be done in conjunction with developing information on the packaging cell line as mentioned previously. For the same reasons, it is important to have this information audited at the same time as the audit on the data for the producer cells. As with the cell line, it is important to know the source and history of how the virus has been handled and passaged. This information can be particularly important if the virus has been carried for many generations in a cell line or medium that is significantly different from the system that will be employed by the company.

One of the critical questions that need to be investigated, deals with the possibility of the reversion of the virus to a replicating state. Most gene therapy virus vectors are engineered to be replication incompetent, with the factors required for virus replication residing in the packaging cell line, but, for purposes of safety, it may be important to know what can happen if the virus were to become replication competent. In many cases the use of double defects greatly reduces the possibility of the virus reverting to a replication state. Even with a single replication defect, the possibility of generating a replication competent virus is very small, but the large numbers of viruses involved creates the possibility that a replication competent virus will, eventually, be encountered.

Given the low probability of generating a replication competent virus, it may be impractical for a company to conduct experiments to deliberately generate such a virus. In these cases it may be useful to study the behaviour of the parent virus and viruses modified in the laboratory without or before the replication defect was introduced. For many viruses, such as retroviruses and adenoviruses, there may be a considerable amount of information already available in the literature, concerning the toxicity or pathogenicity of the replication competent virus. If the company will be generating its own safety data on the product, it may not be necessary to repeat work described in the literature as long as the company does not wish to challenge the conclusions. The QA/QC worker needs to collect this literature information and have it available in the material file for the virus line.

6.   IN-PROCESS TESTING

As will be noted in sections on the viral vector preparations, there are certain situations where the final preparation is made and then must be infused into the patient in a relatively short time. In these situations, the quality control testing of the in-process material becomes especially important as the quality of the manufacturing process becomes an important factor for providing assurance that the product will be able to meet standards of safety and efficacy.

6.1   Special situations

In situations where the final product must be administered before the completion of testing, the analyst should be concerned with the process validation studies and the on-going monitoring of process parameters that are related to the tests that may not be completed before the product is presented to the patient. For gene therapy products, these parameters are those related to the safety of the final product, and this, in turn, is related to sterility and purity issues.

6.2   In-process controls

The subject of in-process controls is covered in other chapters, so this discussion will mainly be concerned with the in-process quality control tests that are related to general monitoring, sterility and purity. Generally, the idea is to provide data on the sterility and purity of the preparations as close as possible to the point where the final preparation is made. However in keeping with the principles of GMP, it is necessary to provide cumulative data that support the conclusions of the ultimate testing.

The earlier paragraphs have dealt with the subject of the purity of the cell preparations and the viruses themselves. This work needs to be done as a matter of qualifying the raw material that will be used in the manufacturing process. Given the nature of the manufacturing process that will be used for producing gene therapy vectors, it is also important to control the sterility and endotoxin content of the starting material. This is because of the nature of the manufacturing process where the medium will support the growth of many microbes and endotoxin could affect cell growth and contaminate the final product.

Even if the starting materials are known to be sterile and to meet the specifications developed for the raw material, the possibility of contamination or mistakes during manufacturing needs to be considered. For this reason the manufacturer needs to identify critical points in the process where samples should be taken to verify the absence of contamination and the proper composition of the product at that stage. In many instances, this can be as simple as checking the pH, A_{260} and A_{280} of the intermediate.

6.3   Sterility testing

In other instances, such as sterility testing, the process will be completed before the test can be completed. This does not mean that the test should not be performed. Rather it should be performed and the results recorded for two reasons. First, if a sterility failure is found, the clinicians should be told so that they will be alert to the possibility of a septicaemia in their patient. In addition the organism could be subjected to screening by rapid identification methods so that the clinicians will know what organism is likely to be the cause of the septicaemia and perform appropriate antibiotic susceptibility testing. Second, as a matter of process monitoring, a sterility test failure should result in a closer scrutiny of the process to determine what may have been the cause of the failure. This should lead to a corrective action especially if there is a history of test failure. Even if a cause for the failure cannot be identified, it should still lead to closer scrutiny of the next process. Companies subject to regulation by the U.S. FDA or working under U.S.P. rules, should note that a sterility test failure constitutes an out-of-specification test result that cannot be ignored without positive proof that the contamination occurred during sampling or in the laboratory. This proof must go beyond the finding of a particular organism or the finding of a negative result upon retesting.

In the case of sterility testing, there is a problem that is created by the reluctance of companies to fully bear the cost of Quality Control testing. This is the practice of taking in-process samples, but not testing them unless a final product sterility test failure is found. This problem is also related to the difficulty that arises when an in-process sample fails in a sterility test, but the sample preceding it and the one succeeding it pass the test. In some companies the test failure is automatically assumed to have been due to laboratory error. However an understanding of the variability and difficulty of detecting low levels of microbial contamination will show that depending on a single sample to prove or disprove a low level of contamination is not a good practice. In addition the possibility of a contaminated preparation producing a single positive result in the middle of negative results is a reasonable possibility. Consequently, good manufacturing practices would argue that if a sample is taken, it should be tested, especially when low levels of contamination need to be detected.

Another problem that often leads to the rejection of a sterility test failure on the grounds of laboratory error arises when the test failure is found to be due to a contaminant that is an organism frequently found in the environment or on personnel. For example, test results are often dismissed on the grounds that the contaminant is found to be *Staph. epidermidus* which is a well-known skin contaminant. The problem here is that unless the analyst has clear data showing how the organism was introduced into the test system, it is necessary to ask how the organism came to be present. The point here is that it is widely acknowledged that the media used in sterility tests do not allow for the growth of all possible organisms. Therefore when an organism is found, the question is not just how the particular organism came to be present in the product, but what else might have been introduced by the same path. Therefore the finding of any organism must act as an indicator showing that the containment of the system was breached. It is this exposure of the system that is important, not the nature of the organism.

6.4   Rapid testing methods

As noted previously, rapid testing methods such as pH, conductivity, refractive index, and UV/Vis absorbance methods are preferred for in-process measurements due to the time limitations that are encountered. These time limitations are created by the need to obtain test results that allow decisions to be made concerning the continuation or adjustment of the manufacturing run without incurring serious delays in processing. Because of the need for rapid testing, in-process assays often sacrifice specificity, sensitivity, accuracy, or precision in exchange for a rapid completion of tests. In general, this is not a problem as long as the utility of the methods has been demonstrated in the process validation studies.

Recently developed analytical methods allow rapid turn around times while maintaining a level of sensitivity, specificity, accuracy or precision that would be acceptable for the testing of bulk or final product lots. At present, many of these methods are not widely accepted or easily validated, but the advantages offered by these procedures are such that they should gain more acceptance in the near future. In particular, assays based on microfluidics or the polymerase chain reaction (PCR) will be major factors.

With respect to the preparation of vector lots, rapid methods for the detection of microorganisms and for the determination of virus titre are gaining in popularity. In particular, the use of HPLC methods for the measurement of virus titre offers the advantage of speed and specificity. The use of chromatography allows the separation of viruses from most proteins and the assay can be performed very quickly compared to the usual procedure for titre determinations. The speed of the assay is such that the worker can make rapid decisions based on the growth of the organism. This allows for in-process adjustments of nutrients and environmental parameters. The problem with this assay is that inactive virus particles will be detected as normal virus in these assays. Thus its utility is mainly for situations where time dependent differences in results are important. Consequently, in the assays on the final product, the normal growth-dependent titre assays are preferred.

Rapid sterility testing methods based on flow cytometry, fluorescence-based enzyme or nucleic acid detection, and other high sensitivity methods are especially useful for situations where small amounts of material are produced or where the product must be infused into a patient within a relatively short time after manufacture. In situations where the product needs to be administered to a patient before results are available from a standard 14-day sterility test, the results from a rapid testing method (most of which require less than a day) would be the best substitute.

As a point of discussion, it should be noted that some workers consider the usual 14-day sterility test to be no better at detecting contaminants than the rapid testing methods that are currently available. The argument has to do with sensitivity and specificity. The 14-day sterility test presumably offers the advantage of being able to detect damaged or defective organisms, but the proponents of rapid methods usually claim that their methods can also detect these organisms as long as they are relatively intact. In other words if an organism is intact enough to be detected in the normal sterility test, it should be intact enough to be detected by rapid testing methods. Sensitivity claims are usually at the single cell level and this, at least theoretically, is the same as the claim for the 14-day culture method.

With respect to specificity, sterility tests usually have the opposite problem. A good test will lack specificity and be able to detect any organism that is present. Rapid testing tests are normally based on reagents that presumably will react with any microorganism, but it is unrealistic to expect that these claims will be comprehensively tested. Similarly, the media used in the 14-day sterility test do not necessarily grow every organism that might be present, especially those that are extremely fastidious. The result of these discussions is the conclusion that rapid testing methods and the 14-day sterility test are about even in terms of being able to provide assurances of sterility. The retention of the 14-day sterility test as a release test for the final product will continue mainly as a result of conservatism, but the inherent advantages of the rapid testing methods will eventually lead to the replacement of the 14-day test in much the same way that LAL-based endotoxin testing has replaced the rabbit pyrogen test.

7. VIRUS VECTOR PREPARATIONS IN GENERAL

The final product testing of virus vector preparations does not need to duplicate many of the tests that were performed on the producer cells or in-process intermediates. In fact, many tests can be performed on end-of-culture cells and not on the final product itself. These would be tests for the presence of replication competent virus and for the genetic stability of the cell line. The usual procedure is to test 5% of the volume of the supernatant and 1% or 10^8 cells whichever is less. As noted previously, what are called "end-of-production" cells are usually production cells that have been grown beyond the point when the culture would have been normally terminated.

The extension of the growth of "end-of-production" cells is normally done in an attempt to increase the sensitivity of the tests. Presumably, the cells will have had a greater opportunity to produce replication competent viruses or to multiply with the modified genome. However this procedure may ignore the fact that many production processes come to an end at a time when the cells have attained a maximum density or the medium has been depleted of an essential nutrient. As a result, cells from an extended culture may be in an altered state. This can be especially important if the cells begin to release unusual amounts or proteases or nucleases that may damage the vectors or the cellular nucleic acids. The analyst needs to evaluate this situation carefully before determining the point at which the tests should be performed.

7.1 Replication competent virus (RCV) assays

Normally, virus vectors are prepared in such as way that the virus lacks one or more genes for replication. The virus must depend on the producer cell line to provide the machinery needed for replication. This is important because of the need to control the spread of the virus within the patient and to avoid infecting other patients and healthcare workers who do not need the particular gene. Besides this, there are situations where the virus is engineered to target certain cells and kill them. In these situations, it is especially important to be able to contain the distribution of the virus.

RCV testing is normally based on the exposure of a susceptible cell line to 10^8 or 1% of the cells (whichever is less) and also to the 5% of the culture supernatant. In some situations, such as those where the vector is doubly defective in elements required for replication, the expected frequency of generation of RCV is so low that one would not expect to ever encounter an RCV. However the test must be performed "just in case" an RCV arises during cultivation.

7.2   Genetic stability

Even if the producer cell line has not been engineered to contain specific sequences for incorporation into the vector, it is still important to know if the end-of-production cells possess the same genetic make up as the cells from the WCB. This is important because the expectations concerning the behaviour of the vector and the effects of its interaction with the patient's tissues are based on the assumption that the vector's genetic make up is known. If the producer cell line "drifts" away from its expected composition, questions must then be raised about the composition of the vector.

Genetic stability is often first determined by an actual sequencing of the portion of the genome that is of concern to the manufacturer. This is normally done during the development process and a suitable primer sequence can be determined for use in PCR-based methods. A PCR test should then be developed and validated. This PCR test should be performed on each lot of end-of-production cells or those from an extended culture. It is important to establish the ability to maintain the appropriate producer cell genome through the product's life cycle. When the PCR is performed, the copy number of the sequence should also be determined.

7.3   Concentration (titre)

The concentration of virus particles may be determined by several methods, including HPLC methods as previously mentioned. Counting of particles by light scattering methods appears to represent the next generation of rapid tests. The problem with the rapid methods is that they do not necessarily measure the potential activity of the viruses. While the simple presence of virus particles may be acceptable for vaccines or in-process testing, the efficacy of viral vectors depends upon their ability to infect cells and express the desired genetic unit in the host cell.

At present, the preferred methods for measuring the concentration of viral vectors are quite varied. The traditional use of plaque-forming or tissue cytopathic effects are valuable in the few cases where replication competent vectors are employed. However these measurements are not valuable in many cases because the vectors are normally designed to be replication defective. The methods actually used often fail to distinguish between active and inactive virus particles. This problem is such that many companies ignore the activity issue for concentration measurements, and address it only when considering potency measures.

Among the procedures that have been used are:

7.3.1   Transmission electron microscopy (TEM)

TEM for determining concentrations requires the use of statistical methods to obtain the best number for the measurement. Simple counting methods with assumed volume relationships are usually highly variable and inaccurate due to sampling effects. These effects arise from practical difficulties when counting large numbers of particles and the problem of properly sampling preparations at low levels of particles. TEM also suffers from the problem that it measures virus particles without providing any indication of their activity.

7.3.2   Light scattering, size-exclusion HPLC and other HPLC methods

These methods have been discussed previously. They offer the advantage of being rapid methods and have sufficient specificity to distinguish between proteins or nucleic acids in general and virus particles. As with TEM, they cannot distinguish between biologically active and inactive viruses.

7.3.3   Fluorescent focus and haemagglutination assays

These procedures, including their inhibition forms, offer the advantage of being able to measure the ability of the viruses to interact with cells, thereby providing some measure of their potential activity. While this, by itself, is no guarantee of biological activity, they are better than no effect at all. These assays are like plaque and cytopathic assays in that the most informative measurements are those performed at or near the limit of detection of the method. Consequently, statistical procedures for both the sampling and calculation of concentrations are needed for the best results.

7.3.4   Polymerase-chain-reaction (PCR) and hybridisation methods

These methods offer the advantage of actually measuring what is presumed to be the pharmacologically active agent, the actual genes that are to be inserted. It is possible for these methods to target sequences that are virus related but not really related to the desired genetic construct. However, the worker usually will prefer a measurement of the actual sequence and also the number of copies of that sequence. Also, while demonstrating the presence of a sequence, this test does not demonstrate the ability of that sequence to produce a clinically effective product. These procedures often require specialised equipment, well-designed laboratories, and good training of analysts.

With certain viruses, such as retroviruses, enzyme measurements can be employed. Assays for reverse transcriptase or other relevant enzymes can offer convenient methods to determine concentrations. These tests do provide some estimate of the viability of the vector. The tests are similar to PCR based methods as they may require specialised equipment and specialised training of employees.

7.4   Identity

Before a product can be released, the identity of the active agent needs to be confirmed. The identity measurements need to be done by methods with high specificity. Common tests are based on immunochemical (ELISA, RIA) or physical (ultracentrifuge, light scattering) determinations. Cell line identity is frequently established by isoenzyme testing using a combination of electrophoresis and enzyme specific staining. A recently developed technique is to employ tests such as PCR methods where concentrations and identities may be derived from the same test.

In addition to the identity testing of the viral vector, it is important to establish the identity of the end-of-production cells. As a simple matter of production quality control, the company needs to show that each production lot of the vector was produced by the correct cell line.

7.5   Purity

The purity of the final product needs to be established. Other substances besides the vector itself need to be monitored and controlled. Specific tests for potential contaminants that may have been introduced by the manufacturing process need to be developed and emplaced. In addition, the presence of chance contaminants (microorganisms for instance) needs to be considered.

7.5.1   Media components

Several components of culture media may create problems if introduced into patients. Material, such as growth factors may possess pharmacological effects that may not be apparent in the cultures, but will affect patients. Insulin is a good example of this problem. It is frequently added to aid in cell growth, and is sometimes added in concentrations that could affect patients if carried through into the final product. Similarly, the use of known allergens creates a need to prove that they have been removed from the final product.

7.5.2   Culture contaminants

These are contaminants introduced as a result of the cultivation process. As an example, the cells in the culture may release DNA into the culture supernatant as a result of cell death or other lytic processes. Since the amount of DNA in a product needs to be held at less than 100 pg/dose[2], a normal procedure is to add a DNAse such as Benzonase™ to an intermediate preparation before the production of the final product. When this is done, testing must be performed to demonstrate the reduction of the DNA content as well as the removal of the nuclease by the process.

A point of confusion arose with respect to the DNA content of final products. This was due to a statement in an FDA document[3] that stated that tests for polynucleotide contamination should have a sensitivity of 10 pg/dose. This sensitivity was taken to be a limit for DNA in products and was clearly in conflict with the limit recommended by the WHO[2]. The situation was resolved in 1997 by an FDA statement[4] that the final product should not contain more than 100 pg/dose of cellular DNA as determined by a method with a sensitivity of 10 pg.

7.5.3   Host cell proteins and other components

In addition to DNA, contamination of the final product by antigenic proteins released by the producer cells can be a problem. The presence of host cell protein was a serious consideration in the early days of working with recombinant products, but these contaminants have become less of a concern as experience has been gained with highly purified recombinant products. In general the cultivation of virus vectors in human cell lines reduces the need to be concerned with host cell proteins. However, cultures that use nonhuman cells and component proteins of nonhuman origin (such as calf serum) create a need for testing the final product for the absence of these potentially antigenic proteins.

In guidance for the testing of bulk lots for potential contaminants and additives, FDA has suggested that a company should be able to demonstrate the removal of host cell proteins using a highly sensitive method[4]. However FDA has stated that these tests might be performed during the process validation study and may not be necessary for lot release. In addition they have considered the possibility that it may not be possible to demonstrate the complete removal of host cell proteins. In the case of serum additives and foreign proteins, FDA guidance[5] has stated that the residual amount in the final product should not exceed 1: 1,000,000 of the original amount added.

Electrophoretic methods and immunological methods such as Western blots and ELISAs are normally used to detect the presence of foreign proteins in final products. In addition, rapid testing methods based on microfluidics have demonstrated sensitivities within the required range. The problem in most cases has been related to the need for a suitable immunological reagent. It has been very difficult to obtain a polyclonal antibody with a broad specificity. In many instances, workers have developed sensitive methods for specific proteins that were known to be present. The clearance of these proteins was then taken as a surrogate for the removal of all host cell or additive proteins.

A related problem is the presence of cell membrane fragments in the final product. These fragments can create the same problems as host cell proteins and also offer the potential for triggering thrombotic events by acting as surfaces for the contact activation of the coagulation cascade. It is important to be able to demonstrate the removal of membrane fragments. If possible, the test for host cell proteins should detect membrane associated proteins so that a single test can cover both potential contaminants.

7.5.4   Endotoxin

The gram negative bacterial endotoxin that is responsible for most pyrogenic side reactions, can be introduced at just about any stage in the manufacturing process. While it is the product of the growth of gram negative bacteria, dead cells also contain the active endotoxin so its presence is not necessarily correlated with sterility problems. It is frequently introduced at the beginning of processing through contaminated raw materials and during processing through the use of contaminated equipment or containers. Once introduced, the endotoxins are difficult to remove since gene therapy products normally cannot tolerate the heating processes or ultrafiltration that are used to reduce endotoxin concentrations in other types of products.

Endotoxin is normally measured by the Limulus Amoebocyte Lysate (LAL) test, which is sensitive, rapid, and easily performed, by well-trained analysts. This test comes in several forms, some of which are amenable to the use of automated or high capacity processes. Test results are normally expressed as endotoxin units (EU) that are based on comparisons that eventually lead back to an international standard. The usual specification is to require a product to contain less than 35 EU per dose. This number is derived from a requirement that a product should not expose a patient to more than 5 EU per kilogram of body weight[6]. The procedure used here is to convert this amount into 350 EU per dose based on the assumption that the average person weighs 70 kg. A ten-fold safety margin is then applied to arrive at the specification of "less than 35 EU per dose". The analyst is warned that this number may need further reduction in the event that a product is directed at a patient population with a low body weight (small adults, infants, or small children) or with a high sensitivity to endotoxin. Also, the original figure is actually given as 5 EU/kg per hour. Since most infusions of gene therapy vectors or cells will take place in less than an hour, there is more support for the ten-fold safety margin.

7.5.5   Sterility

There are three sterility issues. First, there is the usual issue of contamination by microorganisms (bacteria and fungi). Second, is the issue of mycoplasma testing and finally there is the problem of contamination by adventitious viruses.

7.5.5.1   Microorganisms

The main problem with sterility tests for microorganisms arises from the need for what may appear to be a large amount of samples. In this case, we are considering the situation where a final virus vector preparation has been divided into aliquots that will be employed in individual treatments. ICH guidelines call for the testing of 1% of the total vials, but no less than 2 vials, while FDA requests the testing of 40 vials[7]. Since vector preparations yield a relatively small amount of vials compared to normal production lots of biologicals the difference in these sample quantities can be significant.

A second consideration is the length of the 14-day incubation period. Under normal conditions for the final product this is not a problem since the test for mycoplasmas requires 28 days, but with many of the in-process tests the 14-day incubation means that work will proceed without knowledge of the sterility status.

The reader should review the previous statements with regard to sterility testing during in-process testing, as these statements, especially those concerning rapid methods, will also apply to the testing of the final product.

7.5.5.2   Mycoplasma testing

The possibility of contamination by mycoplasmas is an issue that arises due to the handling and prolonged cultivation of the producer cells. Since the MCB should have been tested and found to be free from mycoplasmas, the concern here would be for the introduction of mycoplasmas by exposure to workers or other carriers and the amplification of the mycoplasmas during cell growth. Unfortunately, a considerable amount of evidence has been accumulated to show that contamination of cell cultures by mycoplasmas is a real problem. Also, the testing for mycoplasmas has not been harmonised across regulatory areas[7]. FDA has described what is called the "Points to Consider Method"[5] and this is currently in wide usage, but European methods do differ.

The testing methods are based on multiple cultures and the resulting 28-day culture test creates a major problem when timing the release of product lots. Even when vital staining methods are used, the cultivation for 28 days is often required. As a result of this problem, there is great interest in the development of PCR-based mycoplasma detection methods, and several procedures are being developed or are in use. A kit for PCR-based mycoplasma testing is available from the American Type Culture Collection (ATCC), and workers have claimed sensitivities of 20 – 80 colony-forming-units (cfu) for this kit. The problem, as with all PCR-based methods, has to do with the availability of primers with a broad specificity. Although the primers used for mycoplasma testing are directed at highly conserved ribosomal sequences, regulators have been reluctant to accept this procedure because of the remote possibility that a mycoplasma with a different or mutated sequence may contaminate a preparation. At present, it appears that some organisations are using PCR methods to monitor in-process materials and use the culture methods for the release of final products. It would seem that the reluctance of the regulators will be overcome as experience with and refinements of the methods provide greater assurance of its broad applicability.

In addition, there have been some suggestions that regulatory agencies may allow companies to certify the use of "mycoplasma-free" processes. The resulting validation studies might also follow the "virus removal" requirements and require a demonstration that the processing will significantly reduce the titre of contaminating mycoplasmas.

7.5.5.3   Adventitious viruses

The presence of adventitious viruses is an important concern when establishing the MCB or WCB. It is also important to know if the production culture contains a contaminating virus. For the final product, the concern is based on the same concerns that lead to mycoplasma testing. After all, if the product could become contaminated by mycoplasmas during production culture, would it not be susceptible to contamination by adventitious viruses via the same types of routes? This would be especially important in situations such as the production of adenoviral vectors where the producer cells might also be suitable for the cultivation of a wild-type virus.

Parvoviruses have been detected in finished products, and their source has not always been clear.

In these situations, testing can be very difficult as it is necessary to detect the presence of a low level viral contaminant against a background of a high concentration of live virus. With gene therapy vectors, the advantage is that the vector should not be capable of replicating so it should not contribute to cytopathic or pathological effects. As a result, the test for RCV should be considered as a test for any RCV not just the RCV arising from the vector. Testing can also be performed using three cell lines of the type used during the screening of the cell banks for adventitious viruses. These would be one cell line related to the production cells, one cell line derived from a nonhuman primate, and a human cell line. If the production cell line is a human line, such as 293 cells, two human cell lines representing different tissues or cell types could be employed in addition to the nonhuman primate.

On rapid test that will be useful here would be PCR testing. Primers for an array of viruses could be used to check for specific viruses in final products. The advantage here would be the relative speed of PCR methods compared to culture methods. One criticism that may be directed against PCR methods is that they could detect a virus that is not capable of replicating. The QC laboratory should adopt a strategy where the finding of a critical virus by a PCR method will place a final product on hold, pending a confirmation via a culture method.

7.6   Potency testing

Potency testing of virus vector preparations is often very difficult and, consequently, surrogate tests are frequently employed. The problem is that the normal QC approach is to require the potency test to be closely related to the desired clinical effect. When surrogate methods are employed, there is an assumption that a connection to the clinical effect will be established during the clinical testing of the product. However this may be overlooked in the rush to complete clinical studies.

For example, the potency test for many vector preparations is to determine the virus titre. This has been done on the assumption that the more virus used, the higher the rate of infection, the higher the rate of transduction, and the higher the level of gene expression. But this chain of effects may not be sustainable. What should be done is to expose cells to the preparation and look for evidence of the expression of the gene. Since this has not been done in many cases, there have been instances where the vector was unable to infect patient cells due to the presence of antibodies or other binding proteins, or where the vector bound to the cells but was unable to integrate or insert the gene into the cells. In other cases the gene was inserted, but was not expressed. In one odd case it was theorised that the incorporated gene caused the expression of an antigenic protein on the cell surface, and this resulted in the cell being attacked and destroyed by the patient's immune system. For these reasons it is very important to show that the potency test really does measure the ability of the product to exert a clinical effect.

8.   SPECIAL SITUATIONS AND VECTORS

8.1   *Ex-vivo* gene therapy

When vectors are prepared for *ex vivo* gene therapy the vector preparation should be regarded as a final product in its own right even in situations where the actual final product is a transduced cell preparation that is administered to the patient. The constraints attending *ex vivo* gene therapy can be such that time is not available for fully testing the final product. In these situations the GMP philosophy of building quality into the product is very important as the testing may be incomplete when the product is administered to the patient, and it is necessary to have confidence that the product will pass testing because proper procedures have been followed.

With *ex vivo* gene therapy, there will be a stage where the patient's cells are incubated in the presence of the vector preparation to allow the transduction to take place. All of the components used in the transduction mixture must be considered to be "patient contact" materials and should be subject to release testing before they are employed. If there is further processing (washing and concentration of cells, for instance) the level of scrutiny should be high. The final preparation that is infused into the patient must be considered to be a final product lot. In addition to these tests, if the incubation of the cells is for more than a "short time" (interpreted as three days in many laboratories) a mycoplasma test may also be required.

With respect to the QA/QC of these products, a further difficulty may arise when the final preparation of the cells is performed at the clinical site or at the patient's bedside. This creates a situation similar to that of a shared manufacturing process, and a similar type of QA monitoring is required. The requirement for the monitoring and the data that will be required should be discussed with representatives from the clinical sites, and be included in any contracts between the clinical site and the company.

8.2 Adenoviral vectors

Adenoviruses are present in the environment and many patients will already possess antibodies or an infection by adenoviruses. Therefore the patients should be tested for the presence of these antibodies or viruses. While it may be possible to pursue the treatment even in the presence of antibodies or wild type infections, the potential complications need to be considered.

Adeno Associated virus (AAV) is frequently encountered as a contaminant of adenoviral preparations and infections. Therefore these preparations and the associated cell lines should undergo screening for AAV.

8.3 Genetically modified cells

If the vector is used to transduce cells in an *ex vivo* setting, it will be necessary to view the preparation of modified cells as a product lot. If these cells are derived from a specific individual and are destined for an infusion into a specific individual, each preparation is a separate lot and must undergo lot release procedures. Also the vector lot must be considered to be an intermediate product along with any cytokines, growth factors and culture media that are employed. One of the difficulties here is that the manipulation of the cells usually results in a potential for environmental exposure and a requirement for repeating the array of sterility tests.

The cells and vector preparations at this stage are relatively unstable when compared to the usual biologic product. The freezing of the transduced cells is usually unacceptable due to the reduction in cell viability that accompanies a freeze-thaw cycle. This causes a reduction of the dosage while introducing apoptotic cells and cell debris into the patient. The reduction of the cell dose is undesirable as the usual requirement is for the patient to receive the maximum possible dose. As a result it is frequently necessary to proceed with the infusion of the cells before the results of all the tests are available. In this case, a strict adherence to the rules and philosophy of GMP is very important.

9. STABILITY STUDIES

Stability studies are normally a function of the QA/QC unit of a company. Whether the virus is used for *ex vivo* or *in vivo* therapeutic procedures, stability studies must be conducted on the preparations and their intermediates. Also, with *ex vivo* procedures, the stability of the cells during their exposure to the vector and afterwards needs to be established. Stability studies are more than just a regulatory requirement. The data produced by stability studies has practical importance as well. The manufacturer needs to have a clear idea of how stabile the products and intermediates are. This is especially true with gene therapy where the virus vector is often just as unstable as the cellular preparations and is very sensitive to environmental conditions. As a result stability studies need to be conducted on virus vectors as well as any cellular product[1] that must be held before administration to the patient.

Studies should be conducted both at the storage temperature and at an elevated temperature (usually 15° C higher than the storage temperature). A very low temperature (vapour or liquid phase liquid nitrogen, for example) may also be used to provide a stable control preparation for the studies. Of course in cases where these extremely low temperatures are used routinely, it may not be possible to use an ultra low temperature. It should also be remembered that even when material is stored at an ultra-low temperature, it still needs to be thawed and brought to at least room temperature before administration to a patient or use in a procedure. Therefore the stability study should involve the second condition as well as the first. Consequently, one part of the stability study may be conducted at -80° C and another at 37° C with resulting dating periods of 3 years at -80° C and 3 hours at 37° C.

Studies at an elevated temperature are useful especially in the early development stages in that they allow the extension of dating periods with confidence. These are considered to be accelerated studies due to the expected increase in reaction rates at elevated temperatures. For instance if a product is found to be stable at 40° C and 25° C at three months, and the product is being stored at 25°, the QA/QC department may feel some confidence in extending the product dating to four months.

A second type of study at elevated temperatures, is one that is deliberately conducted to force the degradation of the product. These studies are usually conducted well above normal storage temperatures or those for accelerated studies and result in what are called stress tests. For instance it is common to use 60° C as a stress condition with 40° C as the accelerated condition and 25° C as the normal storage temperature. The role of the stress condition is to cause the degradation of the product so that the QC worker may study the routes and end products of product degradation. It is important to know the chemical and toxic properties of these degradation products. One of the more important roles of stress conditions is to validate the use of assays that are defined as being stability-indicating. Many stability studies have employed methods that were later shown to have no response that was related to the stability of the product. Therefore when reviewing assays used in stability studies, the QA/QC worker should look for the data that support the idea that the assay really does measure product stability.

These studies at higher temperatures are often criticised as being unrepresentative of the actual degradation of the product. The claim is made that the elevated temperatures allow the product to degrade along paths that are different from what would be the normal path under normal storage conditions. This is not necessarily true; thus these claims must be subject to proof. Just as it would be naive to accept the stability of a product without data, it does not make any sense to blindly accept the claim of an alternate degradation path. This proof can only be obtained by comparing the results of a real time, real storage condition study with that of the accelerated study. Unfortunately, this data is often not available during the early stages of the product cycle, and the results of accelerated studies will be the only data that the worker can use.

10. TRANSMISSIBLE SPONGIFORM ENCEPHALOPATHIES (TSE)

In addition to concerns over the possible transmission of Scrapie or Bovine Spongiform Encephalopathy (BSE)[5] , regulators have developed a high level of concern over the potential for the transmission of Creutzfeldt-Jakob Disease (CJD)[8] and other TSE[9,10] by biologic products. Manufacturers of gene therapy products have responded to these concerns by eliminating animal (including human) sourced material and weaning cell lines from a dependence upon serum or serum components. When this cannot be done because of time factors or the growth requirements of the cells, a common procedure has been to use sera sourced from New Zealand or Australia where TSE in animals is not believed to be endemic. As an alternate, the German scoring system[11] has been very useful for assessing the potential infectivity of the products.

While the issue of TSE contamination may not be a problem in the release of final products, it is a major issue with regard to the qualification of raw material. Even with animal-related material that does not come from New Zealand or Australia (the U.S. and Canada are believed to be free from BSE) it is important to be able to verify the source of the product and the quality of its production. The worker needs to remember that CJD transmission by material used as drugs or in medical transplant procedures has been demonstrated. As a result, the use of human-derived material does not relieve the QA/QC worker from having a concern with raw materials.

Therefore QA audits and the QC review of the certificates of origin or analysis are important in the manufacturing of gene therapy vectors.

11.   PHASE IV STUDIES

With certain drug products there are requirements for the extended monitoring of patients after they have received the drug. In fact, in some cases, this requirement for extended monitoring can be a condition for licensure. These are sometimes known as Phase IV Studies and often fall to the QA/QC function for execution and monitoring. Note that if these studies reveal an adverse event, it will need to be reported to the appropriate regulatory authorities following the usual rules for adverse event reporting.

While most of the Phase IV Studies look for adverse events, in the case of newly developed drugs, other subjects may be of interest. With gene therapy, the long-term survival of transduced cells and the long-term presence of the vector are of interest. There may be questions about the development of antibodies against the vector, the transduced cells, or the gene product. These responses may be desirable for gene therapy-based vaccines. However with other therapies, antibody responses may preclude additional treatments of a relapsing patient.

Phase IV testing normally is performed by immunochemical or nucleic acid amplification (PCR) or hybridisation methods. The testing will not have the urgency of in-process testing, but the desire for an early detection of potentially adverse events creates a need for highly sensitive assays.

Gene therapy based on the use of virus vectors creates a need to monitor patients for the potential development of an RCV. A second concern here is the possibility that an initially contained vector could migrate and result in a modification of the patient's germ cells, resulting in a vertical transmission of the vector or gene. If the vector or its genome becomes a permanent component of the patient'' cells, these concerns become a major problem as the exposure of the patient to environmental factors or viral infections increases with time. These exposures have, at least theoretically, a potential to lead to adverse consequences. Thus the monitoring may need to be done over a period of several years, even decades, if the virus or the inserted sequence is not cleared from the patient. Because of the high level of uncertainty surrounding these potential long-term effects, the worker can expect regulators to take a cautious approach until more experience and data are accumulated.

Testing for the incorporation of the viral genome or the therapeutic gene sequence into germ cell lines usually requires an examination of cells obtained by biopsy by PCR-based methods to detect the presence of these sequences. In these studies it is important to distinguish between sequences that have become incorporated into the germ cell genome and sequences that are present due to the presence of the vector in tissue or cell fluids.

Testing for RCV may require a combination of PCR or hybridisation methods and culture methods. The nucleic acid-based methods require knowledge about the specific sequences that are required for the virus to be "replication competent." In addition, the ability to produce synthetic positive sequences or complementary sequences can become a major factor in these tests. Culture methods using susceptible cell lines are sometimes preferred as they, at least theoretically, are just as sensitive as nucleic acid methods and allow a detection of active viruses. The difficulty here is to obtain a cell line that is susceptible to infection by the RCV, and to prepare stocks of the cells for long-term use.

REFERENCES

The following reference documents were employed in the preparation of this chapter. Specific references were not always provided in the text as the author feels that repetitive citations clutter the text and the worker should review all of these documents to have a good idea of how the guidelines apply to a particular product and process. This is especially critical for those in regulatory jurisdictions outside the U.S. where interpretations and nuances of local and international regulations may be different.

1. Guidance for Industry: Guidance for Human Somatic Cell Therapy and Gene Therapy. CBER, FDA, USDHHS, March, 1998. Proposed Approach to Regulation of Cellular and Tissue-based Products. CBER, FDA, UDHHS, February, 1997. Guidance for the Submission of Chemistry, Manufacturing, and Controls Information and Establishment Description for Autologous Somatic Cell Therapy. CBER, FDA, USDHHS, January, 1997. International Conference on Harmonisation; Q5A; Viral Safety Evaluation of Biotechnology Products Derived from Cell Lines of Human or Animal Origin. CBER, FDA, USDHHS, Federal Register Vol. 63, No. 185, 51074 – 51084, September 24, 1998.

2. Acceptability of Cell Substrates for Production of Biologicals. Report of a WHO Study Group, World Health Organization Technical Report Series No. 747, 1987.

3. Points to Consider in the Manufacture and Testing of Monoclonal Antibody Products for Human Use (1987). OBRR, CDB, FDA, USDHHS, June, 1987.

4. Points to Consider in the Manufacture and Testing of Monoclonal Antibody Products for Human Use. CBER, FDA, USDHHS, February, 1997.

5. Points to Consider in the Characterization of Cell Lines Used to Produce Biologicals. CBER, FDA, May, 1993. ICH; Q5D; Guidance on Quality of Biotechnological/Biological Products: Derivation and Characterization of Cell Substrates Used for Production of Biotechnological/Biological Products. CBER, FDA, USDHHS, Federal Register, Vol. 63, No. 182, 50244 – 50249, September 21, 1998.

6. Guideline on Validation of the Limulus Amebocyte Lysate Test as an End-product Endotoxin Test for Human and Animal Parenteral Drugs, Biological Products, and Medical Devices. CDER, CBER, CDRH, CVM; FDA, USDHHS, December, 1987.

7. Cell Bank Validation: Comparing Points to Consider and ICH Guidelines. Lao, M.S., BioPharm 48 – 51, January, 1999.

8. Precautionary Measures to Further Reduce the Possible Risk of Transmission of Creutzfeldt-Jakob Disease by Blood and Blood Products. Letter to All Registered Blood and Plasma Establishments. Director, CBER, FDA, USDHHS, 8 August 1995.

9. Guidelines for Minimizing the Risk of Transmission of Agents Causing Spongiform Encephalopathies Via Medicinal Products. Ad Hoc Working Party on BioTechnology/Pharmacy, Commission of the European Communities. Note for Guidance, 111/3298/91-EN, 1991.

10. Medicinal and Other Products and Human and Animal Transmissible Spongiform Encephalopathies: Memorandum from a WHO Meeting. Geneva, WHO, Bull. WHO: 1997, 75 505-513.

11. Notification of Safety Requirements of 16th February, 1994 for Medicines Containing Animal Substances Deriving from Cows, Sheep or Goats with Respect to the Avoidance of the Risk of Transmission of BSE or Scrapie. Federal Health Office (BGA), 1994. and Notification on the Marketing Authorisation and Registration of Drugs: Measures to Avert Risks Associated with Drugs, Stage II. German Federal Institute for Drugs and Medical Products (BfArM) 1996.

Analytical Assays to Characterise Adenoviral Vectors and Their Applications

ELISABETH LEHMBERG, MICHAEL T. McCAMAN, JOSEPH A. TRAI - NA, PETER K. MURAKAMI, JAMES G. FILES, BRUCE MANN, LINH DO, MEI P. TAN, SPENCER TSE, TAO YU, JEFFREY W. NELSON, JUAN IRWIN, JOHN IRVING, EIRIK NESTAAS AND ERNO PUNGOR, JR.
Berlex Biosciences, 15049 San Pablo Avenue, Richmond, CA 94804-4089, USA

1. INTRODUCTION

Recombinant adenovirus preparations are used for gene delivery in a number of clinical gene therapy approaches. Adenovirus offers short term (transient) gene expression without integration into the host cell genome[1]. The adenovirus is a complex biological system containing several different structural proteins and a linear double stranded DNA molecule held together by non-covalent interactions[2]. The classical assays used to characterise the adenovirus preparations include infectivity analysis (typically a plaque assay format[3]), SDS-PAGE analysis of the adenoviral proteins[2,4,5] and the estimation of virus concentration by lysing the virus in SDS and measuring absorption at 260 nm in the lysate[2].

With the steady increase in the number of adenovirus based programs in clinical development, commercial production is anticipated in the near future. In order to support manufacturing it is essential to significantly improve the analytical definition of the virus product, i.e. to allow verification of product equivalency by chemical, biochemical and *in vitro* biological analysis following changes in the manufacturing process (including scale up, introducing new production sites, introducing changes in the viral vector, etc.). Without improved analytical definition of the virus products, the implementation of such manufacturing changes may require

201

costly and time-consuming human clinical studies to prove product equivalency. Thus, similarly to the therapeutic proteins during the past decade, the recombinant adenovirus products will seek to ultimately reach the "well characterised" (or "well understood") biologic product regulatory status.

In this chapter we give an overview of the approach we took towards developing such analytical support of manufacturing a first generation (E1 deleted) recombinant adenovirus type 5 carrying a growth factor transgene. We discuss some of the key methods we adapted or developed for a systematic physical and physicochemical analysis of both the intact (section 2) and the dissociated virus, i.e. the viral proteome and genome (section 3). The fundamental methods for the *in vitro* biological characterisation of the virus preparations are described in section 4. When discussing our approaches, we focus on the generally applicable aspects (independent of the transgene, serotype, additional deletions in the adenoviral genome, etc.). In section 5, we show examples on how we applied these techniques to solve selected problems.

2.   PHYSICAL AND PHYSICOCHEMICAL ANALYSIS OF THE INTACT ADENOVIRUS

The protein coat surrounding the adenoviral DNA forms a symmetrical icosahedron which has been extensively studied by electron microscopy and crystallography[6,7]. The structural proteins forming the coat (coat proteins) are proteins II, III, IIIa, IV, VI and IX, their copy numbers in the virus are well established by electron microscopy. The remaining structural proteins (proteins V, VII, VIII and X) are localised inside the virus (core proteins) together with the viral DNA. Further structural information is available on the viral proteins (some 3D structures based on crystallography, evidence of proteolytic processing of structural proteins during virus assembly[8,9], evidence of glycosylation on some coat proteins[10], etc.). Besides the known cotranslational and posttranslational processing events, there may be manufacturing related protein modifications (e.g. oxidation, deamidation, proteolysis, denaturation) and other phenomena (e.g. virus aggregation, loss of some of the structural proteins) which can have effects on the surface characteristics of the virus. These surface characteristics, in turn, may affect the *in vitro* properties (e.g. stability) and the *in vivo* performance of the virus (e.g. infectivity, immunogenecity).

We have adapted published methods and developed new procedures to resolve possible differences in surface properties (charge density in ion exchange chromatography, zeta potential in capillary zone electrophoresis, and particle size in size exclusion chromatography) of the intact adenovirus. Our basic requirements from these assays were quantitative recovery of both the virus mass and viral infectivity (assays for virus mass and infectivity, using permissive cells, are discussed in section 4).

2.1 Anion exchange chromatography

The most widely applied analytical method for the determination of both purity and concentration of the adenovirus is analytical anion exchange chromatography developed by Shabram *et al.*[11]. Many other anion exchange resins can be potentially useful for the same purposes, e.g. studies have been performed to separate different serotypes of adenovirus having somewhat different surface properties by anion exchange chromatography[12]. The anion exchange chromatography separates intact adenoviral particles from other components like unassembled hexon, fibre, penton proteins as well as free nucleic acids[11]. Further confirmatory information can be obtained on the purity of the virus peak by monitoring the absorption ratio at 260 and 280 nanometres. This absorption ratio can be easily calculated if an HPLC is used which monitors at least two wavelengths simultaneously. For a pure virus, the typical value of this ratio is 1.2-1.3[11]. A higher value could indicate nucleic acids coeluting with the virus peak, a lower value could indicate proteins coeluting with the virus peak.

For the determination of the particle concentration the integrated peak area at 260 nm is typically calibrated to an adenovirus standard, for which the virus concentration was determined by a suitable, independent method (discussed in section 5).

For our studies we applied the most widely accepted anion exchange method[11]. In short, we used a 1 ml Resource Q column (Pharmacia Biotech), equilibrated the column with 300 mM NaCl in 50 mM HEPES, pH 7.5, eluted with a salt gradient from 300 mM to 600 mM NaCl in 50 mM HEPES, pH 7.5. The flowrate was 1 ml /min and the chromatography was performed at ambient temperature. We only made one modification to the published assay: for the lack of detectable nucleic acid contamination, we did not need to pretreat the purified virus samples with nucleases (like Benzonase or RNase A).

2.2 Size exclusion chromatography

Although size exclusion chromatography is widely applied as purification technique for different viruses (for reference, see[13]) only one application to adenoviruses has been reported[13]. In this case, Toyopearl HW-75F (TosoHaas) combined with a mobile phase of PBS + 2 mM $MgCl_2$ and 2%

sucrose showed signs of nonspecific virus retention resulting in a broad virus peak and poor recovery (15-20%) thus rendering this approach unattractive for analytical chromatography.

We have developed an analytical size exclusion assay for intact adenovirus. Using 4X PBS as mobile phase with a flow rate of 1 ml/min and a G5000 PW XL HPLC column (7.8 mm x 30 cm, TosoHaas), the assay is performed at ambient temperature. Over 90% of the injected virus is recovered (as measured by both infectivity and mass of virus particles) in the void volume of the column. With this method one can separate intact adenovirus from smaller impurities (retained during separation). The assay can be used on crude samples such as cell lysates. For virus concentration estimation the integrated peak area of the virus peak is calibrated to a virus standard with a known virus concentration. Using the peak area at 214 nm for highest sensitivity, the limit of quantitation for our assay is 2×10^9 virus particles/ml (the maximum injection volume was 250 μl).

2.3   Capillary zone electrophoresis

Capillary zone electrophoresis (CZE) has become a routinely applied high resolution separation tool for proteins, peptides and DNA. The utility of CZE to analyse even more complex systems such as viruses, bacteria[14] and red blood cells[15] was demonstrated a decade ago, but has not been exploited until recently[16,17,18]. The most significant problems to overcome in virus analysis by CZE are adsorption of the virus particles to the capillary and the positive identification of the viral peaks in the electropherogram. To our knowledge only the electrophoresis of intact Rhinovirus (approximately 20 nm diameter) has been solved to date by adding detergents to the electrophoresis buffer and identifying peaks by indirect methods such as heat denaturation, enzymatic treatments and biospecific reactions performed before CZE analysis[17].

We developed a capillary zone electrophoretic method for high efficiency separations of intact adenovirus species[19]. In short, we found, that PVA coated capillaries can achieve essentially quantitative recoveries of the adenovirus during separation. The recoveries were determined by measuring both infectivity and DNA content (by performing quantitative PCR of the transgene region, see section 3.2.3.) of collected fractions. For routine studies we use a 57 cm capillary with an internal diameter of 50 μ, the separation voltage is set to 29.5 kV (reverse polarity), the electrophoresis buffer is 25 mM sodium phosphate, pH 7, and the samples are injected by pressure (0.5 p.s.i., 30 sec.). A bubble cell detector is used to monitor absorption at 214 nm. Typical current during separation is approximately 30-35 μA.

3. PHYSICAL AND PHYSICOCHEMICAL ANALYSIS OF THE DISSOCIATED ADENOVIRUS

The adenovirus can be dissociated with several methods into its protein and DNA constituents allowing a detailed analysis of the viral proteome and genome. The proteome consists of approximately 2500 protein molecules/virus (at least 11 structural protein species present in multiple copies) in a molecular weight range from 2000 – 108000[2,4,5,10,20]. The adenovirus particle also contains a single double stranded linear DNA molecule of approximately 36 Kbasepairs with a molecular mass of about 2.3×10^7 Da.

3.1 Analysis of the viral proteome

A fingerprint of the viral proteins (separation of the viral proteins after dissociating the virus) can be obtained by electrophoretic methods, analytical chromatography and even by mass spectroscopy.

The most commonly used technique is to dissociate the virus by boiling the virus sample in the presence of SDS and separating the proteins by SDS PAGE[2]. The separation of the viral proteins upon treating the virus with SDS can also be done by capillary gel electrophoresis (CGE). The proteins separated can be further analysed (mass determination, cotranslational and posttranslational modifications, like glycosylations, phosphorylations, etc.) by well established protein chemistry methods.

An alternative rapid protein fingerprinting technique was developed by two laboratories[21,22] using MALDI TOF MS. In this approach, the virus is placed directly (or after a pretreatment with organic solvents to promote dissociation) into the ionisation matrix of the mass spectrometer. The ionisation dissociates the virus and, at the same time, ionises viral proteins which can be subsequently separated based on their mass/charge ratio in the mass spectrometer.

The above approaches provide useful information on the components of the viral proteome, and can be used for a variety of purposes (including quality control). However, quantitation of the above protein maps and/or further analysis of the separated proteins is limited (for example by the staining selectivity and the interference by SDS in further analysis in SDS-PAGE, by the sample size and sample collection problems in CGE, by differences in the ionisation efficiencies between different proteins and sample collection problems in MALDI TOF MS). We developed a chromatographic fingerprint for the adenoviral proteome using a RP-HPLC procedure to address these shortcomings. Prior to our work, only one published study described a qualitative chromatographic fingerprint for HIV-1 proteins using a RP-HPLC procedure[23].

3.1.1   Reversed phase chromatography of the adenoviral proteome

A detailed description of the RP HPLC assay and its characterisation is given by Lehmberg *et al.*[20]. Briefly, intact virus was injected onto a C4 (Phenomenex) column at 40°C equilibrated in 20% acetonitrile and 0.1% TFA in water. Under these conditions the viral particles dissociated into their structural proteins which subsequently bound to the resin. The proteins were then separated and recovered quantitatively by eluting the column with an acetonitrile gradient from 20% to 60% in the presence of 0.1 % TFA giving a characteristic protein fingerprint. The absorption during elution was monitored at 214 nm and typically also at 260, and 280 nm.

The separated protein peaks (14 peaks) were collected and subjected to MALDI TOF MS and N-terminal sequence analyses for identification.

MALDI TOF MS was performed with sinapinic acid as the matrix. All spectra were taken in the positive ion mode. Aliquots from the RP-HPLC were either spotted directly onto the probe, or concentrated by vacuum centrifugation. Care was taken to not completely dry the fractions. Mass calibration was performed by analysing a set of external or internal standard proteins.

Automated Edman degradation was performed for N-terminal sequencing. For direct sequencing, samples without pre-treatment were loaded to the sample cartridge and analysed. To obtain sequences with blocked N-termini, samples were treated with Chloramine-T to cleave the protein chain by oxidation at the C-termini of tryptophanyl residues. To minimise losses, the reaction mixture was loaded to the sequencer cartridge without separation of the peptides. In most cases we were able to obtain sequences based on the similarity in the amino acid signal intensities in consecutive cycles.

We used the above protein map to solve a variety of analytical problems, including virus concentration determination (as discussed in section 5).

3.2   Analysis of the viral genome

Conventional molecular biology methods (for a general review, see ref.[24]) are well suited for study of the linear, double-stranded DNA genome of recombinant adenovirus species[25,26,27]. For this reason, we only discuss the key approaches we used for the characterisation without going into experimental details.

3.2.1   Quantitation of the adenoviral DNA

DNA concentration is readily measured after disruption of the protein capsid using one of several methods. We have developed a microplate assay for the rapid determination of adenovirus DNA using the fluorescent dye PicoGreen which exhibits >1000 fold fluorescent enhancement upon binding to double stranded DNA[28]. The PicoGreen reagent, an asymmetric cyanine dye, has several important properties desirable for this application including greater fluorescent enhancement upon DNA binding than that of Hoechst 33258, a fluorescent dye widely used for DNA quantitation[29]; compatibility with various sample matrix components[30] and selectivity for double stranded DNA over either RNA or single stranded DNA.

A direct spectrophotometric method does not provide information on the presence of interfering contaminants. Crude samples of virus often contain appreciable amounts of cellular RNA that could lead to a serious over-estimation of viral DNA content. Thus, non-viral nucleic acids must be removed from the preparation before a direct DNA assay can be applied to measure virus DNA. We have shown that an RNAse digestion of the crude samples can effectively remove non-viral nucleic acids prior to the PicoGreen assay[28].

In our studies we used the PicoGreen assay for routine quantitation of the viral DNA and to determine virus concentration (see section 5).

3.2.2   Restriction maps

Viral genome structure can be assessed by restriction mapping of purified DNA. Conventional methods for extracting, precipitating, and purifying viral DNA allow preparation of a reproducible template for restriction site mapping as another tool for genome characterisation. We have studied a spectrum of vectors, including ones with different transgenes, the same transgene in opposite orientations, and vectors whose backbones (Ad5 related sequences) have been shown to differ at certain regions[25]. In each case a unique restriction map fingerprint has been easily developed. Typically, ethidium bromide stained agarose gels require viral samples containing a minimum of 200-400 ng DNA to visualise after digestion. Further sensitivity (reduced sample needs) can be obtained using a more sensitive DNA staining dye such as SyberGold.

We have successfully used such restriction maps to routinely verify product identity.

3.2.3 PCR

Focussed genome structure analysis can be achieved by combining PCR, restriction digestion, and gel electrophoresis methods. Small segments (500 – 3000 basepairs) of viral DNA can be amplified by PCR to facilitate a more detailed characterisation of a viral sample. For example, key viral or transgene sequences can be confirmed as intact and unchanged with increasing passage number as a measure of stability of either product or master virus bank. Restriction mapping of select PCR segments offers confirmation of precise sequence at the (enzyme-recognised) restriction site as well as size verification of flanking segments (controlling for DNA deletion or rearrangement events).

3.2.4 Sequencing

The targeted analysis of viral DNA can be taken one further step beyond the PCR/mapping described above by either direct DNA sequencing of the PCR products or by cloning the PCR reaction end-products into a plasmid for subsequent sequencing. We have made frequent use of TA-cloning vectors designed to ligate readily and directly with PCR products to speed up the cloning process. Mini-screened plasmid DNA can also be easily sequenced, avoiding need of larger scale (and longer timed) efforts for plasmid preparations.

4. IMMUNOLOGICAL AND *IN VITRO* BIOLOGICAL ANALYSIS

4.1 Characterisation of surface antigens

Antibody-based immunoassays provide a straightforward approach to assess the presence of specific surface structures. This can be done with denatured proteins (Western blots) probed with specific anti-sera or monoclonal antibodies. Similarly, the same immunological reagents can be used to detect components of intact virus using either a single antibody (e.g. a FACS-based analysis) or with multiple antibodies (in a sandwich/capture ELISA method). Measuring antibody mediated neutralisation of viral infectivity can be an alternative way to assess the presence or absence of some functional epitopes on the surface of the virus. Existing reagents range from polyclonal sera to purified monoclonal antibodies, from both commercial and academic sources.

We developed assays for antibody mediated adenovirus infectivity neutralisation and for antibody binding. The infectivity neutralisation assays were based on the endpoint dilution infectivity assay (see section 4.2.) performed in the presence of neutralising sera. In the binding antibody assay, adenovirus was bound to 96 well plates (Immulon 4, Dynatech) at approximately 10^8 particles/well by spontaneously adsorbing to the wells at pH 9.1 from a borate buffer. The virus was inactivated by UV irradiation and the wells were blocked. A serial dilution of human serum samples was added to the wells, and after incubation, the human antibody binding was assessed by using a goat anti-human IgG – HRP conjugate. We used this general assay format for different purposes, including assessing human adenovirus binding antibody response in patients in gene therapy trial and comparing adenovirus preparations by evaluating titration curves obtained using a test serum.

4.2   Titering infectivity: endpoint dilution format

Infectivity of a recombinant adenovirus preparation can be determined using different assay formats (plaque, endpoint dilution) and with a variety of readouts (lysis of permissive cells, transgene or viral gene expression, etc.). For a review, see ref[3].

We used an endpoint dilution format assay to determine the infectivity of recombinant adenovirus preparations. Serial two-fold dilutions of the virus were made in 8 replicates on a 96-well cell culture plate containing approximately 5,000 permissive (E1 complementing HEK 293) cells per well. The virus samples were diluted down to "limiting dilutions" for which, on average, no infectious virus particle was present any more in the diluted sample. In this assay format, allowing multiple rounds of infections, one infectious virus particle in the original sample placed in the well causes measurable cytopathic effect (CPE) or full cell lysis in the well. After the appropriate incubation time, the number of CPE positive wells versus virus dilution was fitted to a statistical model[31] (based on Poisson and binomial statistics) to give a "maximum likelihood" estimate of the infectious titer in the original sample in infectious units per mL (IFU/mL). The standard deviation of this infectivity assay is ± 0.2 log10, corresponding to +58% - 37%.

4.3   Testing for replication competent adenovirus (RCA)

Recombinant, E1 deleted adenoviral vectors, currently developed for gene therapy, are amplified in cell culture on permissive mammalian host cell lines transfected with viral E1 sequences. The most frequently used E1 containing permissive cell line is HEK-293[32]. An undesirable by-product of viral amplification is the emergence of replication competent adenovirus (RCA) from recombination events within the infected permissive cell when there is sufficient homology between the virus and host cell genomes, most likely due to sequences flanking the E1 gene. Numerous examples are known of recombinant Ad5 grown in HEK293 cells (which contain viral sequences flanking both sides of the transfected E1 gene) generating RCA that appear to be wild-type by sequence. Several approaches have been taken to reduce the likelihood of forming replication competency, including generating multiple deletions in the adenoviral genome and developing complementary production cells, or producing a new generation of E1 complementary cell lines where the viral sequences flanking E1 have been removed. An example to this approach is the PER.C6 cell line described by Falleaux *et al.*[33].

An RCA can be detected as a CPE in a cell-based assay. This can be readily accomplished by exposing monolayers of non-permissive cells (such as HeLa and A549 cell lines) to high numbers of total virus particles in which RCA may be present at only trace level[34,35]. Given sufficient incubation time, multiple rounds of virus infection will occur.

We developed, optimised and characterised an RCA assay using two rounds of amplification on the same two cell lines (the first on HeLa, the second on A549 cells). If CPE is observed in the secondary (A549) culture, a tertiary A549 amplification and a concurrent PCR amplification of the E1 gene from the cell lysate of the secondary culture are used to confirm RCA. The assay was qualified to detect a single replication competent infectious unit from a background of 3×10^{12} non-replication competent virus particles. The assay is always performed with at least one positive control spiking 1 infectious unit of replication competent virus into the sample to be analysed.

4.4   Transgene expression

Transgene specific assays should be an essential part of the characterisation package of a recombinant virus product. Transgene expression can be tested on at least three different levels: transcription, translation and testing the biochemical / biological effect of the protein expressed as the result of the viral infection. Since these tests are highly dependent on the nature of the transgene, we discuss our general approach without going into assay and assay design details.

We established a quantitative rtPCR assay to assess transcription by quantitating the mRNA carrying the transgene. We use this assay to measure mRNA from both permissive and relevant non-permissive cells.

We adapted a commercial ELISA test to quantitate the transgene protein product. We use this assay to evaluate protein expression upon infection with the recombinant virus in both permissive and relevant non-permissive cells. We also verified molecular mass and immunoreactivity of the expressed transgene (SDS-PAGE and Western blots) and confirmed glycosylation on the protein in a variety of cell lines.

We developed cell based assays to evaluate the biological effect (growth promotion) of the expressed growth factor protein. Among others, we established a format where the viral infection of a non-permissive (primary human) cell line, the expression of the transgene and the growth promotion takes place in the same assay (i.e. without transferring samples, cells, etc.). The readout can either be the expressed transgene (protein) concentration or the growth promotion activity (increase in the number of viable cells).

5. APPLICATIONS

5.1 Measurement of adenovirus concentration: protein and DNA based approaches

The RP-HPLC assay can be utilised for the measurement of the adenovirus concentration based on the quantitation of the individual structural proteins by integrating the peaks of the structural protein at 214 nm[20]. The absorbance of a protein at 214 nm is primarily determined by the number of amide bonds[20]. Since the amino acids have similar molecular masses, the number of peptide bonds is approximately proportional to the injected mass of the protein, independently of its identity. Although modest variations in specific absorbance at 214 nm may arise from differences in higher order structures, these are not anticipated to play significant role under the denaturing conditions applied in our assay (low pH, organic solvent and elevated temperature). Protein standard curves for quantification were generated by injecting different amounts of BSA, GHRF or rDSPA α-1 onto the RP HPLC column, eluting the protein with the same gradient as used for the virus chromatography (see section 3.1.1.) and integrating the protein peak at 214 nm. The protein standard curves were linear between 0.1 and 2.0 μg protein injected. The slopes were 2262 ± 52 area units/μg BSA, 2205 area units/μg GHRF 1-29 and 2413 ± 12 area units/μg rDSPA α-1. The slopes were not statistically significantly different as expected based on the theoretical considerations above. For the quantification of the viral proteins, we used the BSA calibration.

Using the example of protein II with an apparent molecular mass of 107,950 Da (see Table 1) and a copy number of 720 molecules/virus[5,6,10], 1 µg of protein II corresponds to 9.26×10^{-12} [moles of protein II/µg] * 6.022×10^{23} [virus particles/moles of virus] / 720 [moles of protein II/ moles of virus] = 7.72×10^9 virus particles. Similar calculations can be made for all structural proteins of the virus with known copy numbers and molecular masses. In other words, the RP-HPLC assay of the viral proteome can supply multiple, independent virus particle concentration estimates. Using the largest protein peak (protein II) and the current format of the RP-HPLC procedure, the sensitivity of the assay is approximately 2×10^8 virus particles injected.

A similar approach can be taken based on measuring the DNA content of a pure virus preparation[28], utilising the 1:1 stoichiometric relationship between adenovirus DNA and intact particles. Using the values of 35,506 basepairs for the adenoviral DNA and 660 Da / basepair, 1 ng DNA corresponds to 2.6×10^7 adenovirus genomes. For DNA mass measurement we calibrated the PicoGreen method with Bacteriophage λ DNA.

The PicoGreen assay (discussed in section 3.2.1.) has excellent reproducibility, linearity, and sensitivity. In present form, this assay has a limit of quantitation of 10.3 ng/ml viral DNA, predicted to correspond to 2.6×10^8 virus particles/ml[28]. When compared to the benchmark spectrophotometric method described by Maizel *et al.*[2] both the RP-HPLC assay and the PicoGreen assay are 10-20 fold more sensitive.

A side-by-side comparison of the PicoGreen, the RP-HPLC and the spectrophotometric methods is shown in Figure 1. The data show excellent agreement between the virus concentration estimates based on the protein II content and the DNA content over a wide virus concentration range (approximately 3×10^{10} to 3×10^{12} viral particles/ml). Note that the protein based and the DNA based measurements had independent mass calibrations (BSA and Bacteriophage λ DNA, respectively). The spectrophotometric virus concentration estimates[2] showed weaker correlation when compared to the two other methods. The differences were the largest in the lower virus concentrations (approximately 3×10^9 to 10^{11} viral particles/ml corresponding to the absorption range of 0.03 – 0.1 units at 260 nm). The increasing differences from the estimates by the RP-HPLC assay at lower virus concentrations led to an apparent slope of <1 and an apparent positive intercept for the spectrophotometric estimates in this plot. These characteristics are consistent with substantial light scattering in the SDS lysate samples reported by others[3,11].

The observation that contaminant RNA interferes with viral particle count assays based on nucleic acid quantitation is a reminder of the benefits of independent and orthogonal methods for key analytical measurements. In our studies we found that the DNA and structural protein based virus concentration estimates (as shown above) correlated very well over a broad virus concentration range. This strong correlation suggests that the contribution of empty virions (if any) to the protein based estimates was negligible for our purified viral preparations.

5.2 Product comparability: adenovirus type 5 vectors manufactured in different production cell lines

The study discussed below was part of a program to investigate and establish comparability between two recombinant adenovirus type 5 vector candidates carrying the same transgene, manufactured in two different production cell lines: HEK 293[32] and PER.C6[33]. Here we describe some of the physical, chemical, biochemical and *in vitro* biological characterisation performed to compare the HEK 293 product (referred to as Ad5) to the PER.C6 product (referred to as Ad5.1). This discussion is limited to the application of the analytical tools described in sections 2-4.

There were some *a priori* known differences between the two vector candidates besides the difference in the production cell lines. For example, in the Ad5.1 vector we used a modified, shorter version of the promoter driving the transgene expression and we changed the orientation of the transgene expression cassette. Also, we introduced minor DNA differences into the Ad5.1 vector flanking the transgene region to minimise the probability of forming RCA. We anticipated that the *in vitro* studies would also confirm these known differences.

The *in vitro* study discussed below was followed by an *in vivo* comparison program investigating toxicity, biodistribution and efficacy in relevant animal models in order to evaluate if the differences uncovered by the *in vitro* characterisation had any effect on the *in vivo* performances of the vectors.

5.2.1 Physical and physicochemical comparison of the intact vectors

The Ad5 and Ad5.1 vectors were compared in the anion exchange HPLC, the size exclusion HPLC and in the capillary zone electrophoresis assay. The results are shown in Figures 2-4.

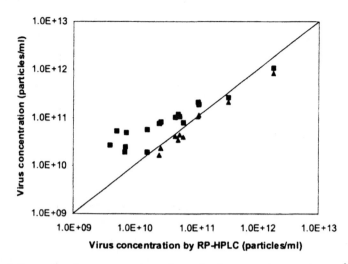

Figure 1. Comparing virus concentration estimates by the spectrophotometric assay[2] (squares) and by the PicoGreen assay (triangles) to the estimates based by the RP-HPLC assay.

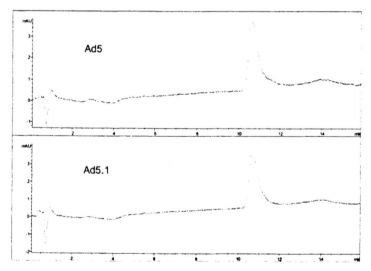

Figure 2. Anion exchange chromatograms of the two purified viral vectors. The absorption was recorded at 260 nm. The method is described in section 2.1. The virus elutes from the column in the peak around 11 minutes.

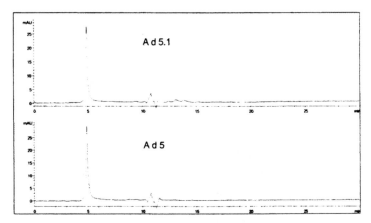

Figure 3. Size exclusion chromatograms of the two purified viral vectors. The absorption was recorded at 214 nm. The method is described in section 2.2. The virus elutes from the column in the peak at 5 minutes. The peaks at 10-12 minutes are the solvent peaks in the included volume.

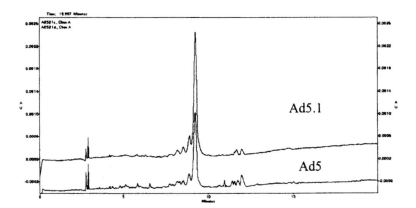

Figure 4. The electropherograms of the two purified viral vectors. The absorption was monitored at 214 nm. The method is described in section 2.3. The virus peaks elute from the capillary between 7 and 10 minutes[19]. The cause for the difference in the electrophoretic mobilities of different virus species, as seen in the smaller 'series peaks' migrating in front of the main viral peak, is yet unknown and is currently under investigation. The distribution between the different virus forms seems to be comparable between the two virus vectors. The peaks at approximately 3 minutes and at 11-14 minutes are caused by the applied solvents.

By the chromatographic and capillary electrophoretic analyses, the intact Ad5 and Ad5.1 vectors were indistinguishable as illustrated in Figures 2-4.

**5.2.2   Physical and physicochemical comparison of the dissociated
           vectors. Proteomic and genomic analysis**

We used the RP-HPLC assay to generate protein maps of the two vectors.
The results are shown in Figure 5.

The proteome fingerprints shown in the RP HPLC chromatograms of the
two product candidates are indistinguishable. For a more detailed analysis,
peak fractions were collected (peak assignments are shown in Figure 6) and
subjected to MALDI-TOF mass spectroscopy and N-terminal protein
sequencing (with or without Chloramine T digestion, see section 3.1.1.). The
results of the N-terminal sequencing and the MALDI TOF mass
spectroscopy are shown in Tables 1 and 2, respectively.

The N-terminal sequencing results show the expected identity between
the two constructs in both the blockade on the N-terminus (only proteins VI
and VII are not blocked) and the internal sequences generated by the
Chloramine T digestion. It is noteworthy that the Chloramine T cleavage
resulted in the same detectable sequences. This shows that the sensitivities of
the tryptophane residues to oxidative peptide chain cleavages are also
indistinguishable between the two constructs. The only differences we
observed in the N-terminal analysis between the two constructs were in two
minor peak fractions. One minor compound, the protein VII precursor
fragment in peaks 2 and 3 was not detectable in the Ad5.1 preparation and
L2 µ was not detected in peak 3 in the Ad5 preparation. These differences
may be explained by fraction collection differences (neighbouring peaks
may not be completely separated in the collected fractions) or by the fact that
the sequencing signals in these minor peaks were very low, <10 picomoles
and close to the sensitivity limit of the N-terminal sequencer instrument.

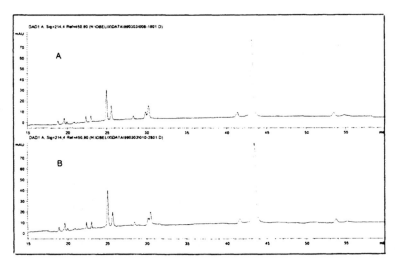

Figure 5. Proteome maps of the Ad5 and Ad5.1 vectors generated by the RP HPLC procedure. The method is described in section 3.1.1.

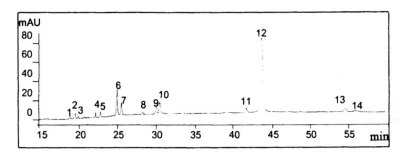

Figure 6. Peak assignments from the RP- HPLC chromatograms for identification and further analysis.

Table 1. N-terminal sequencing of the collected virus protein peak fractions from the RP HPLC (figure 6). For Ad5 all collected fractions were analysed directly and with Chloramine T digestion (see section 3.1.1). For Ad5.1, the peaks for proteins VI and VII were only tested without Chloramine T digestion to verify unblocked N terminus. X = anticipated amino acid was below detection in the corresponding sequencing cycle. No sequences were detected in peak 1.

Peak	Sequences Ad5	Protein ID, position Ad5	Sequences Ad5.1
2	GLRFPSK	VII precursor, from W_{13}	------
	ALTXRLR	Late L2μ, authentic N-term	ALTXRLR
3	GLRFPSK	VII precursor, from W_{13}	------
		Late L2μ, authentic N-term	ALTXRLR
4	QDIGTSTN	VI precursor, from W_{22}	QDIGTSTN
	SYQPQMG	VIII, from W_{11}	SYQPQM
5	PAALVYQ	VIII, from W_{73}	PAALVY
	FRHRVRS	VIII starting at F_{112}	
6	AKKRSDQ	VII, authentic N-term	AKKRSDQ
	VRDSV	VII, from W_{155}	
7	KGRRVKR	V, from W_{66}	KGRRVKRV
	MFRVSAP	V, from W_{270}	MFRVSAP
8	GRSSFTP	C-terminal segment of VIII	GRSSFTP
9	GGDLKTI	III, from W_{119}	GXDLKTII
	VEFTLPE	III, from W_{165}	VEFTXPE
	YLAYNYG	III, from W_{406}	YLAYNYG
10	AFSWGSLW	VI, authentic N-term	AFSWGSLW
	GSL	VI, from W_4	
	NSSTG	VI, from W_{26}	
	QSTLN	VI, from W_{196}	
11	LPPPGFY	IIIa, from W_{390}	LPXXGFY
	DDIDDSV	IIIa, from W_{412}	XDIDDXV
	KTYAQEH	IIIa, from W_{523}	KTYAQEH
12	SYXXISGQ	II, from W_{10}	SYXHISGQ
	DEAXTXL	II, from W_{135}	DEAXTALE
	NXAVDSY	II, from W_{387}	NQAVDSY
	SLDYM	II, from W_{524}	XLDYM
	NFRKDVN	II, from W_{583}	NFRKXVN
	AFTRLKTK	II, from W_{679}	AFTRLKTK
	PGNXXLL	II, from W_{731}	PGNXXLL
	RIPFSSNF	II, from W_{871}	RIPXXSNF
13	AGVRQNV	IX, from W_{22}	AXVRQNVM
14	ILPLLIPLI	Late L2 μ, starting from I_{51}	ILPLLIPLI

Table 2. MALDI TOF analysis of the collected virus protein peak fractions from the RP HPLC. No masses were detected for peak 1.

Peak	Protein ID from N-terminal sequencing	Predicted mass	Measured mass Ad5	Measured mass Ad5.1
2	VII precursor, from authentic N-terminus	2598	3037 + O adducts	3037 + O adducts
	Late L2 μ, authentic N-terminus	8715		2605(minor) 2508 (minor)
3	VII precursor, from N-terminus	2598	2618 2866	2618 2547
4	VI precursor, from authentic N-terminus	26996 3576	3624 + O adducts	3623 + O adducts
5	VIII precursor, from N-terminus to G_{111}	12125	3541* 12038	12037
6	VII, authentic N-term VII	19412	19410	19415
7	V, from W_{66} V, from W_{270}	41447	41461	41441
8	C-terminal segment of VIII	7642	7642	7642
9	III	63293	63377	63524
10	VI	22100	22104	22103
11	IIIa	63502	63736	63823
12	II	108008	107950	108033
13	IX	14458	14369	14369 14431
14	Late L2 μ	2888	2889	2888

The accuracy of the mass spectroscopic mass assignments was ± 0.1% (100 ppm) or better. The MALDI TOF analysis results in Table 2 showed the expected molecular masses for almost all the corresponding major proteins within the accuracy of the assay with the exceptions of proteins III and IIIa. These two proteins show larger molecular masses than predicted from the DNA sequence for both constructs possibly indicating posttranslational processing (for example glycosylation). Both molecules contain consensus N-glycosylation sequences (4 each) and several possible sites for O-linked glycosylation. To our knowledge there is no published report on the posttranslational modifications of these proteins. Published studies only describe some glycosylation of the fibre, protein IV[36]. It is important to note, however, that the observed masses for proteins III and IIIa are statistically indistinguishable between the Ad5 and Ad5.1 vectors possibly indicating similarities in the modifications on these proteins. As shown in Table 2, for some minor peaks we were not able to detect molecular masses consistent with the predictions by the N-terminal sequences. Nevertheless, the detected masses in these peaks (2, 3 and 4) were, for most cases, identical in the two virus vectors.

The analysis of the viral genome was fairly straight forward using the techniques described in section 3.2.1. Without going into experimental details, the characterisation (using 10 restriction digests, direct sequencing of the expression cassette and a variety of PCR based approaches) confirmed the anticipated differences and verified the identity of the rest of the viral genomes.

5.2.3   Immunological and *in vitro* biological comparison

The comparison of the two adenoviral vector products in the antibody binding assay is shown in Figure 7.

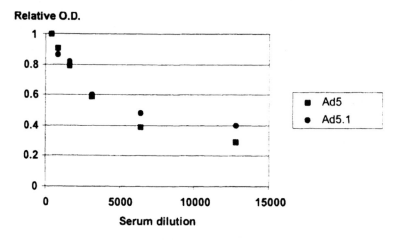

Figure 7. The characterisation of the surface antigens of the intact adenovirus vectors by binding of the viral vectors to serial dilution of a patient serum (from a patient treated with the Ad5 vector) containing anti-adenovirus binding antibodies. The method is described in section 4.1.

According to the data in Figure 7, the two adenovirus vectors were equally recognised by the human serum in the binding antibody assay. Similarly, we have not seen differences between the Ad5 and Ad5.1 vectors in antibody mediated neutralisation assays (data not shown). These findings confirm the observations made in the physicochemical characterisation of the two intact adenovirus constructs: the surface characteristics of Ad5 and Ad5.1 were indistinguishable.

The specific infectivities of the two purified vector constructs were tested using the RP-HPLC assay for mass determination and the endpoint dilution assay for infectivity analysis (section 4.2). The specific infectivities were statistically indistinguishable between the Ad5 and Ad5.1 vectors.

Replication competent adenovirus is frequently detected in the adenovirus produced using the HEK 293 cells. Replication competency was confirmed for the Ad5 vector used in our comparison study at or above 10^{11} particles/assay. The Ad5.1 vector produced in PER.C6 cells did not show replication competency in 20 tested lots at a sensitivity of 3×10^{12} particles.

Transgene expression driven by Ad5 and Ad5.1 was evaluated at the transcriptional level (using quantitative rtPCR in relevant non-permissive cells), and at the translational level (immunologically, using ELISA and Western blotting in both permissive and several relevant non-permissive cells). Without going into experimental details, the quantitative rtPCR and the ELISA assays did not show detectable differences between the two constructs. (Note that transgene expressions were driven by slightly different promoters and the orientation of the transgene cassette was different between the two constructs, as discussed above.) The Western blot analysis was unable to distinguish between the transgene protein products expressed by the Ad5 and Ad5.1 vectors. This observation suggested that even the posttranslational modifications of the expressed transgene products were indistinguishable.

The biological (growth promotion) activity of the expressed growth factor was tested in several relevant, non-permissive cell lines in response to infection with the Ad5 and Ad5.1 vectors. An example of such comparison is shown in Figure 8 using a primary human cell line infected with the virus constructs after growth arrest (exposure to low serum concentration).

The difference observed between the potencies of the two virus products in the experiment shown in Figure 8 is not statistically significant. The difference is less than the error of the infectivity determination by the endpoint dilution assay format (see section 4.2).

6. CONCLUSIONS

We adapted or newly developed an array of physicochemical and *in vitro* biological assays to assemble a comprehensive analytical package for the characterisation of a recombinant adenovirus type 5 gene therapy vector. The aim of our program was to reach a sufficient *in vitro* analytical definition of the virus to allow for demonstration of "product comparability"

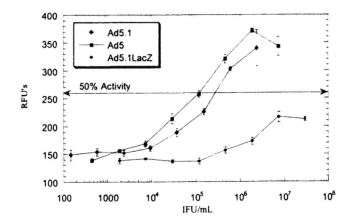

Figure 8. Comparison of the growth promotion activities of the expressed transgene protein products by the Ad5 and Ad5.1 vectors.   An Ad5.1 construct expressing the beta-galactosidase transgene was used as the negative control. Approximately 50000 cells were plated for the assay and infected by the indicated amounts of virus. Viral infectivity was determined independently by the endpoint dilution assay. Viable cell concentration was assessed by Alamar Blue staining and measuring fluorescence (in Relative Fluorescent Units, RFUs)

following process changes (such as optimisation of production process, facility changes, etc.). With this approach we were therefore aiming at bringing the recombinant adenovirus to the regulatory status of "well characterised" (or well understood) biological product.

In this study we gave an outline of our analytical approaches and we used two examples to demonstrate that a meaningful analytical definition of a complex system as a virus is possible with the technologies available today. First we showed how the problem of determining and verifying virus particle concentration can be solved by utilising orthogonal approaches in the analyis of the viral proteome and genome. Then we showed how we applied our analytical package to investigate the comparability of two vector products manufactured using different packaging cell lines.

ACKNOWLEDGEMENTS

We would like to acknowledge the contributions of Francisco J. Castillo, Pearl Chang, Ray-Jen Chang, Moutassem Elsheikh, Wayne Foley, Julia Grey, Joanne Johnson, Jacob Kung, Julie Lai, Dennis Lee, Maria Parkman, and Cynthia Soderblom.

REFERENCES

1. Martin, P.A. and Thomas, S.A., 1998, The commercial development of gene therapy in Europe and the USA., *Hum. Gene Ther.* **9**:87-114

2. Maizel, J. V., White, D. O. and Scharff, M. D., 1968, The polypeptides of Adenovirus 1. evidence for multiple protein components in the virion and a comparison of types 2, 7A, and 12. *Virology* **36**:115-125

3. Mittereder, N., March, K.L. and Trapnell, B.C., 1966, Evaluation of the concentration and bioactivity of adenovirus vectors for gene therapy. *J. Virology* **70**:7498-7509

4. Green, M., and Pina, M., 1964, Biochemical studies on adenovirus multiplication. VI. Properties of highly purified tumorigenic human adenoviruses and their DNA's. *Proc. Natl. Acad. Sci.* **51**:1251-1259

5. van Oostrum, J. and Burnett, R.M., 1985, Molecular composition of the adenovirus type 2 virus. *J. Virology* **56**: 439-448

6. Stewart, P.L., Fuller, S.D. and Burnett R.M., 1993, Difference imaging of adenovirus: bridging the resolution gap between X-ray crystallography and electron microscopy. *EMBO J.* **12**:2589-2599

7. Stewart, P.L., and Burnett R.M., 1995, Adenovirus structure by X-ray crystallography and electron microscopy. *Curr. Top. Microbiol. Immunol.* **199**: 25-38

8. Boudin, M.-L., D'Halluin, J.-C., Cousin, C. and Boulanger, P., 1980, Human adenovirus type 2 protein IIIa, II. Maturation and encapsidation. *Virology* **101**: 144-156

9. Anderson, C.W., 1990, The proteinase polypeptide of adenovirus serotype 2 virions. *Virology* **177**:259-272

10. Burnett, R.M. 1997. The structure of adenovirus. In: *Structural Biology of Viruses* (W. Chiu, R.M. Burnett and R.L. Garcea, eds), Oxford University Press, Oxford, New York, 1997, pp. 209-238

11. Shabram, P.W., Giroux, D. D., Goudreau, A. M., Gregory, R. J., Horn, M. T., Huyghe, B. G., Liu, X., Nunnally, M. H., Sugarman, B. J., and Sutjipto, S. (1997) Analytical anion-exchange HPLC of recombinant type-5 adenoviral particles. *Hum. Gene Ther.* **8**:453-465

12. Blanche, F., Cameron, B., Barbot, A., Ferrero, L., Guillemin, T., Guyot, S., Somarriba, S. and Bisch, D., 2000, An improved anion-exchange HPLC method for the detection and purification of adenoviral particles. *Gene Therapy* **7**:1055-1062

13. Huyghe, B.G., Liu, X., Sutjipto, S., Sugarman, B. J., Horn, M. T., Shepard, H. M., Scandella, C. J., and Shabram, P., 1995, Purification of a type 5 recombinant Adenovirus encoding human p53 by column chromatography. *Hum. Gene Ther.* **6**:1403-1416.

14. Hjerten, S., Elenbring, K., Kilar, F., Liao, J., Chen, A., Siebert, C. and Zhu, M., 1987, Carrier-free zone electrophoresis, displacement electrophoresis and isoelectric focuing in a high performance electrophoresis apparatus. *J. Chromatogr.* **403**:47-61

15 Zhu, A. and Chen,Y.,1989, High-voltage capillary zone electrophoresis of red blood cells. *J. Chromatogr.* **470**:251-260

16. Schnabel, U., Groiss, F., Blaas, D. and Kenndler, E., 1996, Determination of the pI of human rhinovirus serotype 2 by capillary isoelectric focusing. *Anal. Chem.* **68**:300-4303

17. Okun, V.M., Rohacher, B., Blaas, D.and Kenndler, E., 1999, Analysis of common cold virus (human rhinovirus serotype 2) by capillary zone electrophoresis: the problem of peak identification. *Anal. Chem.* **71**:2028-2032

18. Okun, V.M., Blaas, D. and Kenndler, E., 1999, Separation and biospecific identification of subviral particles of human rhinovirus serotype 2 by capillary zone electrophoresis. *Anal. Chem.* **71**:4480-4485

19. Mann, B., Traina, J.A., Soderblom, C., Murakami, P.K., Lehmberg, E., Lee, D., Irving, J., Nestaas, E. and Pungor, E.Jr., 2000, Capillary zone electrophoresis of a recombinant adenovirus. *J. Chromatogr. A.* **895**:329-337.

20. Lehmberg, E., Traina, J.A., Chakel, J.A., Chang, R-J., Parkman, M., McCaman, M.T., Murakami, P.K., Lahidji, V., Nelson, J.W., Hancock, W.S., Nestaas, E. and Pungor E.Jr., 1999 Reversed-phase high performance liquid chromatographic assay for the adenovirus type 5 proteome. *J. Chromatogr. B.* **732**:411-423

21. Traina, J.A., Lehmberg, E., Chakel, J.A. unpublished data

22. Carrion, M.E. personal communication

23. Prior, C., Bay, P., Ebert, B., Gore, R., Holt, J., Irish, T., Jensen, F., Leone, C., Mitschelen, J., Stiglitz, M., Tarr, C., Trauger, R.J., Weber, D and Hrinda, M., 1995, *Pharmaceutical Technology* April: 30-51

24. *Current Protocols in Molecular Biology. Volume 1.* 1998, (Chanda, V.B. ed.) John Wiley & Sons, Inc., New York

25. Bett, A.J., Krougliak, V. and Graham, F.L., 1995, DNA sequence of the deletion/insertion in early region 3 of ad5 dl309. *Virus Res.* **39**:75-82

26. Hierholzer, J.C., Halonen, P.E., Dahlen, P.O., Bingham, P.G. and McDonough, M.M., 1993, Detection of adenovirus in clinical specimen by polymerase chain reaction and liquid phase hybridization quantitated by time-resolved fluorometry. *J. Clin. Microbiol.* **31**:1886-1891

27. Adrian, Th., Wadell, G., Hierholzer, J.C. and Wigand, R., 1986, DNA restriction analysis of adenovirus prototypes. *Arch. Virol.* **91**:277-290

28. Murakami, P., and McCaman, M.T., 1999, Quantitation of adenovirus DNA and virus particles with the PicoGreen fluorescent dye. *Anal. Biochem.* **274**:283-288

29. Cesarone, C.F., Bolognesi, C. and Santi, L., 1979, Improved microfluorometric DNA determination in biological material using 33258 Hoechst. *Anal. Biochem.* **100**:188-197

30. Singer, V.L., Jones, L. J., Yue, S. T. and Haugland, R. P., 1997, Characterization of PicoGreen reagent and development of a fluorescence-based solution sssay for double-stranded DNA quantitation. *Anal. Biochem.* **249**: 228-238.

31. Nielsen, L.K., Smith, G.K. and Greenfield, P.F., 1992, Accuracy of the endpoint assay for virus titration. *Cytotechnol.* **8**:231-236

32. Graham, F.L., Smiley, J., Russell, W.C. and Nairn, R., 1977, Characteristics of a human cell line transformed by DNA from human adenovirus type 5. *J. Gen. Virol.* **36**:59-72

33. Fallaux, F., Bout, A., Van Der Velde, I., Van Den Wollenberg, D., Hehir, K.M., Keegan, J., Auger, C., Cramer, S., Van Ormondt, H., Van Der Eb., A., Valerio, D. and Hoeben R., 1998, New helper cells and matched early region 1-deleted adenovirus vectors prevent generation of replication-competent adenoviruses. *Hum. Gene Ther.* **9**:1909-1917

34. Hehir, K.M., Armentano, D., Cardoza, L.M., Choquette, T.L., Berthelette, P.B., White, G.A., Coutire, L.A., Everton, M.B., Keegan, J., Martin, J.M., Pratt, D.A., Smith, M.P., Smith, A.E. and Wadsworth, S.C., 1996, Molecular characterization of replication-competent variants of adenovirus vectors and genome modifications to prevent their occurances. *J. Virol.* **70**:8459-8467

35. Zhu, J. Grace, M., Casale, J., Chang, A.T.-I., Musco, M.L., Bordens, R., Greenberg, R., Schaefer, E. and Indelicato, S.R., 1999, Characterization of replication-competent adenovirus isaolates from large scale production of a recombinant adenoviral vector. Hu. *Gene Ther.* **10**:113-121

36. Gheesling, K., Haltiwanger, R.S., Hart, G.W., Marchase, R.B. and Engler, J.A., 1990, Relative accessibility of N-acetylglucosamine in trimers of the adenovirus types 2 and 5 fiber proteins. *J. Virology* **64**:5317-5323

Validation of Gene Therapy Manufacturing Processes
A case study for adenovirus vectors

DOMINICK VACANTE, GAIL SOFER, STEPHEN MORRIS, AND CHRIS MURPHY
BioReliance, 14920 Broschart Road, Rockville, Maryland 20850-3349, USA

1.   INTRODUCTION

Specific issues of concern in the validation of gene therapy viral vector manufacturing processes include quality of raw materials, safety testing of cell and viral banks, production and purification of the vector, in-process and final-product testing, and validation of analytical methods. Since most vectors are produced in multi-product facilities, cleaning validation is a major concern due to potential product-to-product cross contamination. Viral clearance also presents a major validation challenge due to the nature of the product. As with any relatively new technology, the testing and validation requirements are still in development, but the basic principles of cGMP and process validation apply. Adenovirus vector manufacturing is presented here to illustrate a logical approach to validation of a gene therapy viral vector manufacturing process.

2.   REGULATIONS   PROVIDING   VALIDATION GUIDANCE

Currently, there are no licensed gene therapy virus vectors in use, and only a few regulatory documents are directly concerned with gene therapy vectors.[1, 2, 3, 4] Some regulatory guidance that applies to other products may provide useful information on validation requirements. It is noted in both European and U.S. publications that other documents should be consulted,

227

including guidances, points to consider, and ICH guidelines. Some relevant topics addressed in these publications are viral safety, cell line characterisation, and validation of analytical methods. Documents intended for specified biotherapeutics and those that address validation of vaccines can also provide some guidance. Guidance documents on chemistry, manufacturing, and controls content and format are available from the FDA's website. They explain what information should be included in a license application.

Most regulations that apply to validation of biological processes and products are broad whereas Code of Federal Regulations (CFR) and Pharmacopeia methods provide more details. There are some overriding safety concerns, however, that dictate some of the compliance and validation requirements. It is important to recognise that a process does not have to be fully validated prior to entering clinical trials. However, retention samples should be taken and stored properly as soon as any product is made, so that as the process is improved one can evaluate the impact of changes.

Many analytical methods will still be in development during Phase I clinical trials, and the process will almost certainly need refinement. However, prior to putting a gene therapy, or any other, product into humans, analytical methods used to assess safety, especially those used for lot release, should be validated. Any equipment related to sterility assurance should also be validated at this time and cell lines and/or banks should be characterised.[5] In preparation for Phase III clinical trials, cell line stability should be validated, and validation of the process capability to remove adventitious agents and known impurities should be conducted. Stability of both process intermediates and product should be evaluated.

The reader is also referred to other chapter(s) in this book that present the latest regulatory issues. Valuable websites include www.fda.gov and www.eudra.org.

3.   GENERAL PRINCIPLES OF VALIDATION

Process validation is documented evidence that a process will consistently produce product meeting pre-determined specifications. Validation takes time and it can be expensive, but in the long run it is good business since it diminishes the likelihood of batch failures. Gene therapy products may be produced in small quantities. In the past most of these products were developed in university hospitals, where familiarity with biopharmaceutical manufacturing and issues such as validation are minimal and funds are often severely limited. Everyone recognises, however, the importance of producing gene therapy products that can be safely administered to patients. Product quality is critical, and consistent, high quality product can only be ensured if a combination of good manufacturing practices and control of raw materials, cell culture, recovery, and purification processes are employed. Analytical methods that can be validated must be used to demonstrate that the process intermediates are consistent, the process raw materials controlled, and the end product adequately tested. Deciding early in development to design validation into the process and taking a logical stepwise approach to validation during clinical trials is the most cost effective strategy. Furthermore, technology transfer can be expedited when a process is well defined, controlled, and documented.

A development report should provide a rationale for the chosen process and control parameters. This same report can be very useful in the future when process changes are made. Since most of the unit operations are performed in multi-product facilities, special attention must be paid to avoidance of cross contamination, validation of cleaning procedures, operator training, and adherence to standard operating procedures. Validated analytical methods are employed for the cleaning validation. For early development products, in particular, the use of disposable equipment is often warranted to avoid cross contamination and reduce the costs associated with cleaning validation.

There are certain activities that are required for validation. It is essential that validation is implemented in accordance with an approved validation protocol. Processes cannot be validated in equipment that has not been qualified. Qualification of equipment generally includes an installation qualification and an operational qualification. Simply put, the installation defines each piece of equipment and/or system, including software that may be used for control and analysis. The operational qualification is performed to ensure that the equipment meets its design specifications and operates as it should. Once equipment is qualified, the process can be validated. Process validation is typically performed in preparation for Phase III clinical trials. Consistency is demonstrated in 3-5 consecutive batches made at pilot or full scale. During the validation runs, state-of-the-art analytical tools are used. Not all of these analytical methods may be required for routine manufacturing.

3.1   Raw materials

The facility and raw material testing and control are some of the corner stones for building quality and consistency into the manufacturing process and, therefore, the product. Programs for testing and control are established prior to initiating validation and product manufacturing. In the manufacturing process all raw materials are controlled, including but not limited to plastic ware, glass vials, salts, buffers, cell and virus banks. Furthermore, as specified by cGMPs, all raw materials must have at least one identity test (21 CFR 211.84). However, for Phase I manufacture of clinical trial materials, less extensive testing may be allowed. To control raw materials, manufacturing operations typically have receiving, inspection, quarantine and release programs under the supervision of the quality control unit with oversight by quality assurance and regulatory affairs. Integral to the program is obtaining supplies from qualified vendors, most of whom also produce the supplies in compliance with cGMPs. With the receipt of each supply, a certificate of analysis is obtained to support the vendor's labelling. After visual inspection of the supply, the material is placed into quarantine until authorised for release for use in manufacturing. Identity and other tests are performed to show the material meets pre-determined specifications.

As an example, sodium chloride would be received and the container and label inspected for proper material, grade, size, seal, any obvious punctures, etc., to provide evidence the material was received as indicated and has not been compromised. Aliquots of the salt would be removed to identify it as a white powder and tests would be performed to show the material consists of sodium and chloride; for example, flame photometry and silver chloride precipitation assays, respectively. Upon passing the inspection and testing, the material would be labelled for release and placed in the appropriate controlled location for released raw materials. Buffer solutions prepared using the salt are also controlled, tested and released.

Complex raw materials such as foetal bovine serum (FBS) typically undergo identity and biological tests to assure quality. For example, in addition to identity tests, FBS would be analysed for viral contaminants and the presence of mycoplasma and other microbial contaminants before release. In addition, cell culture medium would be prepared with an aliquot of the specific lot of FBS for performance qualification or what is more commonly called a growth promotion test. The performance qualification uses the same cells as those in production. Cell growth characteristics of the cells using the new lot of FBS are compared to cell growth with a previously accepted lot of FBS to show the new lot of FBS will meet cell growth requirements for production. The combination of testing and growth promotion assure that the quality of the new lot of FBS is equivalent to previously used lots and that it will perform in a consistent manner in the production process.

Although these examples are just a small part of the overall program they show how a raw material receiving and release program builds quality into the manufacturing process.

3.2   Cell and viral bank characterisation

Cell and virus banks form an important resource for consistent production of the product. Quality is built into the banking system by using qualified raw materials, a cGMP facility, approved procedures and thoroughly characterised banks. Each bank is considered a candidate until it meets all pre-determined specifications outlined in the characterisation and qualification of the bank in production. In practice, several candidate banks are prepared and a final selection made on performance criteria. The following is a description of cell and virus banking for adenovirus products, which is a part of the overall manufacturing quality programme.

Cells used to prepare a cell bank are typically obtained from a research or development laboratory and undergo preliminary testing to assure the absence of mycoplasma and other microbial contamination prior to entry into a cell banking facility. It is also important to know the history of the cells, donor tissue, number of passages to establish the line and method of conversion to continuous growth. For adenovirus vector production, the cell lines currently used are 293 cells[6] and Per.C6[7] These cells are of human origin and contain only a portion of the adenovirus genome, the E1-region, which is integrated in the genome of the cells. The E1-region is necessary for virus replication and is deleted in replication-defective adenoviral vectors, thus the cellular production of the E1 gene products compliment the genes present in the vector for vector production. For the master cell bank (MCB), cells are cultured until sufficient cell numbers are obtained to create the bank, typically, $1 - 2 \times 10^9$ total cells. All the cells for the bank are collected into a single pool, thoroughly mixed, suspended in medium with a cryoprotective agent such as dimethylsulphoxide, then aliquotted into cryovials. A typical bank consists of 100 to 200 vials with each vial containing approximately 1×10^7 cells. The cells are frozen at a controlled rate and stored in the vapour phase of a liquid nitrogen freezer. For extra precaution, after characterisation most banks are split among two or more freezers and if necessary, stored in different facilities to assure safekeeping of the banks over the life span of the product.

Working cell banks (WCB) are prepared from one or more aliquots of the MCB. The cells are propagated and frozen as noted above. Because the WCB is the starting point for production, the WCB is typically larger than a MCB and for example might be in the range of 200 to 500 vials. The combination of a MCB and WCB is considered a two-tiered banking system and allows for a substantial number of production runs. For example, if one vial of a 100 vial MCB is used to make 200 vial WCBs and one vial of a WCB is used to prepare a single lot of product, then it would be possible to produce 20,000 lots before exhausting the MCB and WCB. With the use of a production cell seed train and multiple production lots from each WCB vial, the number of possible lots increases dramatically. Clearly, it would be rare that the two-tiered banking system described above would not survive the useful life of the product.

Characterisation of the MCB and WCB is shown in Table 1. The tests are subdivided into categories based on identity and adventitious virus testing. Because the cells are of human origin, human virus testing is predominant. Characterisation of the WCB involves only a subset of the tests listed in Table 1 because it is assumed that it is prepared under controlled conditions and some of the agents wouldn't be possible contaminants.

Table 1. Testing cell and virus banks for adenovirus vector production

Identity (Genetic, integrity, stability, and isoenzymes for cells only)
Microbial contaminants including Mycoplasmas
Adventitious and endogenous agents
    In Vitro and *In Vivo* virus screening assays
    Electron microscopy
    HIV-1 & HIV-2, Hepatitis B and C viruses, EBV, CMV, HHV-6, 7, 8
    HTLV-I and HTLV-II, Adeno-associated virus, Bovine viruses, Porcine Parvovirus
    Replication competent adenovirus (for virus banks only)

Preparation of virus banks presents a particular challenge to the manufacturer because the size of the master virus bank (MVB) and working virus bank (WVB) can vary widely depending on the efficacious dose found in clinical trials and the scale and efficiency of manufacture. Further, because adenovirus is a lytic virus, killing the cells during its growth cycle, preparing large quantities requires a stepwise approach of sequential pools of infected cell harvests. For Phase I clinical trials, an MVB resulting from the infection of 5×10^9 cells, typically yields sufficient virus for preparation of multiple lots needed for a dose escalation trial with a high dose of 1×10^{11} virus particles. MVBs and WVBs are prepared as crude virus from the clarification of a lysed virus-infected cell harvest. This method maximises the yield and the protein in the medium acts as a cryoprotectant during long term storage. However, purified virus banks can successfully be made and are acceptable.

Testing of the MVB and WVB is shown in Table 1 and is similar to the cell bank characterisation where human virus testing is predominant. However, in addition to these tests, a test unique to adenovirus vector production is the replication competent adenovirus (RCA) testing, which is necessary because replication-defective adenoviral vectors have the potential to form RCA during production due to recombination of the vector sequences with the E1-region found in the cellular genome. When 293 cells were the most likely cell substrate for production, RCA results compiled from a wide variety of adenoviral products indicated there was approximately a 25% chance of detecting at least 1 RCA per dose at doses of $1x10^8$ to $1x10^9$ PFU or up to ~2 x 10^{10} virus particles[8]. Per.C6 cells have been designed to minimise the formation of RCA; however at the time of this writing, results from a large number of large-scale production runs have not been published. Nevertheless, calculations based on molecular genetics suggest the level of RCA to be much lower than the experience with 293 cells.

3.3   Vector production

The production of adenoviral vectors involves the propagation of cells from the MCB or WCB to build a cell seed train to sufficient number of cells to meet production needs. Production needs vary and have been projected by one group to be as high as 10^{16} to 10^{18} total virus particles per year[9]. Using typical yields per cell and cell density as an estimate of volume, the production volume needed to produce these quantities of virus could be up to 10,000 litres per year with projected commercial needs significantly exceeding this level. For production of most Phase I clinical trial materials, cells are grown in an attachment-dependent manner. Production methods range from production in plastic ware (such as dishes, 500 cm^2 trays, Nunc^{TM} cell factories and roller bottles) to use of the Costar Cell Cube^{TM} and microcarriers in bioreactors. Improvements have been made to provide easier scale-up by adapting the cells to growth in suspension and in most cases they are adapted to suspension growth in serum-free medium[10]. Serum-free growth avoids some of the regulatory issues encountered with the use of fetal bovine serum. However, typically the supplements used in the medium are derived from bovine or human sources, which carry some of the same or similar regulatory issues as FBS. The other advantage to serum-free medium is that the medium has defined proteins and known concentrations that may provide feedstreams that are more amenable for development of downstream processing procedures. The upstream process also would be part of the overall validation program and the following is a description of the basis for virus production and details the starting point for recovery and purification of adenovirus vectors.

Adenovirus is very cell-associated with most of the virus remaining within the cell membrane even though they exhibit extensive cytopathic effect (CPE), however at very late stages of virus growth the virus is released into the culture medium. Some production methods take advantage of this biology by infecting cells with virus from the MVB or WVB at a high multiplicity of infection, that is, in the range of 2 to 10 infectious virions per cell. This leads to a synchronous infection of all the cells in the culture. After

36 to 48 hours, the cells typically show extensive CPE and detach from plastic surfaces or microcarrier surfaces. When using plastic ware in a small-scale mode of operation, the cells are harvested and concentrated using low-speed centrifugation. Virus is released from the cells by freezing and thawing the cells, then the virus purified by caesium chloride density ultracentrifugation. These methods in most cases are not amenable to large-scale operations; thus, there has been significant effort to develop scalable downstream methods.

In a large-scale operation, there are, in general, two approaches to the initial processing of the infected cells to release the virus: 1) allow the infection to go to completion until the cells lyse and most of the virus is found in the culture medium, or 2) disrupt the intact cells to release the virus. Cell disruption is performed either mechanically using a microfluidiser to break open the cells or chemically using detergents to solubilise the cell membrane. Specific methods are established during process development to ensure that the initial processing does not affect the integrity of the product or its stability. In either case, the feedstream at this stage is a complex mixture of intact virus, viral nucleic acids and proteins, cellular proteins and nucleic acids, and culture medium components. Ultrafiltration technology and chromatographic procedures are the primary techniques used for purification of adenovirus vectors at large-scale. The next section contains information on the validation of the recovery process and chromatographic procedures for purification in cGMP manufacture.

3.4   Vector recovery and purification

3.4.1   Purity, yield, impurities, and contaminants

During process validation, the yield and purity of each step is measured and recorded, generally while operating at the outer limits of the specified control parameters (worst case). Consistency of the removal of impurities is documented. Impurities may include in-process materials such as cell culture additives, solubilising agents, as well as host cell proteins and nucleic acids. Viral clearance validation (or evaluation) studies will need to be performed prior to licensure. (see Conclusions). Control over microorganisms is demonstrated by bioburden and endotoxin assays. In general, assays are developed to detect the product and product- and process-related impurities in process intermediates and the final product.

3.4.2 Processing times and stability

The entire recovery and purification operations must be carried out within a defined time frame. In addition to adhering to the time specified for each unit operation, holding times for process intermediates must be defined. The stability of process intermediates is demonstrated by using validated assays such as infectivity determinations and other stability indicating assays.

3.4.3 Cleaning validation

Holding times for equipment prior to its cleaning are also specified, and the ability of the cleaning protocol to remove impurities to a defined level after the maximum holding time is validated. The type and concentration of the cleaning reagents are defined along with the SOPs for their preparation, contact time and temperature, cleaning frequency, and storage conditions prior to and after cleaning. These factors are all held within specifications during cleaning validation. Safety margins should have been designed into the process to ensure adequate cleaning with a worst case soil. Validation of removal of cleaning agents is also part of the entire cleaning validation. Cleaning validation requires the use of both specific and broad-spectrum methods. For the manufacture of viral vectors, PCR can be a useful tool for cleaning validation. Further information on cleaning validation can be found in a few books[11, 12].

3.4.4 Case study

The following case study for recovery of an adenoviral vector illustrates many of the most important validation principles. Figure 1 illustrates the overall process. After production of the vector at a 10-liter scale in CellCube™, microcarrier, or suspension culture, the cells are disrupted by a microfluidiser, clarified by depth filtration, concentrated and diafiltered by hollow fiber ultrafiltration. The DNA from the 293 cells in which the vector is cultured is enzymatically degraded. An anion exchanger is used as a capture step, and this is followed by a size exclusion chromatography step that is used for polishing and buffer exchange. A final concentration and formulation step will complete the process.

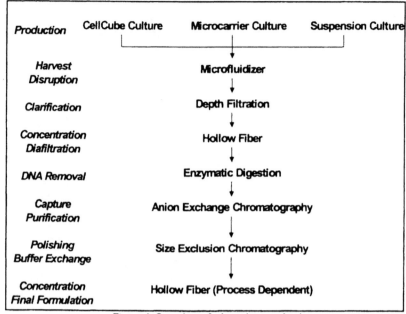

Figure 1. Overview of adenovirus production

3.4.4.1   Cell disruption and feedstream clarification

Most cell disruption processes are controlled by pressure and in some cases by heat. There may also be other parameters depending on the methodology. When validating the cell disruption step, one must determine the percentage of broken cells, product quality and product recovery. In the case of the microfluidiser, the percentage of viable cells remaining is determined by trypan blue staining.

Microfluidisation was chosen for this example because it offers a simple flow-path. It processes both cells and supernatant, which is advantageous for adenovirus since the virus is found both within cells and in the supernatant. With freeze-thawing that had been used previously, adenovirus in the supernatant is lost. The microfluidiser has a high throughput and product quality is shown to be equivalent to or better than that produced by 3 freeze-thaw cycles (see Figure 2). This kind of information is put into the development report to justify the selection of the microfluidiser as the method for cell disruption.

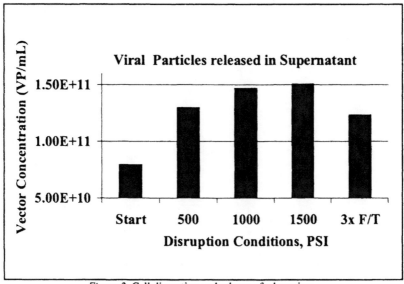

Figure 2. Cell disruption and release of adenovirus

The microfluidiser provides a closed, sterilisable system, but cleaning validation is an issue that needs to be addressed. This system is dedicated for use with one product, and PCR can be used as one of the assays for cleaning validation. The feedstream is clarified in line with a depth filter. There is only one pass, which further simplifies the validation of this step as compared to validating freeze-thaw cycles.

3.4.4.2   Product concentration and diafiltration

Hollow fiber ultrafiltration concentrates the product and is used to exchange the buffer into a solution suitable for the next step, which is digestion of DNA. For the ultrafiltration, one validates product recovery and purity using appropriately designed assays. For the buffer exchange, validation includes determining the number of volumes it takes until the correct pH and conductivity are reached. In development it was found that ultrafiltration using a high molecular weight cutoff is a good technique for this phase of the process because it does not shear adenovirus (see Figure 3). This step reduces the total protein, most of which comes from cell culture, by 60%. Ultrafiltration is a controllable, scalable process that is readily validated. The consistency of protein reduction during 3 pilot scale runs is shown in Figure 4.

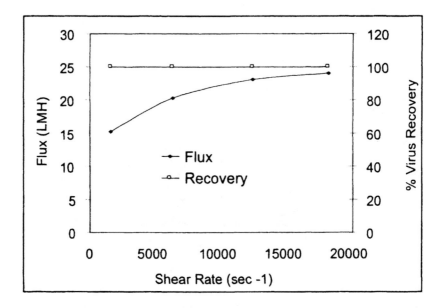

Figure 3. Effect of shear rate on the recovery of adenovirus using hollow-fibre ultrafiltration

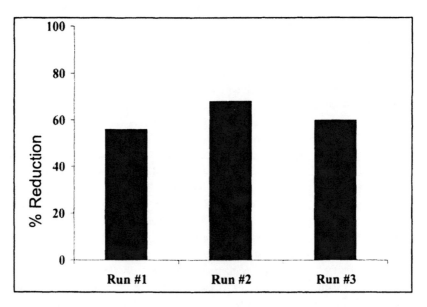

Figure 4. Use of hollow-fibre ultrafiltration to reduce the total protein content of the process feedstream

3.4.4.3   DNA removal

DNA is removed to reduce viscosity and increase the capacity of the first chromatography column for the product. The critical issues for validation of this step include amount of DNA digestion, product recovery, and removal of the nuclease. The nuclease and a validatable assay to determine its removal from the product feedstream are commercially available. In development, conditions for optimal nucleic acid digestion were established, which include magnesium chloride concentration, pH, ionic strength, time and temperature of the enzymatic digestion, and mixing speed. During validation, it is shown that these conditions lead to consistent DNA removal and product recovery in subsequent steps.

3.4.4.4   Purification

When adenovirus vector is grown in medium containing FBS there is a heightened concern for safety, which makes the purification process critical to the safety of the final product. In the anion exchange capture step, the highly negatively charged adenovirus binds, along with any residual contaminating DNA. Impurities such as BSA from the serum pass through the column. During validation of this capture step, we measure product recovery and purity, as well as the impurity profile. We control column geometry, linear flow, sample concentration and volume, pH, and conductivity. Ranges that provide the requisite product intermediate are used. In the development report, we describe screening of several anion exchangers, the criteria for our selection, and why we chose this step, which provides a 1000-fold purification and concentrates the virus for the next step.

A size exclusion chromatography column that provides a 3-fold purification and buffer exchange for the next processing step follows the capture anion exchange step. Again, we measure conductivity, pH, column geometry, linear flow, and sample load. In size exclusion, sample volume is a critical parameter and our specification for volume is within a much narrower range than in the capture anion exchange step. Within the established ranges for these parameters, we validate product recovery and purity and impurities profile. Figure 5 illustrates the type of data that can support the development report. It shows the purity measured as number of viral particles per mg protein plotted against four different chromatographic resins. This justifies our selection of resin number 4 and a flow rate of 15 cm/hr.

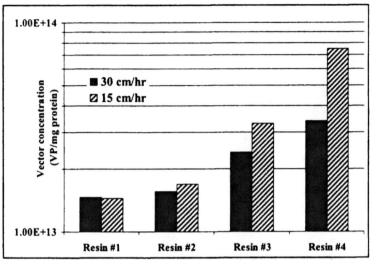

Figure 5. Purification of adenovirus using size-exclusion chromatography

3.4.4.5   Overall process validation considerations

Figure 6 shows the consistency of the process during three runs. There are no steps that present control problems.

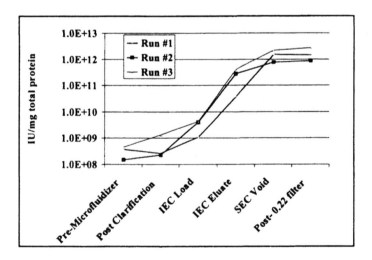

Figure 6. Reproducibility of the adenovirus manufacturing process

3.4.5 Process control systems

As indicated above, facilities and raw material control programs form the foundation on which quality and consistency are built into the process and therefore the product. Validation of the process also assures consistency and sets the guideposts for in-process testing. Process control systems are all of the analyses within a process that are used to assure the process is performing as previously determined. The testing programs are used as a series of checks throughout the process and for release of the product. The control systems include in-process analysis, product characterisation, lot release testing, and product stability programs. Validation of lot release tests and other tests provide another level of quality assurance to the overall manufacture of the product.

Acceptance criteria for each of these analyses, an important part of the overall control strategy, are established during product development and validation studies. The material at each stage must conform and be within the acceptable limits set for each respective analysis to be acceptable for use. As defined in the ICH guideline on specifications[13], "Specifications are critical quality standards that are proposed and justified by the manufacturer and approved by regulatory authorities as conditions of approval." The guideline further states that specifications are not intended to establish full characterisation but are used to ensure safety and efficacy of the product. Furthermore, in-process testing may be recorded as action limits or reported as acceptance criteria.

The major areas of testing are the drug substance and the drug product. The drug substance is the bulk drug, which in this case includes the unprocessed bulk harvest or in the case of adenovirus, the infected cell harvest and the purified bulk product. The drug product is the dosage form of the product or the final formulated product in the container/closure system. Tests are designed to determine identity, potency, quantity, purity, product- and process-related impurities, and contaminants. For example, for adenovirus vectors, product-related impurities could be excess penton in the product and process-related impurities could be cellular DNA and host cell proteins. Contaminants are those materials that are adventitiously introduced into the manufacturing process, such as other viruses, mycoplasma or other chemicals not intended to be a part of the process.

Furthermore, adenoviruses contain proteins, hexon and penton, that make up the virion coat that are toxic in some testing systems. This is especially true when testing the infected cell harvest. In those cases, an approach could be taken which is similar to testing used for some vaccines, that is, the use of production control cells, which are uninfected cells that are carried through the manufacturing process and then used for testing purposes (21 CFR 630.13).

An example of the control system for an ion-exchange column is shown in Table 2 and for the bulk drug substance is shown in Table 3. In addition to the tests listed in Table 3, a more thorough characterisation of the virus might be performed. Analyses using liquid chromatography coupled to mass spectrometry or tandem mass spectrometry and carbohydrate analysis of individual protein species would provide more detailed information of the product's characterisation. Although informative, at this time this type of analysis of adenovirus vectors is beyond that required as part of the specifications.

Table 2. Process control for ion-exchange chromatography

Volume of eluate
Conductivity
Virus concentration
Particle determination
Infectivity determination
Yield

Table 3. Process control for the bulk drug substance

Bioburden
Bacterial endotoxins test
Virus concentration
Particle determination
Infectivity determination
Particle-to-infectivity ratio
Genetic identity
Protein identity
Residual host cell protein
Residual host cell DNA
Residual nuclease
Purity
Potency
Particle size distribution
Protein concentration
pH
Overall yield

4. CONCLUSIONS

4.1 Today

Validation should be designed into gene therapy processes. Validation of a gene therapy vector process not only enhances patient safety, it also prevents failed batches. Validation should be planned for and implemented in a logical, timely manner. Key elements of a validation plan for manufacture of a gene therapy viral vector include cell and viral bank characterisation, vector production, and vector purification. Some guidance can be found in U.S., European, and ICH documents, but it is also necessary to take a case by case approach in which dosage, patient population, and indication are considered.

4.2 Future

New technologies are being developed that are increasing the sensitivity of analytical tools used for validation. Among these techniques is PCR. Although there are no licensed gene therapy products, and thus it is difficult to predict what future regulatory requirements will be, it seems logical that somehow it will be necessary to demonstrate that no harmful viral agents are co-purified with a viral vector. PCR should enhance our ability to perform this type of validation study. Cleaning validation continues to be a major concern and PCR can be of significant value. As with any relatively new product, it is important to stay current with the latest developments and guidance documents from regulatory authorities.

ACKNOWLEDGEMENTS

The authors would like to thank Erik Wilhelm for technical support and Michael Wiebe for discussions and critical reading of the manuscript.

REFERENCES

1. U.S. FDA, CBER, Guidance for Human Somatic Cell Therapy and Gene Therapy, March 1998.

2. CPMP, Gene Therapy Products-Quality Aspects in the Production of Vectors and Genetically Modified Somatic Cells, December, 1994.

3. CPMP, Annex, Safety Studies for Gene Therapy Products, 1998.

4. CPMP, Note for Guidance on the Quality, Preclinical and Clinical Aspects of Gene Transfer Medicinal Products, 2001.

5. U.S. FDA, Draft Guidance for Industry: CMC Content and Format of INDs for Phases 2 and 3 Studies, Including Specified Biotechnology-Derived Products, 1999.

6. Graham F, Smiley J., Russell W., and Nairn R. (1977), Characteristics of a human cell line transformed by DNA from human adenovirus-type 5. *J. Gen. Virol.*, **36**: 59-72.

7. Introgene, Leiden, The Netherlands.

8. Vacante D, Spaltro J, and Dusing S. (1997), Replication Competent Viruses: Assays and Issues. Forum 1997, Gene Therapy Conference sponsored by the NIH and FDA/CBER, Bethesda, Maryland.

9. Chang S (1998), Emerging Vectors and Technology Symposium. American Society of Gene Therapy, Seattle, Washington.

10. Iyer P, Ostrove J. and Vacante D. (1999). Comparison of manufacturing techniques for adenovirus production. *Cytotechnology* **30**:169-172.

11. PDA Biotechnology Cleaning Validation Subcommittee, 1996, *Cleaning and Cleaning Validation: A Biotechnology Perspective*, PDA, Bethesda, MD.

12. Sherwood, D. (2000) Cleaning: Multiuse Facility Issues. In *Biopharmaceutical Process Validation* (G. Sofer and D. Zabriskie, eds), Marcel Dekker, New York, pp.235-250.

13. CPMP/ICH/365/96. Note for Guidance on Specifications: Test procedures and acceptance criteria for biotechnological/biological products. Step 4, Consensus Guideline, 10 March 1999.

Gene Delivery

AKSHAY ANAND
Dept. of Immunopathology, Post Graduate Institute of Medical Education and Research, Chandigarh, India

1. INTRODUCTION

Gene therapy - no other medical treatment has ever been given such an over whelming advance credit. What is the reason for this euphoric belief in success? It is the visionary concept of treating human genetic diseases in the most straight forward, direct and effective way: by supplementing or replacing defective genes by their normal counterparts causal therapy, in contrast to symptomatic therapy, should be possible. We have witnessed an explosive growth of the field in terms of numbers of approved clinical trails and in the wide range of disorders that are under treatment during the past 5 years. The majority of the trials are aimed at treating disease affecting large patient populations, such as cancer and AIDS rather than rare single-gene defects. Although substantial progress on the way to efficient gene therapy in humans has been made, a number of problems, particularly in the area of specific and efficient gene delivery need to be solved before gene therapy can be safely and effectively applied.

The technical possibility to transfer any cloned gene into any cell of an organism demands a fundamental distinction: is the gene transferred into a specific somatic cell type - as is the case in human gene therapy, where the affected individual and not its offspring is treated and where integration of foreign DNA has to be excluded - or is the gene deliberately introduced into the germline of a laboratory animal, which is the experimental approach to study the function of a specific gene during development, and to establish precise animal models for human genetic diseases.

At the most basic level, gene therapy can be described as the intracellular delivery of genetic material to generate a therapeutic effect by correcting an existing defect or providing cells with a new function. Initially, only the inherited genetic disorders were in focus but now a wide range of diseases, including cancer, neurodegenerative disorders vascular disease and other acquired diseases are being considered as plausible targets.

The ultimate goal of gene therapy is the alleviation of disease upon a single administration of an appropriate therapeutic gene.

The genetic materials considered for use are intended to replace a defective or missing gene, augment the functions of the genes present, instil a specified sensitivity to a normally inert prodrug or to interfere with the life cycle of infectious diseases.

1.1   Germline gene therapy

Currently Germline gene therapy is considered to be ethically unacceptable, even though it has the potential to eradicate many hereditary disorders. The technology is relatively simple and requires no targeting as genetic abnormalities can be corrected by direct manipulation. With increasingly sophisticated techniques being developed in the field of transgenic animals, it is now possible, using homologous recombination, to replace old genes for new. The results of the human genome project and establishment of gene function will open up new possibilities for genetic intervention that could be passed down through generations. It is this persistency that frightens many ethicists, and so to date all gene therapy applications have considered only somatic gene manipulation.

Until such time as gene therapy for many illnesses is commonplace, with all the hurdles to routine use of gene therapy overcome and side effects identified and dealt with, germline gene therapy will not be attempted. Ultimately, however, the question of germline gene therapy will have to be re-evaluated and we may see many of these crippling inherited diseases disappear from the gene pool. This type of scenario serves to underline the untapped potential of gene therapy. For the first time we are at a point where it is within our grasp to rid an individual and their progeny of the misery of an inherited disorder.

1.2   Somatic gene therapy

Somatic gene therapy involves the insertion of genes into diploid cells of an individual where the genetic material is not passed onto its progeny. The transfer is currently achieved by use of various vectors like lipids and virus-mediated systems. To date, there are three divisions of somatic gene therapy.

1.2.1   *Ex vivo* delivery

This system employs delivery of genetic material is after explantation, cultivation and manipulation *in vitro*, followed by subsequent re-implantation.   The *ex vivo* route is attractive mainly due to minimal complicating immunological problems and enhanced efficiency of vector delivery *in vitro*.   Cells can be manipulated to incorporate DNA-encoding proteins dysfunctional or lacking in the host.   An example of this is the transduction of T lymphocytes to express adenosine deaminase (ADA), the enzyme known to be defective in the severe combined immunodeficiency (SCID) ADA disorder.   Alternatively, in the case of familial hypercholesterolemia, secretion of protein (e.g. ApoAI/ApoE) could provide benefit by lowering serum cholesterol.

The protein expressed at therapeutic levels for extended periods reduces the need for readministration of purified protein or drugs to control the effects of the disease.   This contributes significantly to reducing healthcare costs and improving the quality of life of many sufferers.   There are, however, major constraints on the use of *ex vivo* therapy in that only some disease states are suited to the approach.   In addition, at present only a small percentage of re-implanted cells remain viable.

1.2.2   *In situ* delivery

Administering the material directly to the desired tissue is currently the major area of clinical interest, as many of the current delivery systems lack effective targeting.   One disease that has shown some success with this strategy has been cystic fibrosis (CF).   This disorder results from mutations in the cystic fibrosis transmembrane conductance regulator (CFTR).   As a consequence, mucociliary action is impeded leading to bacterial colonisation of the airways.   The CFTR gene has been delivered using lipid and adenoviral vectors to defined sites in the respiratory tract. Results have demonstrated a 20-30% correction in CFTR mediated ion transport at these sites.

This form of delivery is also used in the treatment of cancer. A study in a rodent model used direct administration of a retroviral vector expressing the suicide gene herpes simplex virus thymidine kinase (HSVTK) directly into small intracerebral gliomas.   Upon treatment with the prodrug gancyclovir, 75% of the tumours showed regression.

However, low efficiency of transduction is a continuing problem. This is a particularly important consideration in cancer therapy, as a single malignant cell remaining would re-establish the tumour.

1.2.3   *In vivo* delivery

Potentially the most useful, administration of material systemically is probably the least advanced strategy at present. The main reason for this is insufficient targeting of the vectors to the correct tissue sites.   The liver also rapidly clears adenoviral vectors together with lipid/DNA complexes after administration. Current retroviral vectors activate complement resulting in their neutralisation, and all of the viral vectors used to date stimulate strong immune responses upon repeated dosing.   However, improvements in targeting and vector development are addressing these problems.

All gene therapy applications depend on efficient delivery of genetic material across the cell membrane and ultimately to nucleus of the cell.

The major obstacle to harnessing the potential of gene therapy remains inefficient intracellular delivery. Significant work is underway to develop efficient gene delivery vehicles using modified virus particles or chemically synthesised entities. More recently, a possible aid to the low efficiency of gene transfer has been the identification of the unusual trafficking properties of one of the herpes simplex virus tegument proteins, VP22. This protein has the capacity to spread (either alone or as a fusion peptide) to the nuclei of surrounding cells from the cells in which it is expressed.

Table 1. Ligands used in gene delivery system

Ligand	Comments
Lung surfactants	Lung epithelial cells
Asialoglycoprotein	Asialoglycoprotein receptor (ASGPR) for hepatocyte delivery
Tetra-antenary galactose	ASGPR
Lectin: wheat germ agglutinin	Myoblasts delivery
Other lectins	Tumour delivery
Antibody: ani-CD3, anti-CD4, anti-CD7	T-cell delivery
Antibody: anti-secretory component	Transcytosis into epithelial cells
Antibody against truncated carbohydrate	Tumour delivery
Antibody against thrombomodulin (lung epitope)	Lung, tumour delivery
Insulin	Broadly distributed receptor
Transferrin	Broadly distributed receptor

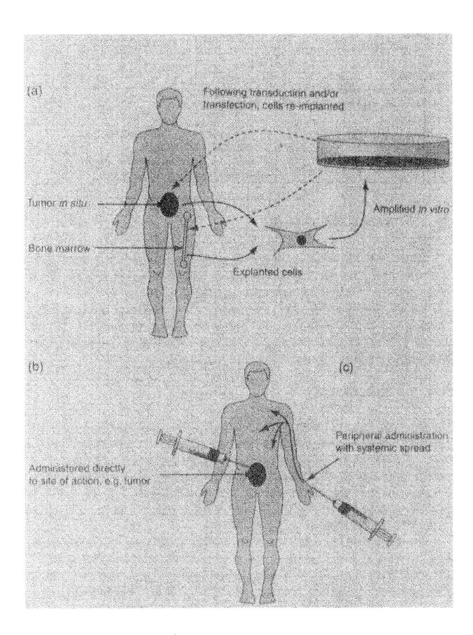

Figure 1. Diagrammatic representation of the different ways of applying gene therapy vectors. (a) *Ex vivo* gene therapy, (b) *in situ* gene therapy, (c) *in vivo* gene therapy.

Table 2. Agents that enhance cellular entry of DNA complexes

Ligand	Comments
Psoralen-inactivated Adenovirus	No viral gene expression required for DNA delivery
Rhinovirus	RNA virus
Membrane active peptide GALA	Membrane disruptive
Influenza HA peptide and variants Adenovirus	Membrane disruptive
Chicken adenovirus CELO	Naturally defective in human cells
Gramicidin S	Amphipathic, cyclic peptide
Chloroquine	Functions to elevate pH, osmotically lyse endosomes; useful in a few cells lines

2.   VIRAL VECTORS

Vectors are the vehicles by which DNA is directed to the target cell. DNA delivery lies at the heart of gene therapy and its successful application depends on the continued development of gene delivery vectors.

Vectors can be divided into two group: those that employ viruses and those that don't. The Gene Delivery Groups working at various Institutes are involved in the development of non-viral and viral vectors, including those based on synthetic polymers.

2.1   Retroviruses

The RNA viruses which replicate through a DNA intermediate, where the viral DNA integrates randomly into the host genome are called Retroviruses. Most recombinant retroviral vectors are based on the murine leukaemia virus and were one of the first vectors developed for use in gene therapy. They are also the most popular vector for gene transfer in clinical trials, being used in over in 60-70% of human gene therapy trials. Murine leukaemia virus (MuLV) has been the most widely used.

The capacity of retroviral vectors to carry therapeutic genes is relatively small, with a maximum insertion size of around 8 kb. They are able to target dividing cells with a high degree of efficiency of gene transfer and a moderate level of gene expression. They have a wide host range and, as a result of their integration into the host genome, they allow long term expression of the transgene.

Safety is one of the main limitations of the use of retroviruses. Recombinant retroviral vectors are generally replication incompetent. However, it has been shown that recombination can occur in producer cell lines leading to the production of replication competent viruses and potential infection in the host. Another potential safety concern is the possibility that the random integration of the retrovirus genome into the host genome might lead to insertional mutagenesis of a cellular oncogene leading to tumour formation. There are other potential drawbacks that may limit the usefulness of retroviruses in gene therapy. For example, retroviruses are generally inactivated by the complement cascade, although recently the development of new packaging cell lines is trying to circumvent this problem.

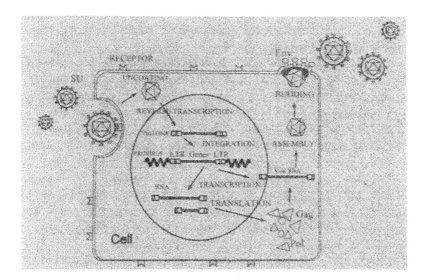

Figure 2. Retroviral lifecycle. Retroviral infection is initiated by an interaction between the viral envelope proteins and cell membrane-associated proteins which act as receptors, leading to internalisation of the virus and release of the core into the cell cytoplasm. The viral genetic information, in the form of an RNA molecule, is reverse transcribed into a double-stranded DNA form, using the virion-associated reverse transcriptase, which then integrates into the host cell DNA as a provirus. This integration requires the virally encoded integrase enzyme. The provirus has the structure LTR–*gag-pol-env*-LTR. LTR (long terminal repeats) result from duplication of terminal sequences as a consequence of reverse transcription, and carry the transcriptional regulatory elements of the virus. The three genes encode the viral core (*gag*), reverse transcriptase and integrase (*pol*) and envelope (*env*) proteins. Transcription of the provirus gives rise to RNA that code for the viral proteins and that provide the new viral genetic information. The viral RNA is then packaged by the viral Gag proteins into cores by a specific interaction and the newly formed cores bud out of the cell, taking part of the envelope protein containing host cell membrane with it.

One of the biggest disadvantages to using murine leukaemia retroviral vectors, however, is the fact that their infectivity is limited to dividing cells. For this reason retroviruses are generally used in an *ex vivo* gene transfer capacity where the target cells can be removed from the patient, stimulated to divide *in vitro* then transfected with the retrovirus and re-administered to the patient. There are a number of techniques which can be used to try to get retroviruses to infect outside their normal host range. One such method involves the co-infection with an adenovirus, which has been shown to allow retroviral vectors to transduce cells that were normally resistant to infection. A second method involves the use of an HIV-based vector to produce retroviral vectors capable of transducing non-dividing cells. The Human Immunodeficiency Virus (HIV) is able to infect non-dividing cells such as lymphocytes and as such a number of attempts have been made to produce gene transfer vectors using the HIV gag, pol and env genes. Despite the potential advantages of an HIV-based vector these types of vectors should be used with caution as there is some evidence that the HIV Env protein may be neurotoxic and even immunosuppressive.

Current methods of gene transfer have a rather low efficiency, especially *in vivo*. Therefore, one tries to achieve the highest possible levels of expression in the few cells that do take up foreign DNA. One approach is to use self-amplifying expression vectors. These vectors are based on the (+) - strand RNA viruses (alphaviruses) Sindbis virus and Semliki Forest virus. In these vectors, the viral capsid protein coding sequences are replaced with the gene of interest. After introduction into the target cells, the viral replication proteins will replicate the recombinant genome. The increased levels of mRNA generate very high transgene expression levels. Furthermore, spread throughout large cells (muscle, neurons) is much better compared to conventional expression cassettes. Self-amplifying vectors can be introduced into target cells as RNA, DNA or virions[1].

The potent tools for stable delivery of a gene of interest into the body are myoblasts, the proliferative cell type of skeletal muscle tissues as they become an integral part of the muscle into which they are injected, in close proximity to the circulation. The recent development of improved tetracycline-inducible retroviral vectors allows for the control of recombinant gene expression levels. The combination of ex vivo gene transfer using myoblasts and retroviral vectors provides a powerful toolbox with which to develop gene therapies for a number of human diseases[2].

The efficient insertion of genes by retroviruses is often complicated by transcriptional inactivation of the retroviral long terminal repeats (LTRs) and by the production of replication-competent retroviruses (RCR). Solutions to these are being found in modular vectors, in which the desirable features of different vector system are combined. Examples of synergistic vectors include virosomes (liposome/virus delivery), adeno-retro vectors, and MLV/VL30 chimeras. As gene delivery systems become increasingly complex, methodology is also needed for precise assembly of modular vectors. Gene self-assembly (GENSA) technology permits seamless vector construction and simultaneous, multifragment assembly[3].

The reversal of brain damage in the beta-glucuronidase-deficient mucopolysaccharidosis type VII mouse, an animal model of human lysosomal storage diseases has been studied using a HIV-based vector[4].

2.2   Adenoviruses

Belonging to the family Adenoviridae which is divided into genera Mastadenovirus and Aviadenovirus, Adenoviruses represent the second most suitable choice of gene delivery vector for gene therapy clinical trials after the retroviral vectors. Adenoviral vectors are based on a family of viruses that cause benign respiratory tract infections in humans and there are 42 serotypes of adenovirus known to infect humans. They are non-enveloped isometric particles approximately 45-50 nm in diameter with an icosahedral surface (capsid) and a DNA-containing core. Viral replication occurs without integration into the host genome, leading to only transient expression of the transgene.

Adenoviral vectors for use in gene therapy are typically based on serotype 5, with the majority of the El a and El b regions deleted to prevent virus replication. The E3 region can also be deleted to provide additional space for the insertion of up to 7.5 kb of exogenous DNA. This is larger than the amount that can be inserted into retroviral vectors, but is still too small to be of use in all gene therapy applications.

The main advantages of adenoviral vectors is that the efficiency of transduction is high, as is the level of gene expression, although this is only transient and can deteriorate rapidly within a few weeks.

One of the study has tried to optimise efficiency, and evaluate the duration, of catheter-based adenoviral vector-mediated pulmonary artery gene transfer in newborn pigs has been carried out by Badran *et al*[5]. The major clinical interest in adenovirus vectors stems from their broad host range and high infectivity *in vitro* and *in vivo*, exhibiting tropism for most cells in the human body, and the ability to infect quiescent as well as dividing cells.

One of the main disadvantages of adenoviral vectors is that cell-specific targeting is difficult to achieve as the virus has no envelope to attach cell-specific ligands to, as can be achieved with retroviruses. Furthermore, the adenovirus receptor is virtually ubiquitous and consequently systemic administration is likely to lead to adenoviral uptake in cell types other than the target cell thereby reducing the specificity of the gene therapy. As adenovirus genome replicates extra chromosomally, for long term expression, repeated administration is a requirement.

There are also important safety concerns over the use of adenoviral vectors for gene therapy. Aside from the potential generation of replication competent virus, there is also the risk of provoking an inflammatory response in the patient, particularly when repeated administrations are given. It has been demonstrated that although repeat administration of adenovirus is possible, the gene transfer becomes progressively less efficient. More, antibodies are produced against the viral and transgene products and a relatively mild inflammatory response was observed in the lungs of CD-1 mice. Studies in non-human primates have also revealed evidence of an inflammatory response following exposure of lungs to adenoviral vectors. It was demonstrated that lungs treated with high-dose (1010 plaque forming units) showed inflammatory cells present in the peribronchial and perivascular regions as well as alveolar accumulation of neutrophils and macrophages. An increase in the bronchoalveolar levels of the cytokine interleukin 8 was also detected. These inflammatory responses were not observed in the low dose studies which correlate to the proposed dose for phase 1 clinical trials in humans. However, the effect of repeat administrations of this dose of adenovirus was not investigated. Adenoviruses have also been implicated in causing cardiotoxicity and brain damage, as well as causing neurogenic and pulmonary inflammation at high doses and over time. It has, however, been successfully used.

There has been development of specific adenoviral gene delivery system with monoclonal antibody (mAb) AF-20 that binds to a 180 kDa antigen highly expressed on human hepatocellular carcinoma (HCC) cells. A bifunctional Fab-antibody conjugate (2Hx-2-AF-20) was generated through AF-20 mAb crosslinkage to an anti-hexon antibody Fab fragment. Uptake of adenoviral particles and gene expression was examined in FOCUS HCC and NIH3T3 cells by immunofluorescence; beta-galactosidase expression levels were determined following competitive inhibition of adenoviral CAR receptor by excess fibre knob protein. The chimeric complex was rapidly internalised at 37 degrees C, and enhanced levels of reporter gene expression was observed in AF-20 antigen positive HCC cells, but not in AF-20 antigen negative NIH 3T3 control cells. Targeting of recombinant adenoviral vectors to a tumour associated antigen by a bifunctional Fab-antibody conjugate is a promising approach to enhance specificity and efficiency of gene delivery to HCC[6].

Despite many apparent drawbacks, the adenovirus remains a popular vector for gene therapy due to its high gene transfer efficiency and high level of expression in a wide variety of cell types. Some effort has been made to modify the inflammatory and immunogenic properties of the adenovirus capsid, but so far little progress has been made. Next generation adenovirus vectors are currently under a rapid phase of development involving the complete elimination of all adenoviral coding sequences retaining only cis-acting elements essential for adenoviral replication and packaging. The so called pseudoadeno-virus (PAV) or gutless vectors may be rescued and propagated to high titres upon trans-complementation of the adenovirus genes from an adenovirus helper genome(s) co-introduced into helper permissive cell lines.

2.3 Adeno-associated and other viral vectors

The adeno-associated virus belonging to Parvoviridae (AAV) is a vector that combines some of the advantages of both the retroviral and adenoviral vectors. It is a single stranded DNA parvovirus that is able to integrate into the host genome during replication, thereby producing stable transduction of the target cell. The virus can also infect a wide range of cell types, including both dividing and non-dividing cells. AAV vectors are also not associated with any known human disease and show high efficiency transduction. They can, however, only carry a fairly small therapeutic gene insert (around 5 kb).

AAV vectors have not been studied to the same extent as adenoviral or retroviral systems, however they appear to be associated with fewer safety risks than the other viral systems. This is due to the elimination of all sequences coding for viral proteins, thereby greatly reducing the risk of an immune reaction against the vector. There remain, however, the potential problems of insertional mutagenesis and the generation of replication competent virus.

Other viral vectors have also been developed for use in gene therapy. These include the herpes simplex virus, the vaccinnia virus and syndbis virus. However, these vectors have not been widely studied and it is not clear what advantages they may hold over retroviral, adenoviral or AAV vectors. Recently, limited-replicating viral vectors, such as the Onyx virus, have been developed. The Onyx virus is an adenovirus that only replicates in p53 deficient cells. The advantage of these vectors is that their replication is limited mainly to tumour cells and that permitting the replication of the virus produces more efficient transfection of the tumour cells. Studies are going on to demonstrate the usefulness and safety of replicating vectors.

Viral vectors based on the nonpathogenic human adeno-associated virus, when coupled with the strong, rod photoreceptor specific opsin promoter, offer an efficient and nontoxic way to deliver and express ribozymes in photoreceptor cells for long time periods of time. Effective ribozyme-mediated therapy also demands careful *in vitro* analysis of a ribozyme's ability to efficiently and specifically distinguish between mutant and wild type RNAs[7].

Approximately 90% of cervical carcinomas are causally linked to infections with high-risk human papillomaviruses (HPVs), whose oncogenicity has been assigned to the continued expression of two early genes, E5 and E7, whose inhibition, therefore, provides a suitable goal for future tumour therapy. Using recombinant adeno-associated virus type 2 (AAV-2) vectors, the monocyte chemoattractant protein-1 (MCP-1) gene encoding MCP-1 which indirectly represses E6/E7 gene expression and is consistently absent in tumourigenic HPV-positive cervical carcinoma cell lines was expressed. The effect of these therapeutic genes on tumour formation was analysed in nude mice after ex vivo gene transfer into a PHV16-or HPV18-positive cervical carcinoma cell line. This suggests that transfer of therapeutic genes mediating a systemic effect via recombinant AAV-2 vector offers a promising approach for the development of gene therapies directed against papillomavirus-induced human cancers[8].

Recombinant adenovirus vectors are highly efficient at *in vitro* and *in vivo* gene delivery. The *in vitro* infection of a mouse colon adenocarcinoma cell line MCA-26 with the adenovirus AdV-LacZ can reach a maximal 75% of infectivity at an MOI of 1000. Intratumoural injection of AdV-LacZ (2X10(9)pfu) has resulted in substantial gene transfer in nearly 70% of MCA-26 tumours. In mice bearing large tumours, the treatment of tumours with pCI-TNF-alpha delivered by the gene gun did not induce significant tumour inhibition. These results indicate that the adenoviral delivery of TNF-alpha gene is more efficient than the particle-mediated gene gun device, and that adenovirus-mediated cytokine gene therapy may be a useful approach in the clinical management of human solid tumours[9].

Although most research in the field of somatic gene therapy has investigated the use of recombinant viruses for transferring genes into somatic target cells, various methods for nonviral gene delivery have also been proposed. Both type of gene delivery systems have advantages and drawbacks. Schematically, viral vectors are particularly efficient for gene delivery, whereas nonviral systems are free of the difficulties associated with the use of recombinant viruses but need to be further optimised to reach their full potential. In order to bridge the gap between viral vectors and synthetic reagents (some specific features of the viral vector systems) can advantageously be taken into account for the design of improved nonviral gene delivery systems. Indeed, although nonviral systems differ fundamentally from viral systems, one possible approach towards enhanced artificial reagents aims at developing 'artificial viruses' that mimic the highly efficient processes of viral infection[10].

2.4   Herpes simplex virus-based vectors

Herpes simplex virus (HSV) belongs to the herpesvirus family, a diverse group of large DNA viruses, all of which have the potential to establish lifelong latent infection[11]. HSV consists of 110 nm diameter particles comprising an icosahedral nucleocapsid, surrounded by a protein matrix, the tegument, which in turn is surrounded by a glycolipid-containing envelope[12]. The HSV genome consists of a linear double-stranded DNA molecule of 152 kb in length encoding 81 known genes, 38 of which are essential for virus production *in vitro*[12]. HSV has a wide tropism, infecting virtually any human cell, and is capable of nuclear delivery, infecting dividing as well as quiescent cells[12]. Herpesviruses infect and persist in cells of the nervous system, hence herpesvirus-based vectors may provide a unique strategy for gene transfer to cells of the nervous system[13].

Table 3. The major properties of the four most reported gene therapy viral vectors

Virus	Retrovirus	Adenovirus	Adeno-associated virus (AAV)	Herpes simplex virus (HSV)
Structure	Enveloped	Nonenveloped	Nonenveloped	Enveloped
Maximum insert capacity (kb)	9-10	7-8 helper-independent, 37 helper-dependent	4, 5	Large capacity Up to 150: gutless-helper-dependent
Infects quiescent cells	Type C (e.g. MLV) : No Lentiviruses (e.g. HIV) : Yes	Yes	Yes	Yes
Integrating Theoretical titres	Yes	Yes	Yes	No
Genome size (kb)	10^8-10^8	10^{11}-10^{12}	10^9-10^{10}	10^8-10^9
Particle size (nm)	~ 10 Pleomorphic	~ 36 60 – 90	~ 4.7 (AAV-2) 18-26	152 (HSV-1) ~110

Current HSV-based vectors involve replacement of one or more of the five immediate early (IE) genes whose functions are *trans*-complemented by packaging cell lines[12,14]. Titres of 108-109 viral particles per ml are possible with replication-competent HSV vectors, although the possible titres drop as more genetic sequence is deleted (~106 particles/ml). Single IE gene deletion HSV vectors have been demonstrated to efficiently direct reporter gene expression in both myoblasts and differentiated myotubes in culture to similar degrees[12]. *In vivo* application to newborn mouse muscle was reported to result in significant numbers of positive fibres, but low level transduction was achieved in adult myofibres, believed to relate to HSV particles poorly penetrating the basal lamina[12]. The persistence of expression in both newborn and adult mice was, however, very limited which was believed to stem from cytotoxic effects resulting from expression of the remaining IE genes. Although preliminary results have indicated that the cytotoxic effects are reduced upon deletion of further IE genes, a great deal of research is required to refine HSV vectors including mechanisms of permeabilising the basal lamina[12]. The major interest comes from the possibility of a gutless HSV vector, which has the potential of accommodating up to 150 kb of insert DNA[15].

Chimeric vectors have been developed utilising the favourable aspects of retroviral integration and adenoviral infectivity combined to achieve stable genetic transduction and efficient delivery *in vivo*.

3.   POLYMER VECTORS

Synthetic vectors are designed to overcome many of the problems associated with viral vectors. Much of the research has been directed toward the development of vectors based on cationic lipids. Research has been directed to the development of vectors based on synthetic polymers. Scientists are investigating a variety of polymer-based delivery systems, but a common feature of all the vectors is the use of a cationic polymer (such as polylysine) that is designed to bind to and condense DNA into a particle with a diameter of around 50 nm. The overall aim is to develop a fully functional gene delivery vector as efficient as a virus.

The main requirements of an ideal gene delivery system are listed below:

3.1   Requirements of an ideal vector

1. Must be capable of extended circulation in the bloodstream
2. Must be small enough to gain access to tissues and cells
3. Must have flexible tropisms for applicability in a range of disease targets
4. Following endocytosis, must be capable of escaping endosome-lysosome processing
5. Must be able to deliver DNA into the nucleus and allow gene transcription

3.2   DNA condensation and complex formation

DNA is an extremely large molecule. The chromatin of a typical human chromosome would be about 1 mm in length if it were extended and would therefore span the nucleus more than 100 times. In its normal state it is condensed by various cellular proteins, such as the histone proteins, that allow such a large molecule to fit inside the cell nucleus. Even a plasmid DNA molecule of around 6 kb in size has an extended structure several hundred nanometers in diameter. In order to efficiently deliver plasmid DNA into cells it is necessary to condense them to a much smaller size.

Endocytosis of plasmid DNA can be achieved if the DNA can be condensed to 100 nm or less in diameter, while access to tumour tissue from the vasculature is limited to particles less than 70 nm in diameter. It is, therefore, important to condense plasmid DNA to as small a size as possible to facilitate gene transfer, and indeed the size of the condensed DNA complex may be one of the most important factors for successful *in vivo* gene delivery.

Polylysine was used to condense DNA as early as 1969 and was initially used merely as a model for the interaction of biopolymers such as DNA and histone proteins. DNA is a highly negatively charged polymer due to the repeating phosphate groups along the polymer backbone. The interaction with cationic polymers such as polylysine is, therefore, an electrostatic one. The exact parameters that govern the ability of cationic polymers to condense DNA are still under study, although it is generally accepted that neutralisation of the charges on the DNA molecule, followed by hydrophobic collapse, as water is displaced from the DNA structure, may play an important role.

In 1987, Wu & Wu proposed polylysine-condensed DNA as a gene delivery vector. Since then it has been used as the basis of most cationic polymer-based vectors. The size of the polylysine/DNA complexes varies according to the molecular weight of the polymer used and the formation conditions such as the charge ratio between the polymer and DNA, the salt concentration and the temperature. However, complex sizes of 50-300 nm diameter can generally be achieved, while careful choice of the polymer molecular weight can produce complexes with an average diameter of less than 30 nm. Careful control of mixing conditions and salt concentrations has been reported to produce complexes as small as 12 nm in diameter.

One of the major advantages of using cationic polymers to condense DNA is that very large genes can be used. Viral vectors are limited by the amount of genetic material that can be inserted into the viral genome. This limitation does not apply to polymer-based vectors, and even DNA molecules as large as 45 Kb have been successfully condensed. This permits the incorporation of gene regulatory regions such as locus control regions that may afford better control of gene expression.

Once DNA condensation has been achieved and a satisfactory complex has been produced it is possible to incorporate functional groups such as targeting moieties, nuclear localisation or targeting peptides, endosomalytic peptides and hydrophilic shielding.

3.3   Nuclear targeting

Delivery of DNA into the cytoplasm of the target cell does not represent the final stage of gene delivery. The DNA needs to be transported into the nucleus and expressed. it is thought that less than 1% of the plasmid molecules that reach the cytoplasm eventually reach the nucleus. For efficient nuclear delivery it may be necessary to incorporate nuclear targeting or nuclear localisation signals into the vector. It has been suggested that certain components of polymer/DNA complexes may possess intrinsic nuclear homing capabilities, for example polylysine and the fusogenic peptide INF7 have both been implicated as having a possible role in increasing the efficiency of nuclear delivery of the complexed DNA. It has also been shown that certain sequences of the plasmid DNA may lead to import of the DNA into the nucleus.

In order to improve the efficiency of nuclear delivery it may be necessary to incorporate specific nuclear localisation signal (NLS) peptides into the polymer/DNA complex. All proteins are synthesised in the cytoplasm and those proteins that are required in the nucleus must be imported from the cytoplasm, through the nuclear pores and into the nucleus. To achieve this nuclear proteins contain short peptide sequences termed nuclear localisation signals that allow selective transport of these proteins to the nucleus. It is possible that the attachment of these peptide sequences to the polymer/DNA complexes may result in much greater transfection efficiency. There are several candidate NLS peptides that to might be suitable for incorporation into a gene delivery vector, and many of these peptides contain lysine and arginine motifs. The NLS from the SV40 large T antigen was one of the first NLS peptides to be discovered and has an amino acid sequence of PKKKRKV. Since then many other sequences have been identified that may prove useful additions to a synthetic, polymer-based gene delivery vector.

Work is on-going to identify novel nuclear targeting peptides using phage display. Scientists are trying to exploit viral nuclear delivery mechanisms, for example through the attachment of adenovirus hexon protein to polylysine/DNA complexes.

4.   NON-VIRAL VECTORS

The limitations of viral vectors, in particular their relatively small capacity for therapeutic DNA, safety concerns, difficulty in targeting to specific cell types have led to the evaluation and development of alternative vectors based on synthetic, non-viral systems.

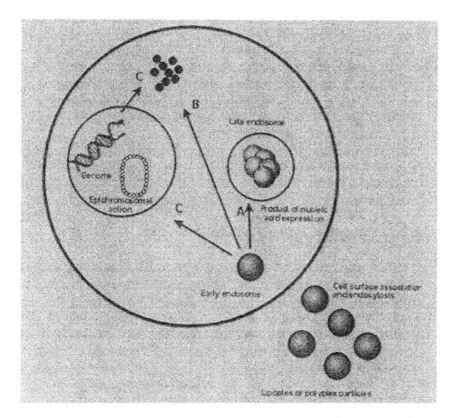

Figure 3. Schematic diagram to show the process by which chemical non viral delivery systems deliver nucleic acids to cells. Lipoplex (or polyplex) particles which have not succumbed to aggregation and/or serum – inactivation (as appropriate), associate with the cell surface and enter usually by endocytosis. The majority of lipoplex (or polyplex) particles in early endosomes become trapped in late endosomes (Path A) and the nucleic acids fail to reach the cytosol. A minority of lipoplex (or polyplex) particles in early endosomes are able to release their bound nucleic acids into the cytosol. Path B is followed by RNA which may act directly in the cytosol. Path C is followed by DNA which must enter the nucleus in order to act. The diagram is drawn making the assumption that plasmid DNA has been delivered which is expressed in an epichromosomal manner.

Apart from synthetic polymers there are numerous systems for the non-viral delivery of DNA. The main alternatives to viruses are described below and include liposomes, naked DNA, liposome-polycation complexes and peptide delivery systems.

4.1 VP22

The herpes simplex virus I (HSV-I) tegument protein VP22 has the exciting property of spreading from the cytoplasm of the cells in which it is expressed to the nuclei of surrounding cells. The mechanism by which this function occurs is currently unknown. However, this propensity remains when fusion molecules are generated between VP22 and a cargo partner including reporter genes such as the *aequoria victoria* green fluorescent protein, and genes currently under examination as therapeutic agents such as p53. As such it heralds the prospect of improving the numbers of cells receiving therapeutic molecules, even under conditions whereby the transfer of genetic material is quite inefficient.

Despite the difficulties with the current delivery systems many diseases have reached the stage of clinical trial using these vehicles. It will be interesting to see how these fare in a clinical setting. This will give us an insight into the factors that need to be addressed to improve these vectors.

4.2   Liposomal gene delivery

Negatively charged, or classical liposomes have been used to deliver encapsulated drugs for some time and have also been used as vehicles for gene transfer into cells in culture. Problems with the efficiency of nucleic acid encapsulation, coupled with a requirement to separate the DNA-liposome complexes from "ghost" vesicles has led to the development of positively charged liposomes. Cationic lipids are able to interact spontaneously with negatively charged DNA to form clusters of aggregated vesicles along the nucleic acid. At a critical liposome density the DNA is condensed and becomes encapsulated within a lipid bilayer, although there is also some evidence that cationic liposomes do not actually encapsulate the DNA, but instead bind along the surface of the DNA, maintaining its original size and shape.

Cationic liposomes are also able to interact with negatively charged cell membranes more readily than classical liposomes. Fusion between cationic vesicles and cell surfaces might result in delivery of the DNA directly across the plasma membrane. This process bypasses the endosomal-lysosomal route which leads to degradation of anionic liposome formulations. Cationic liposomes can be formed from a variety of cationic lipids, and they usually incorporate a neutral lipid such as DOPE (dioleoylphosphatidyl-ethanolamine) into the formulation in order to facilitate membrane fusion. A variety of cationic lipids have been developed to interact with DNA, but perhaps the best known are DOTAP (N-1(-(2,3-dioleoyloxyl)propyi)-N,N,N-trimethylammoniumethyl sulphate) and DOTMA (N-(1-(2,3-dioleoyloxy)-propyi)-N,N,N-trimethylammonium chloride). These are commercially available lipids that are sold as *in vitro* transfecting agents, with the latter sold as Lipofectin.

There have been several studies on the *in vivo*, systemic use of liposome/DNA complexes. The factors controlling the transfection efficiency of liposome/DNA complexes following intravenous administration are still poorly understood. Complexes formed between the cationic lipid DOTMA and DNA are rapidly cleared from the bloodstream and were found to be widely distributed in the body and expression was detected mainly in the lungs but also in the liver, spleen, heart and kidneys. Similar results were found when DOTAP-based liposomes were used and it was found that the main factors controlling transfection efficiency were the structure of the cationic lipid and the ratio of the cationic lipid to DNA. The type of helper lipid used was also important as the addition of DOPE was found to reduce the *in vivo* transfection efficiency of DOTAP/DNA complexes.

The transfection efficiency of liposome/DNA complexes *in vivo* has been shown to be relatively low, especially when compared to viral vectors. One study has suggested that the *in vivo* transfection efficiency of adenoviruses is around 200 times greater than that observed with liposomes. One explanation for the relatively poor transfection efficiency of liposome/DNA complexes is that they are susceptible to disruption by serum proteins. A variety of proteins are known to bind to liposomes *in vitro* and *in vivo* and may cause membrane destabilisaton. There are now serious efforts being made to develop liposomal vectors that are resistant to serum disruption. Novel cationic lipids are also being developed to try to improve the transfection efficiency of liposome/DNA complexes. Targeting of the liposomes to specific cell types has also been investigated as a means of improving the transfection efficiency.

There have been several clinical trials of liposome/DNA complexes, although almost all of these have been involved in the treatment of cystic fibrosis. Most protocols involve the use of DC-chol/DOPE liposomes directly instilled onto the nasal epithelium of CF patients. The effect of gene expression on CFTR function was determined and the presence of the gene in the target cells was determined by PCR (polymerase chain reaction) and Southern blot analysis. The results from these trials are described here.

Initial clinical studies found no evidence of any safety problems with the use of liposome/DNA complexes. This is surprising as it is well documented that the liposome/DNA complexes used in clinical trials are directly cytotoxic *in vitro*. Furthermore, studies in mice and macaques have demonstrated that exposure to high doses or to repeat doses of liposome/DNA complexes results in histopathology and gross lung pathology, suggesting that these vectors may not be as clinically safe as previously thought.

Some data has shown that cationic liposomes were generally more effective at transfecting genes than were micelles of the same lipid composition thus suggesting a role for the bilayer structure in facilitating transfection. In addition, the transfection efficiency of liposome-delivered genes was highly dependent upon the lipid composition, lipid/DNA ratio, particle size of the liposome-DNA complex, and cell lines used.

4.3   Liposome / polycation / DNA (LPD) complexes

The limitations of liposome mediated gene delivery have led to the development of novel lipids in an effort to improve these vectors. A different approach to improving the transfection efficiency of cationic liposomes has been developed. The approach involves the use of polycations such as polylysine in the formation of liposome/DNA complexes. Polylysine and the liposome preparation DC-chol/DOPE were mixed with DNA in order to form liposomelpolycation/DNA or LPD complexes. These complexes showed higher transfection efficiency than the corresponding liposome/DNA complexes without the presence of polycations and also showed enhanced resistance to degradation by nucleases. A further advantage of the use of polylysine was that the resulting complexes were much smaller than complexes formed with liposomes alone. Complexes formed between DNA and DC-chol/DOPE were found to have an average diameter of 1.2 μm, possibly due to an aggregation of the liposome/DNA complexes. When the polycations polylysine or protamine were incorporated into the complexes the average diameter was reduced to around 100 nm, a size that may allow more efficient cellular internalisation.

Systemic therapy with plasmid DNA complexed with cationic liposomes merits further development as an alternative method for anti-inflammatory treatment of arthritis. A novel formulation of cationic liposomes containing the novel cytofectin ACHx was used for delivery of an anti-inflammatory cytokine gene, IL-10, to mice with established collagen induced arthritis. A single intraperitoneal injection of human IL-10 expression plasmid complexed with liposomes 2 to 4 days after the onset of arthritis was sufficient to give significant and prolonged amelioration of arthritis for 30 days[16].

Transgene expression in lymphoid cells may be useful for modulating immune responses in, and gene therapy of, cancer and AIDS. Although cationic liposome-DNA complexes (lipoplexes) present advantages over viral vectors, they have low transfection efficiency, unfavourable features for intravenous administration, and lack of target cell specificity. The use of a targeting ligand (transferrin), or an endosome-disrupting peptide, in ternary complexes with liposomes and a luciferase plasmid, significantly promoted transgene expression in several T-and B-lymphocytic cell lines. The highest levels of luciferase activity were obtained at a lipid/DNA (+/-) charge ratio of 1/1, where the ternary complexes were net negatively charged. The use of such negatively charged ternary complexes may alleviate some of the drawbacks of highly positively charged plain lipoplexes for gene delivery[17].

Cationic lipids are under active consideration as gene delivery vehicles for cystic fibrosis. Initial studies have shown cationic lipids to be effective agents of gene transfer to epithelial cells *in vitro*. Instillation of these vectors into animal models has led to widely different degrees of transfection of the airway epithelia. Newer generations of cationic lipids, with improved transfection efficiencies, appear to result in mostly alveolar gene delivery. Aerosol delivery of cationic lipid-DNA complexes has resulted in variable transfection of the airways in animal models. Initial human clinical trials using intranasal instillation have shown variable low levels of expression, accompanied by little toxicity. Recent developments in the formulation of cationic lipid-DNA complexes have resulted in an ability to aerosolise high concentrations of the complex, and should permit an evaluation of the efficacy of these delivery vehicles when aerosolised into cystic fibrosis patients[18].

4.4   Direct injection of naked DNA

Direct injection of free DNA into certain tissues, particularly muscle, has been shown to produce particularly high levels of gene expression, and the simplicity of this approach has led to its adoption in a number of clinical protocols. In particular, this approach has been applied to the gene therapy of cancer where the DNA can be injected either directly into the tumour or can be injected into muscle cells in order to express tumour antigens that might function as a cancer vaccine.

The direct injection of naked DNA can also be used to treat genetic diseases, particularly of tissues that are available for direct injection such as the skin . Direct injection of skin from patients suffering from the genetic skin condition lamellar ichthyosis caused by a loss of transglutaminase 1 (Tgase1) expression was performed in a study by Choate and Khavari (1997). It was shown that some skin regeneration was possible following repeat injections of plasmid DNA encoding the Tgase1 gene, furthermore the restoration of Tgase1 expression occurred in the correct location in the suprabasal epidermis. Further analysis, however, revealed that the pattern of expression was non-uniform and failed to correct the underlying histological and functional abnormalities of the disease.

Although direct injection of plasmid DNA has been shown to lead to gene expression, the overall level of expression is much lower than with either viral or liposomal vectors. Naked DNA is also unsuitable for systemic administration due to the presence of serum nucleases. As a result direct injection of plasmid DNA seems destined to be limited to only a few applications involving tissues that are easily accessible to direct injection such as skin and muscle cells.

Somatic cell gene transfer is a potentially useful strategy to alter lung function. In a study, the rat haem oxygenase-1 (HO-1) gene was delivered to the lungs of neonatal mice via transpulmonary injection. A bidirectional promoter construct coexpressing both HO-1 and a luciferase reporter gene was used so that *in vivo* gene expression patterns could be monitored in real time. HO-1 expression levels were also modulated with doxycycline and assessed *in vivo* with bioluminescent light transmitted through the tissues from the coregulated luciferase reporter. As a model of oxidative stress and HO-1 mediated protection, groups of animals were exposed to hyperoxia. After gene transfer, elevated levels of HO-1 were detected predominantly in alveolar type II cells by immunocytochemistry. With overexpression of HO-1, increased oxidative injury was observed. Furthermore, this model demonstrated a cell-specific effect of lung HO-1 overexpression in oxidative stress. Specific control of expression for therapeutic genes is possible *in vivo*. The transpulmonary approach may prove useful in targeting gene expression to cells of the alveolar epithelium or to circumscribed areas of the lung[19].

A DNA and RNA chimeric oligonucleotide was designed to induce host cell mismatch repair mechanisms and correct the chromosomal mutation to wild type. Direct skeletal muscle injection of the chimeric oligonucleotide into the cranial tibialis compartment of a six-week-old affected male dog, and subsequent analysis of biopsy and necropsy samples, demonstrated *in vivo* repair of the GRMD mutation that was sustained for 48 weeks. Reverse transcription-polymerase chain reaction (RT-PCR) analysis of exons 5-10 demonstrated increasing levels of exon 7 inclusion with time[20].

Preclinical studies involving intramuscular injection of plasmid into animals have revealed at least four significant variables that effect levels of gene expression (i.e., >fivefold effect over controls), including the formulation, injection technique, species and pretreatment of the muscle with myotoxic agents to induce muscle damage. The uptake of plasmid formulated in saline has been shown to be a saturable process, most likely via a receptor-mediated event involving the T tubules and caveolae. Pharmacokinetic studies have demonstrated that the bioavailability of injected plasmid to muscle cells in very low, due to rapid and extensive plasmid degradation by extracellular nucleases. One group has developed protective, interactive, (PINC) delivery systems designed to complex plasmids and to (I) protect plasmids from rapid nuclease degradation, (ii) disperse and retain intact plasmid in the muscle and (iii) facilitate the uptake of plasmid by muscle cells. PINC systems result in up to at least a one log increase in both the extent and levels of gene expression over plasmid formulated in saline. They have combined the PINC delivery systems with two different muscle-specific expression plasmids. After direct intramuscular injection of these gene medicines, it was shown both local myotrophic and neurotrophic effects of expressed human insulin-like growth factor (hIGF-I) and the secretion of biologically active human growth hormone (hGH) into the systemic circulation[21].

4.5   Peptide mediated gene delivery

The use of cationic polymers as gene delivery vectors has been investigated by several groups. A related approach is to use naturally occurring or synthetic peptides as gene delivery systems. This approach is based upon the observation that the functionally active regions of proteins such as enzymes, receptors and antibodies are relatively small, typically consisting of around 10-20 amino acids. Synthesising peptides based upon functional regions of DNA binding proteins or a variety of viral proteins is an approach that has been used to replace the use of whole proteins (such as histone Hl) or large, polydispersed polymers (such as polylysine) as gene delivery vectors.

A number of peptide sequences have been shown to be able to bind to and condense DNA. One such sequence is the tetra-peptide serine-proline-lysine-lysine located in the C-terminus of the histone Hl protein. Rational design of peptide sequence has also been used to develop completely synthetic DNA binding peptides. For example, peptide sequences based on the tetra-amine spermine were developed by Goftschalk et al (1996) who showed that the peptide tyrosine-lysine-alanine-lysine8-tryptophan-lysine was very effective at forming complexes with DNA.

DNA binding peptides can also be synthesised that can be coupled to cell specific ligands, thereby allowing receptor mediated targeting of the peptide/DNA complexes to specific cell types. One such approach has been to synthesise a cationic peptide based on 16 lysines and an RGD peptide sequence. The RGD sequence has been shown to facilitate binding of the peptide/DNA complex to the integrin receptor on Caco-2 cells *in vitro*.

Synthetic peptides have also been developed that enhance the release of the peptide/DNA complexes from the endosome following endocytosis. Without a means of escaping the endosome, the endosomal-lysosomal pathway presents a major barrier to transfection via receptor mediated endocytosis.

Current methods of gene transfer have a rather low efficiency, especially *in vivo*. Therefore, one tries to achieve the highest possible levels of expression in the few cells that do take up foreign DNA. One approach is to use self-amplifying expression vectors. These vectors are based on the (+)-strand RNA viruses (alphaviruses) Sindbis virus and Semliki Forest virus. In these vectors, the viral capsid protein coding sequences are replaced with the gene of interest. After introduction into the target cells, the viral replication proteins will replicate the recombinant genome. The increased levels of mRNA generate very high transgene expression levels. Furthermore, spread throughout large cells (muscle, neurons) it is much better compared to conventional expression cassettes. Self-amplifying vectors can be introduced into target cells as RNA, DNA or virions.

A novel formulation of cationic liposomes containing the novel cytofectin ACHX was used for delivery of an anti-inflammatory cytokine gene, IL-10, to mice with established collagen induced arthritis. A single intraperitoneal injection of human IL-10 expression plasmid complexed with liposomes 2 to 4 days after the onset of arthritis was sufficient to give significant and prolonged amelioration of arthritis for 30 days) Preliminary experiments suggested that the therapeutic effect was IL-10 dose-dependent. The distribution of the human IL-10 DNA after injection was widespread, including the inflamed paws. Human IL-10 mRNA was also detected in the paws 24 h after injection. IL-10 protein was below the level of detection in paws and serum but was detected in some tissues up to 1 0 days after injection. The target cell of transfection was demonstrated to be the macrophage. These results suggest that systemic therapy with plasmid DNA complexed with cationic liposomes merits further development as an alternative method for anti-inflammatory treatment of arthritis.

A major advantage of synthetic peptide based DNA delivery systems is its flexibility. By design, the composition of the final complex can be easily modified in response to experimental results *in vitro* and *in vivo* to take advantage of specific peptide sequences to overcome extra-and intracellular barriers to gene delivery. The extreme heterogeneity which greatly complicates both the kinetics of DNA-poly (L-lysine) interaction and the thermodynamic stability of the final DNA complexes is avoided. Other unique features include the absence of biohazards related to the viral genome as well as the production of the viral vector and the absence of limitations one the size of the therapeutic genes that can be inserted in the recombinant viral vector. In principle, if the gene can be cloned into an expression plasmid, it can be delivered as a synthetic DNA complex. Since these synthetic delivery systems are composed of small peptides which may be poorly antigenic, they hold the promise of repeated gene administration, a highly desirable feature which will be important for gene targeting *in vivo* to endothelial cells, monocytes, hepatocytes and tumour cells[22].

Lipidoc glycosides with amino alkyl pendent groups have been synthesised and evaluated for *in vitro* DNA transfection activity. The first representative of this new class of cationic lipids has shown good gene delivery and low toxicity to HeLa and 3T3 cells[23].

4.6 Poly (ethylenimine) – mediated transfection: a new paradigm for gene delivery

Poly (ethylenimine) (PEI) is a synthetic polycation that has been used successfully for gene delivery both *in vitro* and *in vivo* due to, in theory, a form of protection that is afforded to the carried plasmids. In a recent study the stability of PEI/DNA complexes was demonstrated using deoxyribonuclease (Dnase) 1 and Dnase 2, various levels of pH, and increasing exposure times. DNA that was complexed with PEI was not degraded when exposed to at least 25 units of either enzyme for 24 h while uncomplexed forms of the same plasmid were digested when exposed to .010 Units of Dnase 1 for 0.05 h or 003 Units of Dnase 2 for 1 h. For further comparison, the stability of complexes made with poly (L-lysine) PLL) and DNA was examined and found to be lower than that of PEI/ DNA complexes; PLL – complexed DNA was digested on exposure to 1.25 Units Dnase 1 for 3 min. Cells were transfected with PEI/ DNA complexes and, by using a pH indicator and optical recording techniques, it was found that the normal lysosomal pH value of 5.0 was not altered, bringing into question PEI's hypothesised lysosomal entry. Confocal microscopy showed that PEI/ DNA complexes and lysosomes do not merge during transfection (although PLL/DNA complexes do). The lack of lysosomal involvement in PEI – mediated transection is surprising because it goes against the conventional wisdom that has attempted to explain how PEI functions during transfection. PEI forms a stable complex with DNA, which moves from endocytosis to nuclear entry without significant cellular obstacles[24].

Local delivery of therapeutic molecules represents one of the limiting factors for the treatment of neurodegenerative disorders. *In vivo* gene transfer using viral vectors constitutes a powerful strategy to overcome this limitation. A study was carried out to validate the lentiviral vector as a gene delivery system in the mouse midbrain in the perspective of screening biotherapeutic molecules in mouse models of Parkinson's disease. A preliminary study with a LacZ-encoding vector injected above the substantia nigra of C57BL/6j mice indicated that lentiviral vectors can infect approximately 40,000 cells and diffuse over long distances. Based on these results, glial cell line-derived neurotrophic factor (GDNF) was assessed as a neuroprotective molecule in a 6-hydroxydopamine model of Parkinson's disease. Lentiviral vectors carrying the cDNA for GDNF or mutated GONF were unilaterally injected above the substantia nigra of C57BL/6j mice. Two weeks later, the animals were lesioned ipsilaterally with 6-hydroxydopamine into the striatum. Apomorphine-induced rotation was significantly decreased in the GDNF-injected group compared to control animals. Moreover, GDNF efficiently protected 69.5% of the tyrosine hydroxylase-positive cells in the substantia nigra against 6-hydroxydopamine-induced toxicity compared to 33.1% with control mutated GDNF. These data indicate that lentiviral vectors constitute a powerful gene delivery system for the screening of therapeutic molecules in mouse models of Parkinson's disease[25].

5.   CONCLUSIONS

The choice of vectors in gene delivery is crucial in the success of genetic interventions. Vectors employed range from retroviruses, adenoviruses, to Herpes simplex virus based types varying in their genome size and maximum insert capacity, however, they should embody capacity of extended circulation in the blood stream, small size, flexible tropisms and ability to deliver DNA into nucleus and allow gene transcription. Apart from other non viral vectors like synthetic vectors designed to overcome problems of viral vectors, liposomal and direct injection of naked DNA is being evolved. The application of PEI as a synthetic polycation and lentiviral vectors constitute another potential delivery system which has gained ground lately. The ultimate success of gene therapy hinges on the design of gene delivery vectors based on the growing knowledge of the gene uptake mechanisms.

ACKNOWLEDGEMENTS

I am grateful to my colleagues and my wife for their help and patience. I am particularly grateful to Prof. S. Prabhakar for critical analysis of the manuscript and encouragement. Acknowledgements are also due to Mr. Lalit Kumar for efficient and flawless typing.

REFERENCES

1. Wolff JA, Herweijer H (1997), Self-amplifying vectors for gene delivery, *Adv Drug Deliv Rev,* **27**(1): 5-16.

2. Ozawa CR, Springer ML, Blau HM (2000) A novel means of drug delivery: myoblast-mediated gene therapy and regulatable retroviral vectors. *Annu Rev Pharmacol Toxicol,* **40**: 295-317.

3. Solaiman F, Zink MA, Xu G, Grunkemeyer J, Cosgrove D, Saenz J, Hodgson CP (2000): Modular retro-vectors for transgenic and therapeutic use. *Mol Reprod Dev,* **56**(2 Suppl): 309-15.

4. Bosch A, Perret E, Desmaris N, Trono D, Heard JM, (2000): Reversal of pathology in the entire brain of mucopolysaccharidosis type VII mice after lentivirus-mediated gene transfer. *Hum Gene Ther,* **11**(8): 1139-50.

5. Badran S, Schachtner SK, Baldwin HS, Rome JJ, (2000), Optimization of adenoviral vector-mediated gene transfer to pulmonary arteries in newborn swine. *Hum Gene Ther,* **11**(8): 1113-21.

6. Yoon SK, Mohr L, O'Riordan CR, Lachapelle A, Armentano D, Wands JR (2000): Targeting a recombinant adenovirus vector to HCC cells using a bifunctional fab-antibody conjugate. *Biochem Biophys Res Commun,* **272**(2): 497-504.

7. Hauswirth WW, LaVail MM, Flannery JG, Lewin AS (2000): Ribozyme gene therapy for autosomal dominant retinal disease. *Clin Chem Lab Med,* **38**(2): 147-53.

8. Kunke D. Grimm D, Denger S, Kreuzer J, Delius H, Komitowski D, Kleinschmidt JA (2000): Preclinical study on gene therapy of cervical carcinoma using adeno-associated virus vectors. *Cancer Gene Ther,* **7**(5): 766-77.

9. Wright P, Braun R, Babiuk L, Little-van den Hurk SD, Moyana T, Zheng C, Chen Y, Xiang J (1999), Adenovirus-mediated TNF-alpha gene transfer induces significant tumor regression in mice. *Cancer Biother Radiopharm,* **14**(1): 49-57.

10. Navaroo J, Oudrhiri N, Fabrega S, Lehn P (1998), Gene delivery systems: Bridging the gap between recombinant viruses and artificial vectors. *Adv Drug Deliv Rev* **30**(1-3): 5-11.

11. Efstathiou, S. and Minson, A.C,. (1995) Herpes virus-based vectors. *Brit. Med. Bull.,* **51**, 45-55.

12. Huard, J., Krisky, D., Oligino, T., Marconi, P., Day, C.S., Watkins, S.C. and Glorioso, J.C. (1997) Gene transfer to muscle using herpes simplex virus-based vectors. *Neuromusc. Disord.,* **7**, 299-313.

13. Hermens, W.T. and Verhaagen, J. (1998) Viral vectors, tools for gene transfer in the nervous system. *Progress Neurobiol.*, **55**, 399-432.

14. Brehm, M, Samaniego, L.A., Bonneau, R.H., DeLuca, N.A. and Tevethia, S.S.(1999) Immunogenicity of herpes simplex virus type 1 mutants containing deletions in one or more of the genes: ICP4, ICP27, ICP22, and ICP0. *Virology*, **256**, 258-269.

15. Frenkel, N., Singer, O. and Kwong, A.D. (1994) Minireview: The herpes simplex virus amplicon-a versatile defective virus vecto. *Gene Ther.*, **1**, S40-S46.

16. Fellowes R, Etheridge CJ, Coade S, Cooper RG, Stewart L, Miller AD, Woo P (2000): Amelioration of established collagen induced arthritis by systemic IL-10 gene delivery.

17. Simoes S, Slepushkin V, Gaspar R, Pedroso de Lima MC, Duzgunes N (1999): Successful transfection of lymphocytes by ternary lipoplexes. *Biosci Rep*, **19**(6): 601-9.

18. Scheule RK, Cheng SH, (1998): Airway delivery of cationic lipid: DNA complexes for cystic fibrosis. *Adv Drug Deliv Rev.* **30**(1-3): 173-184.

19. Weng YH, Tatarov A, Bartos BP, Contag CH, Dennery PA (2000): HO-1 expression in type II pneumocytes after transpulmonary gene delivery. *Am J Physiol Lung Cell Mol Physiol*, **278**(6): L1273.

20. Bartlett RJ, Stockinger S, Denis MM, Bartlett WT, Inverardi L, Le TT, Man Nt, Morris GE, Bogan DJ, Metcalf-Bogan J, Kornegay JN (2000): *In vivo* targeted repair of a point mutation in the canine dystrophin gene by a chimeric RNA/DNA oligonucleotide. *Nat Biotechnol*, **18**(6): 615-22.

21. Rolland AP, Mumper RJ, (1998): Plasmid delivery to muscle: Recent advances in polymer delivery systems. *Adv Drug Deliv Rev.* **30** (1-3): 151-172.

22. Sparrow JT, Edwards V, Tung C, Logan MJ, Wadhwa MS, Duguid J, Smith LC (1998): Synthetic peptide-based DNA complexes for nonviral gene delivery. *Adv Drug Deliv Rev,***30**(1-3): 115-131.

23. Bessodes M, Dubertret C, Jaslin G, Scherman D (2000): Synthesis and biological properties of new glycosidic cationic lipids for DNA delivery. *Bioorg Med Chem Lett*, **10**(12): 1393-5.

24. Godbey WT, Barry MA, Saggau P, Wukk, Mikos AG: Poly (ethylenimine) mediated transfection: A new paradigm for gene delivery of *J. Biomed Mater Re*; **5**(3): 321-8.

25. Bensadoun JC, Deglon N, Tseng JL, Ridet JL, Zurn AD, Aebischer P (2000), Lentiviral Vectors as a Gene Delivery System in the Mouse Midbrain: Cellular and Behavioral Improvements in a 6-OHDA Model of Parkinson's Disease Using GDNF. *Exp Neurol*, **164**(1): 15-24.

Regulatory Issues in Gene Therapy
Good science – good sense

[1]NANCY CHEW AND [2]JOY A. CAVAGNARO
[1]*President Regulatory Affairs, North America LLC, P.O. Box 72375, Durham, NC 2772, USA;*
[2]*President, Access BIO, Leesburg, VA 20177-1400, USA*

1. BACKGROUND AND OVERVIEW

The U.S. Food and Drug Administration (FDA) has defined gene therapy as a medical intervention based on modification of the genetic material of living cells. The cells may be modified *ex vivo* for subsequent administration or may be altered *in vivo* by gene therapy products given directly to the subject[1]. The NIH defines human gene therapy as the deliberate transfer of recombinant DNA or DNA derived from recombinant DNA into a human subject (1984 FR 17844, 17847). Gene therapy products falling under the scope of the 1994 CPMP guidelines are defined as medicinal products developed by means of one of the following processes: recombinant DNA technology or controlled expression of genes coding for biologically active proteins in prokaryotes and eukaryotes including transformed mammalian cells.

The first gene therapy trial – for adenosine deaminase deficiency-induced severe combined immunodeficiency disease – was initiated in the U.S. in the fall of 1990[2]. Ten years later, although all patients receiving this therapy were not helped, the first patient treated was still alive and thriving[3]. By early 2000, more than 4,000 patients were enrolled in gene therapy studies. Much research and progress in the field of gene therapy has occurred since the first trial, not only in basic research but also in product and clinical development as well as regulatory oversight and guidance in order to develop a knowledge base for this new class of products. Over the past decade an increasing number of countries, mainly in Europe and more

273

recently Japan, have established specific areas of expertise and have become prominent in the global community of gene therapy research and development.

The scope of gene therapy encompasses a variety of techniques including alteration or supplementation of the function of a mutated gene by providing a copy of the normal gene; directly altering and/or repairing a mutated gene; and providing a gene that adds a missing function(s) or one that regulates other genes. Because most human gene therapy trials have focused on safety rather than efficacy, it has been suggested that it may be more accurate to refer to the application of this technology at the current time as "gene transfer" rather than "gene therapy," at least until there is more evidence of therapeutic benefit.

The success of gene therapy ultimately depends upon not only the delivery of genetic material into the target cells, but also the level and persistence of expression of the gene once it reaches the target site. Both of these requirements pose considerable technical challenges. A variety of vectors have been developed to make gene delivery into target cells possible. Most of these vectors are disabled viruses which work by delivering genes into certain cell types in much the same way as ordinary viruses infect cells. Recently several non-viral vector systems have been used.

The regulation of gene therapy requires state-of-the art knowledge of new science and technology to anticipate risks and devise methods to address them. Science-based regulation requires regulators with expertise both in the use of laboratory animal models to develop information about gene therapy products and in the science of drug development and applicable laws, regulation and policy. Effective regulation requires that regulators recognise the careful balance of potential but unproven benefits against real or theoretical risks. Responsible regulation requires a commitment by regulators to surveillance and compliance activities, sensitivity to public concerns including safety and ethics, and the support for and facilitation of active and ongoing scientific exchange among academia, industry and regulatory groups. Regulations continually evolve as regulatory experience grows from the generation of increasingly more and better data.

The protection of human subjects is fundamental to all FDA-regulated clinical research (21 CFR 50, 56, and 312.120). When filing an Investigational New Drug (IND) application with the FDA, investigators must follow NIH Guidelines, register with the NIH Office of Biotechnology Activities (OBA), and obtain Institutional Review Board (IRB) and Institutional Biosafety Committee (IBC) approvals. The IRB reviews protocols to assess risks to the subjects and adequacy of the informed consent process. The IBC assesses risks posed by the research study to the individual and the community.

For gene therapy products, public involvement in the decision-making process occurs at every level of research. The Congress and the public, especially patient advocacy groups, have become established stakeholders in the FDA regulatory process. Public participation in social and ethical decisions, the openness and availability of research information, and the control that is exerted by the research process itself as well as federal oversight exceeds that known for any other therapeutic product class.

1.1   Regulatory history

The FDA and NIH have oversight of human gene transfer research. They each make unique and complimentary contributions to the scientific evaluation of the safety and efficacy of preclinical studies and clinical trials. The regulatory authority for gene therapy is the responsibility of the United States Food and Drug Administration (FDA), which bases its decisions on the broad principle of regulating products, not technology. FDA scientists at the Center for Biologics Evaluation and Research (CBER) make scientific evaluations of the information and decide whether individual protocols should proceed. Components of scientific review include regulatory research, testing, and compliance activities that include inspections, education, and training. The NIH's oversight of human gene transfer research is embodied in the activities of the Recombinant DNA Advisory Committee (RAC) and the NIH Guidelines for Research Involving Recombinant DNA Molecules (NIH Guidelines[4]). The NIH/RAC conducts public review and discussion of the scientific, safety and ethical issues raised by specific protocols.

In the early 1970s, biomedical leaders came forth to contain emerging social and ethical issues. The scientific community was concerned that SV40 might be able to transform human as well as monkey cell lines. Plans to introduce SV40 DNA into *E. coli* were put on hold. A group of leading scientists published a letter in the journal *Science* calling for a temporary halt to rDNA research[5]. At the June 1973 Gordon Conferences in New England, ardent discussions about the occupational safety of laboratory personnel working with rDNA brought forth a request that National Institutes of Health (NIH) and Institute of Medicine (IOM) appoint a committee to study the subject. An article appeared in the prestigious Proceedings of the National Academy of Science describing the potential biohazards of recombinant DNA molecules[6]. The federal government's response to this request marked the beginning of the regulation of recombinant DNA research in the United States.

NIH formed the Recombinant DNA Advisory Committee (RAC) as a direct consequence of this outcry. The RAC has played a major role in the federal regulation of biotechnology research in the United States ever since. A scientific conference, organised by NIH in 1975, was held in Asilomar, California, where leading academic researchers met to discuss potential hazards of biological research. This conference, which also attracted legal scholars and journalists as well as scientists, is widely recognised as the pivotal event that initiated U.S. public policy debate on the use of the products of biotechnology. Participating scientists published a summary statement immediately after the Conference[7] and a full report soon thereafter[8].

These events are particularly notable for two reasons. First, scientists themselves called for a moratorium on research and requested governmental involvement in its control. Second, immediately after the Asilomar Conference the public joined the scientific community and the regulators in policy-setting debates.

NIH Guidelines for Research Involving Recombinant DNA Molecules (NIH Guidelines) were first published 23 June 1976. They applied only to federally funded research. During the late 1970s, Congress considered passing legislation to regulate biotechnology. Industrial and academic scientists alike were concerned that the legislation would be so restrictive that American scientists would leave the country to continue their research elsewhere. Congressional interest waned, and the control of research was passed to federal regulatory authorities.

There have been many revisions to the NIH Guidelines since they were originally published. In 1978, a public hearing forced NIH to revise the Guidelines to increase the requirement for public participation in local Institutional Biosafety Committees and to require major decisions be forwarded to the RAC. FDA notified the regulated industry that any product developed using rDNA procedures must comply with the Guidelines.

In 1990, the guideline was amended to include further guidance on clinical gene transfer research. In 1991, FDA issued the first guidance document on gene therapy entitled "Points to Consider in Human Somatic Cell Therapy and Gene Therapy." Since then several FDA Points to Consider documents and Guidelines on manufacturing and testing have appeared[9-13]. Over the past 10 years the various guidance documents were continually revised as a better understanding of the actual level of risks for the various procedures became better understood[4,10,12].

In 1995, the Director of NIH proposed to eliminate the redundant RAC review and approval of gene transfer clinical research on the grounds that an institutional review board (IRB) and the FDA review were sufficient. The proposed process of "consolidated review" was intended to eliminate an unnecessary duplication of efforts. In 1997, following much debate, NIH issued a compromise that left the RAC review process intact while eliminating the requirement of RAC "approval" of gene transfer protocols, novel or otherwise. This change did not alter the requirement for all NIH funded investigators (or investigators at institutions receiving NIH funding) to register their protocols adhere to NIH Guidelines, and participate in public discussion if requested by the RAC.

In November 1999, NIH issued a proposal to amend its Guidelines, so that "serious adverse events" during gene transfer trials would be reported immediately to NIH, and this information would automatically be considered "public." At the same time, in a letter to gene therapy IND sponsors and principal investigators, FDA outlined a process for submission of gene transfer INDs and subsequent adverse events reports to CBER, and how that process related to the required NIH submissions[14]. Currently there are ongoing discussions to better define what must be reported when a serious adverse event occurs. The main concerns about the safety reporting process are timelines and adequacy of information. Safety reporting procedures must ensure that all pertinent information regarding the scientific and ethical conduct of human gene transfer trials proceed to review and analysis in a timely fashion.

2. THE HIGH PRICE OF PROGRESS

Over the past two decades progress in gene transfer clinical research has been much slower than anticipated or hoped. Nevertheless, incremental steps forward have been made. The rationales and strategies for treating particular diseases are varied and the most promising findings have been described for the treatment of primary immunodeficiencies and haemophilia.

From 1989 through the first half of 2000, 270 new gene transfer INDs were submitted to the FDA, with 55 submitted in 1999 alone[15]. An increasing number of clinical programs progressed from establishing adequate safety profiles and proof-of-concept determinations to optimising dose levels and refining regimens.

The significant progress in gene transfer studies was called into question following the death of an eighteen-year-old patient enrolled in an adenoviral vector gene transfer clinical protocol in the fall of 1999, nine years after the first patient was treated. The death was presumed to be the first death directly related to the gene transfer. The event prompted concern and an immediate investigation of the processes by which gene transfer trials had been reviewed, conducted, and monitored by FDA and NIH up to that point. Ardent debate over the exact cause of death impelled intense investigation by both FDA and NIH. As the results of inspections became available, concerns were raised about the conduct of clinical research that extended beyond this single patient, trial and sponsor.

The problem areas identified by FDA inspectors included:

- failure to submit information to FDA and IRB (*e.g.* animal data, adverse clinical events, annual reports);
- failure to amend protocols and informed consent forms;
- lack of SOPs for conduct of trials, training, and case report forms;
- lack of documentation of informed consent; and
- inadequate protocol adherence (*e.g.* entry criteria, stopping rules, testing).

As a result of these findings, gene therapy research was suspended at major medical institutions while Congress and the President of the United States called for more careful oversight of clinical trials by regulatory authorities[16]. Ironically, 24 years after the Asilomar Conference was convened to discuss the potential hazards of biological research, the concern over research in gene therapy was due to the conduct of research, not the technology itself.

The regulatory action following a year 2000 review of selected gene transfer studies resulted in an increase in enforcement of compliance with adverse event reporting requirements and a more rigorous oversight of clinical trials by both sponsors and FDA. NIH also called for increased safety evaluation, stricter reporting guidelines and independent, third party review and tightening of measures to protect human subjects.

3.   GENE TRANSFER BASICS

The methods used for gene transfer and expression usually involve a gene, a delivery system known as a vector, and an expression system that can be part of the vector. Gene transfer can replace a defective gene and thereby correct genetically inherited disease. Gene transfer also can modulate cell functions directly to either treat or cure disease (Gene Therapy), to prevent diseases (*e.g.* DNA vaccines), or to act as gene markers for other therapeutic or diagnostic purposes.

There are several kinds of gene therapy products currently in research and development, including:

- naked nucleic acid (*e.g.*, natural or enzymatically synthesised nucleic acid ligated into appropriate plasmids);
- complexed nucleic acid (*e.g.* formulated with salts, proteins or other polymers; encapsulated in lipids or coated on gold particles); and
- replication deficient viral vectors (*e.g.*, murine leukaemia virus, HIV, adenovirus, AAV, herpes simplex virus, poxvirus).

Products also may be produced within genetically modified somatic cells containing and functionally expressing the transferred genetic material (*e.g.* fibroblasts, myoblasts or other autologous, allogeneic or xenogeneic cells).

4. QUALITY ISSUES

From a regulatory perspective, the particular biological and safety characteristics of a vector must be well characterised before introducing it into human subjects. Product testing requirements include demonstration of product safety, assessment of product characterisation and maintenance of product lot consistency.

The most studied delivery systems are the viral vectors which include animal or human viruses, pathogenic or not (*e.g.*, retroviruses, SV40, adenovirus, AAV, herpes simplex virus, vaccinia virus, and lentiviruses). The criteria for use of viral vectors include setting acceptance limits of that take into account dose, type of viral agent, assay methods used, and safety testing.

The non-viral DNA vectors currently under investigation include DNA compacted, cationic and anionic lipids, polymers, peptide, polymer particles, or DNA non-condensed, DNA administered with the use of special devices *e.g.* gene guns or electroporation methods to increase efficiency of delivery.

The inherent features of viral vectors and non-viral vectors make them attractive to specific clinical indications. Characteristics include both advantageous features as well as safety-related concerns. The key features and/or concerns contributing to the selection of a particular gene transfer method include: insert size limit; chromosomal integration; therapeutic protein expression; vector localisation; ability to infect many cell types; efficiency of gene transfer; expression of viral proteins; immunogenicity; ability to produce high titres; ease of manipulation; stability; and experience.

5. SAFETY ISSUES

The design of preclinical safety assessment programs should consider appropriate selection of a relevant model, the manner of delivery including the dose, and the route of administration and the treatment regimen. Relevant model choice should include selection of both species and physiological state. Treatment regimen includes frequency and duration of dosing. The requirements will likely be very similar to those for other biotechnology-derived pharmaceuticals.

The unique concerns and potential risks associated with viral delivery systems are insertional mutagenesis, a theoretically possible germ line transduction, replication competent retroviruses, and/or homologous recombination *in vivo*.

Among the concerns and risks associated with non-viral delivery systems the most important are related to the toxicity of the delivery system itself and/or the route of administration.

The rationale for introduction of the gene transfer agent into human clinical trials is based upon demonstration of the efficiency and feasibility of gene transfer as well as a demonstration of safety. In both viral and non-viral systems there is the potential concern of the lack of an effect/therapeutic benefit due to low transduction efficiency.

Preclinical studies are necessary to answer specific questions to aid in subsequent clinical development. For first-in-human dose (FHD) studies the information should include, recommendation of a safe starting dose, dose escalation scheme and regimen (route, frequency and duration), identification of appropriate monitoring parameters and an understanding of the relationship of toxicities of vector and the transgene product.

Pharmacodynamics studies help provide useful information regarding the most appropriate model for subsequent safety testing, optimisation of route of administration, and dose selection. "Proof-of-concept" may be best achieved in an animal model that mimics the human deficiency or disease state. As is true for other biotechnology-derived pharmaceutical products, the non-human primate should not be considered *a priori* as the most relevant model.

The purpose of toxicity testing is to obtain information regarding single and repeated dose exposure to the vector product, including the delivery system. This information should include identification of target organs for subsequent monitoring in the clinic, information of level of expression, and persistence and reversibility of any adverse effects.

In vivo toxicology studies are generally needed for first generation vectors/new molecular entities. Programs should include establishment of safety pharmacological profiles, dose/response relationships, toxicology profile (single and repeat dose), and vector distribution. The goals of the vector biodistribution determination are to assess dissemination of vector to the germline as well as distribution to other non-target tissues. The studies to date suggest that the risk of foreign gene transfer to germ line cells and future progeny is low.

In vivo safety studies also are needed if adverse effects are known to occur with other products within the same vector class and/or transgene product, or when the route of administration is changed during the development of a product, or if there are significant changes in formulation.

6. EFFICACY ISSUES

The potential risk factors in gene transfer clinical research include the highly innovative nature of the therapy and the lack of precedence with respect to clinical trial design in the area of surrogate markers of activity or safety. The practical risks are the inexperience of sponsors, sponsor/investigators and investigators in clinical trial design, analysis and monitoring, and insufficient knowledge or understanding of regulatory practices. Finally, the economic environment in the field of biotechnology, in which gene therapy research is conducted may give rise to potential financial conflicts of interest.

The characteristics inherent to gene transfer products provide both challenges and opportunities to clinical trial design and analysis. These characteristics include the fact that the ultimate active form is not the product administered; the complex transition from delivered form to active form; the novel time course for activity, the complex correlation of active form dose with dose of product administered, and unique immunogenicity considerations.

The lessons learned from the rigorous regulatory compliance investigations conducted during year 2000 provide the opportunity to enhance the successful completion of clinical trials and evaluation of clinical data. Enhancements may include:

- designing studies based on earlier preclinical or clinical findings;
- developing and using detailed protocols with eligibility criteria and comprehensive evaluation plans;
- ensuring that sub-investigators are informed of the study follow-up requirements prior to enrolling patients;
- using unique study identification numbers;
- including stopping rules and independent safety monitoring;
- making protocol revisions comprehensive by incorporating each amendment within the body of the protocol such that the clinical experiment is summarised in a single document;
- using Case Report Forms (CRFs) or other documentation measures to permit data verification;
- ensuring that all Adverse Event reporting requirements are met (include not only FDA but IRBs, IBCs, and NIH);
- ensuring that patients are informed of accumulating important safety data;
- including follow-up outcome assessments in study protocols;
- providing comprehensive analysis of important outcome data (safety, bioactivity assessments, and gene expression);

- completing study reports;
- maintaining verifiable records.

7.   COMPLIANCE ISSUES

Inadequate oversight of manufacturing, design and analysis of preclinical studies, and oversight of clinical trials, reporting of adverse events, and record keeping all were identified as areas of concern in the field of gene transfer research and early development during the rigorous examination of gene therapy research that took place in the U.S. during the year of 2000. Applications of "Current" Good Manufacturing Practices (cGMPs), Good Laboratory Practices (cGLPs) and Good Clinical Practices (cGCPs) are viewed as a compliance continuums in which adherence to principles are essential during the initial stages of clinical research and full compliance in all areas is expected during later stages of clinical development.

7.1   The principles and practice of cGMP

The GMP continuum applies to manufacturing processes and facilities, laboratory controls, and the materials used, including containers and closures. Compliance with cGMPs is expected throughout clinical studies. Adequate documentation (traceability) and facility design and layout, sterility assurance, and QC/QA oversight are required for FHD studies. Development of defined in-process controls and full process and assay validation are necessary for final product development. The elements of GMP include: appropriate facility design to control operation, adequate documentation/records, production and process controls, quality control/assurance, validation, personnel training, and certification and monitoring, and environmental monitoring.

In March of 2000, FDA issued a request for manufacturing and testing information to all sponsors of INDs or master files using or producing gene therapy products[17]. The request asked for submission of information on all lots of material ever used in animal or human studies; authorisations to cross-reference other submissions; lot release data and characterisation testing for each lot used in clinical trials; and a summary of product manufacturing quality assurance and quality control programs.

7.2   The principles and practice of cGLP

The key elements of a GLP study include: a study protocol, study director, trained personnel, standard operating procedures (SOPs), adequate facilities (*e.g.* animal housing and care and calibration and maintenance of equipment) and a quality assurance unit. Initial pharmacology and/or proof-of-concept studies are generally not expected to be GLP-compliant; however, key elements of GLP should be considered especially in cases where animal models of disease are used to support safety. "Pivotal" toxicology studies should be conducted in full compliance with GLP. However, if there are specific aspects that cannot be performed under full compliance, they should be prospectively identified. For example, some aspects of safety assessment studies may not be fully compliant especially in cases where the methodology is evolving (*e.g.* PCR testing for biodistribution). Other areas which fall on the compliance continuum include biochemical and non-animal based in vitro systems, assay systems in support of process change or process development, QC testing for bioactivity and comparability, and animal disease models.

7.3   The principles and practice of cGCP

A unified standard for Good Clinical Practice has been agreed by the parties to the International Committee on Harmonisation (E6)[18]. The key elements of Good Clinical Practice include the protection of human subjects, scientific soundness, trained personnel, and adequate documentation/records. Protection of human subjects includes the use of an Institutional Review Board or Ethical Committee (which reviews and approves the study protocol and an informed consent statement, and continues to review the ongoing study), and maintenance of the confidentiality of patient records. Scientific soundness includes a well written protocol and sufficient nonclinical and clinical information to support it.

In March of 2000, FDA outlined a plan to enhance the conduct clinical trials of gene transfer studies[17]. The plan included the proposed review of safety monitoring plans, an increase in inspections of sites during IND phase not only "for cause", and an increased outreach and opportunity for training initiatives. In a letter to Sponsors[17], FDA requested a 2-3 page summary for each clinical trial be submitted documenting that procedures were in place to ensure adequate monitoring of the clinical investigations, to demonstrate the trial(s) were being conducted in accordance with regulatory requirements and GCPs; that the protocol and the rights and well-being of human subjects were protected; and that data reporting was accurate and complete.

NIH and FDA also outlined a proposal to develop procedures to further assure that oversight of the conduct of clinical trials is appropriately independent[19].

8.   GLOBAL ISSUES

Over the past decade an increasing number of countries, mainly in Europe and more recently Japan, have established specific areas of expertise and have become prominent in the global community of gene therapy research and development.

In 1997, U.S. sponsors originated more than 90% of all gene therapy trials, and 80% of the patients were in the U.S. The situation changed rapidly during the ensuing two years.

The European marketing authorisation for Gene Therapy products falls under the scope of part A of the Annex to Council Regulation 2309/93. In 1994, a CPMP guideline on gene therapy products that addresses quality aspects in the production of vectors and genetically modified cells was finalised.

In the United Kingdom, Gene Therapy Advisory Committee (GTAC) established a standing subgroup whose remit was to examine the potential clinical and ethical implications of new technologies. The subgroup met for the first time in November of 1997 to examine *in utero* gene therapy. The GTAC works with agencies, which have responsibility in the field of gene therapy (including local research ethics committees) and other agencies that have statutory responsibilities such as the Medicines Control Agency (MCA), The Health and Safety Executive and the Department of the Environment. The GTAC committee provides advice to UK Health Ministers on developments in gene therapy research and their implications. Their focus of review is the identification of risk level, the determination of the safety of production process, the exclusion of dissemination in the environment, and the patient's safety in clinical trial.

Other committees in Europe responsible for oversight of gene therapy include:
–   the CGG (Commission du Genie Genetique), CGB (Commission du Genie Biomolecularie) and Viral Security Committee in France;
–   the Komission fur Somatische Genterapie in Germany; and
–   the GGO (Ministry of Environment) and CCMO (Ministry of Health) in the Netherlands.

9.   CONCLUDING REMARKS

Considerable progress has been made in the discovery and early clinical development of a variety of gene transfer products. Success has been enabled by the availability, validation and implementation of new technologies not only for product manufacture, but also for testing and evaluation. New challenges will need to be overcome to ensure that products will also be successful in later clinical development and ultimately for marketing authorisation. These new challenges include improvements in delivery systems, better control of in vivo targeting, increased levels of transduction and duration of expression of the gene (or silencing of expression), and manufacturing process efficiencies that enable reduction in production costs.

Authorities and institutions responsible for regulatory oversight have recognised the need for and have instituted the necessary programs to support the continued progress in the field of gene transfer research leading to safe and effective therapy. Regulatory initiatives fostering cooperative approaches towards the ultimate goal of approval and availability of these new medicines include:

- increasing training programs and conferences;
- providing reviewer templates;
- increasing 'not for cause' inspections during the IND phase;
- providing additional guidance on an "as needed" basis to reflect current scientific knowledge.

Towards this shared goal, sponsors and investigators should be vigilant in:

- providing a history of manufacturing in facilities used to manufacture preclinical and clinical materials;
- providing lot release and characterisation of all cell banks, viral banks and final product;
- supporting QA/QC programs for manufacturing, monitoring plans and oversight;
- providing timely reporting of animal safety information and yearly update reports of all key data and throughout the entire clinical development program; and
- ensuring protection of patient rights and well-being in the clinical trials in which they participate.

Academia, industry and regulatory bodies need to constantly share information on proposed development strategies and assessments in order to ensure that public health is not put at risk.

REFERENCES

1. Application of Current Statutory Authorities to Human Somatic Cell Therapy Products and Gene Therapy Products. *Federal Register*, 1993: 53248-51.

2. Blaese RM, Culver KW, Miller AD, Carter CS, Fleisher T, Clerici M, *et al.* T lymphocyte-directed gene therapy for ADA-SCID: initial trial results after 4 years. *Science* 1995; **270** (5235): 475-80.

3. Anderson WF. Editorial: The Best of Times, the Worst of Times. *Science* 2000; **288**: 627-9.

4. NIH. Guidelines for Research Involving Recombinant DNA Molecules. 1999.

5. Berg P, Baltimore D, Boyer HW, Cohen SN, Davis RW, Hogness DS, *et al.* Letter: Potential biohazards of recombinant DNA molecules. *Science* 1974; **185**(148): 303.

6. Potential biohazards of recombinant DNA molecules. *Proc Natl Acad Sci USA* 1974; **71**(7): 2593-4.

7. Berg P, Baltimore D, Brenner S, Roblin R, Singer MF. Summary Statement of the Asilomar conference on Recombinant DNA molecules. *Proc Natl Acad Sci USA* 1975; **72**(6): 1981-4.

8. Berg P, Baltimore D, Brenner S, Roblin R, Singer MF. Asilomar conference on recombinant DNA molecules. *Science* 1975; **188**(4192): 991-4.

9. Establishment Registration and Listing for Manufacturers of Human Cellular and Tissue-Based Products. *Federal Register* 1998 Thursday May 14, 1998; 26744-26755.

10. CBER. Guidance for Human Somatic Cell Therapy and Gene Therapy. 1998: 1-30.

11. CBER. Guidance for Industry: Guidance for Human Somatic Cell Therapy and Gene Therapy. 1998: 1-25.

12. CBER. Guidance for Industry: Supplemental Guidance on Testing for Replication Competent Retrovirus in Retroviral Vector Based Gene Therapy Products and During Follow-up of Patients in Clinical Trials Using Retroviral Vectors. 1999.

13. CBER. Guidance for Industry: Content and Format of Chemistry, Manufacturing and Controls Information and Establishment Description Information for a Vaccine or Related Product. 1999: 30.

14. Zoon KC, CBER. Dear Gene Therapy IND Sponsor/Principal Investigator: CBER, 1999:2.

15. Zoon KC. Approaches to the Regulation of Biotech Products in the USA. 5th Interlaken Conference, 2000.

16. Amy Patterson M.D., Director of NIH Office of Biotechnology Activities, Testimony. *Subcommittee on Public Health Committee on Health, Education, Labor and Pensions.* Washington, D.C.: NIH, 2000.

17. Siegel JP, CBER. Dear sponsor of an IND or master file using or producing a gene therapy product: CBER, 2000: 1-3.

18. ICH. Harmonized Tripartite Guideline for Good Clinical Practices. 1996; E6.

19. NIH, FDA. New Initiatives to Protect Participants in Gene Therapy Trials. *HHS News*, 7 March 2000, 2000: 1-3.

Regulatory Aspects in Gene Therapy
Special highlights on European regulation

ODILE COHEN-HAGUENAUER
Euregenethy Coordinator; Laboratoire TGOM and Service d'Oncologie Médicale, Hôpital Saint-Louis, 1, avenue Claude Vellefaux, 75 475 PARIS Cedex 10, FRANCE

1.   INTRODUCTION

The field of gene therapy represents one of the most challenging therapeutics for the next millennium. As such the concept might have been oversold ahead of time. Nevertheless, in some instances, potential clinical efficacy is currently showing in the picture with reports of significant successes in e.g., inherited severe combined immuno-deficiencies, haemophilia, arteritis obliterans and even cancer[1,2,3,4,5]. Potential applications of gene therapy are extremely large extending from monogenic hereditary diseases to acquired and multifactorial disorders. Therapeutic gene transfer addressing such a large panel of conditions is currently being investigated, bringing up therapeutic options in diseases where none had been available to date. The development of recombinant DNA technology has induced in the public fears and speculation regarding its potential risks. In fact, the report of undue accidents in the United States has resulted in a broad debate on the subject of gene therapy oversight brought up to US Senate and government[6,7,8,9,10,11].

2. WHY IS THERE A NEED FOR REGULATION OF GENE THERAPY?

The implementation of gene therapy involves technological approaches which might not be devoid of potential side effects just as with many conventional therapeutic means addressing severe conditions[12,13]. There are different levels where quality and safety need to be ascertained: the patient, the carers and the environment. Safety and regulatory aspects of gene therapy can be envisaged along three lines: 1st/ Experimental and preclinical research; 2nd/ Manufacture of gene therapy products; 3rd/ Clinical trials and development. The regulation of Gene Therapy is intended to assess for these risks in order to minimise them, and also delineate a margin where safety is most likely secured.

Appropriate evaluation of risk/benefit of gene therapy intervention is needed as this field adapts to evolving knowledge. There is no doubt that a significant complexity results from a number of critical levels of concern: gene discovery; gene regulation; gene delivery systems with potential biohazards; manufacture of biotechnology products and finally, efficacy and safety outcome following treatment of a patient. With respect to basic principles involved in the protection and respect of the human person submitted to biomedical research and clinical trials, gene therapy is no different from other fields of medicine. The main guarantee of patient well-being is sound knowledge, specific expertise and wise judgement in evaluating the risk/benefit ratio, rather than specific regulations.

3. REGULATION OF GENE THERAPY: CURRENT STATUS

3.1 Pre-clinical research

Regulation applying to preclinical research mainly relates to the use of Genetically Modified Organisms. In order to harmonise positions within EEC, three Directives were adopted by the Council: *(1) Containment of genetically modified organisms (GMOs)* to protect workers and environment during the production process. This is governed by the application at the national level of the EC Directive (90/219/EC, DG XI; revised 98/81)[14,15,16,17] on contained use of genetically modified microorganisms *(2) The potential adverse consequences of the deliberate release of GMOs* (eg recombinant viruses) into the environment. This is also governed nationally by the application of the EC Directive (90/220/EC, DG XI; currently being revised)[18] on the deliberate release of genetically modified organisms *(3) Directive 90/679/EC on the protection of workers*[19] from the risks related to exposure to biological agents at work.

The member states of the European Union have to implement the provisions of these Directives into National laws. Difficulties have been encountered; in some countries, implementation is still in process while the initial texts have either been revised or currently are being updated. In fact, the Directives are flexible enough so that each national Authority may interpret them in a way that reflects the existing national procedure. This has resulted in a significant level of heterogeneity from one country to the next.

There are elements within existing regulations that apply to the development of gene therapy approaches[20,21,22,23,24,25] (review [26,27]). Biohazard related to gene transfer technologies needs to be carefully assessed and regulation adapted to evolving knowledge. Although GMOs Directives were not initially meant at addressing clinical trials in humans, they do apply to the manufacture of gene therapy products.

3.2 Manufacture of a clinical grade gene therapy product

3.2.1 Regulatory framework

The manufacture and distribution of a product are matters of critical importance. Strict adherence to the principles of Good Laboratory Practice (GLP) is mandatory whatever the level of concern: Quality of Products-Safety Testing-Efficacy Testing from production to clinical assessment in patients. Many of the principles laid down in existing Community Guidelines on Biotechnology Products, such as the requirements for cell banks, genetic stability and safety testing, are applicable to the products designed for gene therapy.

In the USA, FDA is the relevant body dealing with manufacturing facilities and marketing of products. A first document had been released in October 1993 and revised in March 98 which is intended to clarify its regulatory approach and to provide guidance to manufacturers of products intended to be used in somatic cell therapy or gene therapy. FDA premarket approval is required for biological products.

3.2.2 Product safety (see section 4 for details)

Safety aspects related to the product mainly address quality which is a prerequisite for administration in human. These mainly deal with Quality.

Strict adherence to both Good Laboratory Practice (GLP) and Good Manufacturing Practice (GMP) is compulsory. Principles governing quality of biotechnology products derive from those governing biologics in general. These processes involve living organisms; thus variability is of major concern.

3.3   Clinical developments-clinical trials

3.3.1   Levels of concern

The safety and efficacy of any gene therapy procedure should be tested in clinical trials of appropriate design. This question receives much public attention. Safety / efficacy trials in human subjects-regulation is covered by Ethical Committees and Scientific Review. Any Clinical trial, whether intended for market authorisation or performed in the context of clinical research programmes, should stick to Good Clinical Practice (GCP) in compliance with worldwide consolidated guidelines on GCPs as of 1996 (CPMP/ICH/135/35 and FDA E6)[28]. The text of the 'Amended proposal for a European Directive on the implementation of Good Clinical Practice in the conduct of clinical trials' has reached consensus approval by the European Council and is currently being reviewed by the European Parliament (COMC1999/193 final)[29]. Taking into account recent problems that have occurred in the USA, the nature of protocol violations that have been sorted out by the FDA are consistent with misadherence to Good Clinical Practice (GCPs)[10,30,31].

Gene therapy raises questions that are scientific, ethical and social. They arise at the level of the single patient, at the level of healthcare providers and at the level of the environment in general. A fundamental prerequisite is that of safety; whether concerning the patient or his environment. Whatever the technological strategy, non-propagation and non-transmission of the gene transfer delivery system is mandatory. The ethical issues surrounding gene therapy are no different from those related to other new therapeutic approaches. At the individual level of a patient: potential benefits must be shown to outweigh potential risks. The rationale for any gene therapy proposal must therefore take into account the possibility of any alternative treatment strategies. There may in some instances be no alternative to gene therapy. Public health considerations relate in particular to the possible spread of any helper virus either to healthcare providers, to family or to the general population and environment.

3.3.2 Scientific and ethical review

3.3.2.1 Scientific review

In the United States, The Recombinant DNA Advisory Committee (RAC) of the National Institutes of Health (NIH) has been dealing for years with Ethical and Social concerns, scientific evaluation and public discussion. Interaction with the Public has been an issue of great concern and widely encouraged since all meetings of the RAC are open to the public. RAC currently reviews only those protocols involving new technologies or original strategies (Guidelines 64 FR 25361-Appendix M)[32]. It holds consultative and not statutory authority which belongs to FDA. There is no such federal established authority at the level of the European Union; and certainly no forum where scientific debates would be kept open to the public. Some Member States have settled on an ad hoc review process.

3.3.2.2 Ethical review

Local ethics committees

Review of a proposal for a clinical trial involving human subjects is compulsory worldwide. This is being processed by a local ethics committee. In the USA, a proposal will be considered by the FDA and the RAC only after the protocol has been approved by the local Institutional Biosafety Committee (IBC) and by the local Institutional Review Board (IRB) in accordance with Department of Health and Human Services (DHHS) *Regulations for the Protection of Human Subjects*. The IRB and the IBC may, at their discretion, condition their approval on further specific deliberation by the FDA and the RAC. The same processes are being followed Europe-wide although at the level of each local institution. The International Bioethics Committee has released guidances and recommendations which may help local institution better define the overall context of such applications.

International Bioethics Committee

In 1993, the Executive Board of UNESCO stressed "the need to develop information exchanges and to carry out extensive consultations within an intercultural framework, in order to precisely identify the issues involved in the control of the human genome". These decisions led Mr Federico Mayor, Director-General of UNESCO, to create the International Bioethics Committee (IBC) - the first institution of its kind in the world - so as to examine the ethical, socio-cultural and legal questions raised by research in genetics and their applications. The action engaged by the IBC is three-fold.

First, it is conceived as a forum for the exchange of ideas and debate on the ethical, legal, social, and, more broadly, cultural implications of genetic research and their applications. Second, the IBC promotes action aimed at enhancing the participation of the public in this debate. Third, the General Conference of UNESCO has adopted a Universal Declaration on the human Genome; the terms of implementation of which are being debated and updated at regular intervals.

European Group on Ethics

The European Group on Ethics in science and new technologies was established in 1999 with an extended and more official remit as compared to the former Group of 'Advisers on the Ethical Implications of Biotechnology', a body created by the European Commission at the end of 1991. The former Group of Advisers also chaired by Noëlle Lenoir marked the start of its second term of office with a press conference to present its opinion on gene therapy (No 4, 13 December 1994), based on the report by Luis Archer, Professor of molecular genetics and Members of the Portuguese National Ethics Council.

3.3.3   Clinical assessment

Once the therapeutic intervention will have been performed, long term patient follow-up is important in particular at these early stages in the development of gene therapy products where knowledge needs to accumulate from experience gained. Ideally patients should be monitored for: - survival of genetically-modified somatic cells; - biodistribution or anatomical location of the genetically-modified somatic cells; - monitoring of the therapeutic product released over time; - toxicity and potential adverse effects including the development of antibodies to the therapeutic protein, the genetically-modified somatic cells, the vector or those cell antigens modified as a consequence of the presence of genetically-modified somatic cells and/or vectors (especially where repeated administration is intended); - whenever viral vectors are used, patients should be monitored for viral vector-shedding as this may result in infections spread to clinical personnel, relatives and other close contacts.

3.3.4   Clinical development

Current status in various countries demonstrates a high level of heterogeneity. To date the implementation of Gene Therapy clinical trials in Europe is being regulated at two distinct levels, without overlap[33,34,35,36,37]. 1°/ Early phases of clinical trials are currently being reviewed at the national level; the process is not harmonised from one country to the next; 2°/ Marketing authorisation is being covered by the centralised procedure through the European Medicines Evaluation Agency (E.M.E.A.), entering into force on 01.07.95, and involving a decision binding on all Member States of the European Union (Regulation (EEC) Nr. 2309/93)[38]. The CPMP (Committee for Proprietary Medicinal Products) has adopted guidance to support marketing authorisation of gene therapy products (December 1994: III/5863/93 Final)[39]. These 'Notes for Guidance' (NfG) are currently being revised; a draft was open for comments until June 2000[40].

Current regulations differ widely in European countries since early phases are being reviewed at the national level by three to four bodies which have to be consulted. The process is not harmonised from one country to the next. From the user's standpoint, this is an unworkable and detrimental situation, even for the world's biggest pharmaceutical companies. In the case of rare disorders and Academia-driven clinical trials, patients would be forced to travel to a foreign country.

With the help of national officers, the Euregenethy network has published a record of regulatory procedures in each European country in the form of a newsletter[36]. In order to rationalise these data, a summary table using common main headings has been prepared[41]. This contributes to put emphasis on the scientific data and rationale which constitutes the basis for a clinical trial application. In fact, the pre-clinical safety and efficacy requirements are universal and thus should not vary from one country to the next.

Current CPMP-NfG drafts on the Quality, Preclinical and Clinical aspects of gene transfer medicinal products are mostly intended to support marketing authorisation. Requirements, including GMP-related, should best be upgraded as the study phase progresses. Before the development of a product reaches the mature market-phase, the implementation of these guidelines should proceed on a case by case basis, according to the phase of the clinical study, to the disorder of interest (e.g., rare diseases). A consultation with the Committee on Orphan drugs most recently hosted at the EMEA will certainly be of major interest[42].

Specific guidelines are also available from US-FDA (and from NIH-RAC where applicable)[32,43,44,45]. In the USA, FDA is the relevant authority towards marketing authorisation. Guidelines have been released in 1998 dealing with the Application of current statutory authorities to human somatic cell therapy products and gene therapy products. In general, somatic-cell therapy and gene-therapy products based on viral vectors meet the statutory definition of biological products and are subject to appropriate regulation. Some products could also be regulated as drugs; e.g., gene therapy products such as chemically synthesised products meet the drug definition. Some products may even contain a combination of Biological products and drugs or devices. Technical requirements are in principle less stringent during the early phases of clinical investigation. The status of the product is that of Investigational New Drug (IND). Clinical trials are being conducted under INDs. Later developments require <Product License> which involves license of both product (Product License Application: PLAs) and Manufacturing facilities (Establishment License Application: ELAs).

4.   TECHNICAL -SCIENTIFIC AND SAFETY- ISSUES

This section is intended to provide highlights in (i) attempts at increasing the efficacy and (ii) improving the safety of gene transfer vectors. There is indeed a critical need for improving gene delivery technologies as a pre-requisite for achieving therapeutic gene transfer. Whatever the delivery system, non-propagation and non-transmission are compulsory in order to contain biohazard. This has been the focus of a special session towards a better definition of the risks and on elaboration of possible standards during the Euregenethy multidisciplinary forum (June 200).

4.1   Progress with replication incompetent viral and non viral vectors

4.1.1   Retrovirus vectors

Safety considerations in the long-term follow-up of interventions using Mo-MuLV retroviruses arise from their identified ability to integrate into the genome of transduced cells at random sites. Jan Nolta from the Children's Hospital (Los Angeles) presented biosafety data collected after 7 to 18 months follow-up in 412 immune deficient bnx mice co-transplanted with transduced human haematopoietic and mesenchymal stem cells[46]. These studies document that there is very low risk from RCR or from insertional mutagenesis since all assays remained negative. In addition a similar study has been initiated with lentivectors in 35 mice. No helper virus has been detected so far. David Parkinson from Novartis-US reported reassuring extensive safety data collected in the context of the Phase III trial in Glioblastoma multiforme. The company decided to put the protocol on hold several months ago because of lack of demonstrated benefit to patients. Keith Humphries, from Terry Fox Labs (Vancouver) presented successful *ex-vivo* gene modification of haematopoeitic stem cells using mouse models and oncoretroviruses. Constructs accommodating elements limiting variegation have now been shown to drive sustained expression of globin proteins to a level approaching that required for therapeutic use[47]. Alain Fischer's group in Paris has reported on a trial in X-linked SCID-1 patients. Bone-marrow cells were genetically modified *ex vivo* and re-implanted in 5 infants under the age of one year. The vector is derived from a well-known MuLV backbone. A strong selective advantage of genetically corrected T-cells accounts for striking clinical efficacy. Four children have left their bubbles and are at home without any further treatment. These data represent the first clear proof of efficacy of a gene therapy trial[1].

4.1.2   Lentivirus vectors

The most significant progress reported in the recent past by Inder Verma and his colleagues, Luigi Naldini and Didier Trono, has been the paradoxical use of a potent human pathogen the HIV virus[48]. The packaging system, using the VSV-G envelope protein to generate pseudotyped virions, extends the natural virus cell tropism. Lentivectors have been shown to successfully transduce mitotically quiescent cells, such as neurons, photoreceptors from the retina, undamaged liver, muscle, macrophages and also human haematopoietic stem cells with reconstitutive potential in NOD/SCID mice[49,50,51,52]. These properties extend to third generation self-inactivating (SIN) vectors[53] which have been deleted of a great majority of virus sequences and in particular those encoding for key accessory proteins[54,55]. Regulated and/or inducible expression has been obtained with defined promoter sequences cloned inside SIN vectors[56]. In addition HIV cis-elements accounting for increased karyophilic properties of helper virus have recently been identified to improve recombinant vectors-mediated transduction efficiency[57]. These features make lentivectors plausible candidates for overcoming a significant number of limitations. Notable progress has been achieved with regard to safety and in developing assays for the detection of helper virus breakout[58]. Vectors derived from SIV are compatible with *in vivo* safety evaluations in non-human primates as emphasised by Jean-Luc Darlix[59] Production scale-up, reproducibility, quality and safety controls will require the development and use of stable packaging cell-lines[60].

4.1.3   Adenovirus vectors

As a consequence of the report of the major adverse event which occurred at the Institute for Human Gene Therapy (IHGT) (University of

Pennsylvania) in September 1999 where a young asymptomatic patient died, the field is currently under extreme scrutiny. The IHGT will no longer sponsor clinical trials or produce clinical grade adenovirus stocks, as former NIH-sponsored adenovirus-NGVL.

While toxicity data related to the use of adenovirus vectors seems homogenous with a main emphasis on liver and immuno-toxicity, virus-shedding reports in patients are quite uneven from one study to the next. Bernard Escudier[61] reported detailed data pertaining to adenovector- and putative RCA- shedding in clinical trials (NSCLC) sponsored by the company Transgene. Virus shedding was found from blood, stools and sputum in the first 2 days following intra-tumoural injection of 10^7 to 10^9 pfu (at most), but no infectious particles were detected in urine at any time. Subsequent analyses for up to several weeks remained negative. Such studies will help to establish reasonable standards for monitoring. Michael Grace[62] from Shering-Plough-USA presented some adenovirus shedding data from the clinical experience with SCH 58500 which includes RCA in clinical batches. Cobra-Ltd (Staffordshire, UK) currently has substituted regular 293-based adenovirus vector packaging cells with PERC6 cells which results in significant decrease of RCA-contaminants in clinical batches. Pharmaco-toxicologic characteristics have been analysed in animal models, defining routes, dosages and preparations leading to toxicity[63].

Heterogeneity in the manufacture, processing and unit/dose definitions of adenovirus vectors have led companies form a task force with a view of harmonising these data (see below).

4.1.4   Adeno-associated vectors

Evidence for efficacy is also emerging in haemophilia, with AAV-mediated factor IX gene transfer into muscle[4]. Taken together these trials demonstrate the validity and true potential of gene therapy by providing appropriate treatment in a timely and safe manner, once sufficient pre-clinical data have been accumulated to ensure that the benefits for the patient can reasonably be expected to outweigh the potential risks.

4.1.5   Non-viral vectors

Seppo Yla-Herttuala from Kuppio (Finland) has reported on clinical improvement in arteritis obliterans by means of intra-vascular injection of non-viral vectors carrying pro-angiogenic factors like VEGF[5]. This application represents the third area where clinical improvement has been achieved.

4.2   Replication competent viral vectors

The fact that conditional replication is retained as part of vector efficiency raises additional safety concerns and a requirement for specific monitoring in patients[64]. Strikingly, the first evidence for clinical efficacy in cancer has been obtained with such vectors as shown by David Kirn (formerly with ONYX, USA)[3]. The empty adenovirus, ONYX O15 conditionally replicates in p53 deficient tumour cells[65]. This study has now reached Phase III in head and neck cancer. This comes as a fourth application where improvement has been achieved. Rabkin and colleagues in collaboration with the company Neurovir (Vancouver) have reported the extensive preclinical safety study performed in non-human primates with a replication competent Herpes simplex derived vector[64]. A Phase II trial is ongoing in glioblastoma multiforme, without occurrence of serious adverse events[66].

4.3   Viral safety

In those cases where knowledge has been accumulating over years worldwide, the release of standards could be worked out step by step in particular for virus-derived vectors (e.g., retrovirus see FDA-CBER, supplement November 1999)[45]: (i) in case of inadvertent wild-type virus breakout while making-use of an otherwise replication-defective vectors; (ii) applying to the level of replication competent viruses contaminating replication-defective virus stocks which could be considered to qualify (or not) for clinical use; (iii) extending to controlled replication-competent virus-vectors (e.g., adeno or herpes virus-derived) and lentivectors. Again, there is a critical requirement for flexibility as knowledge evolves.

In the case of adenovirus, in particular, the Companies-driven current initiative is aiming at standardising both Adenovirus production and dose definition (Oct 5th meeting at the Williamsburg BioProcessing Foundation: <<Adenoviral vector testing: product characterisation and methods standardisation>>; a report is available upon request from Beth Hutchins, at Canji-Inc and *Molecular Therapy* 2000; 2: 532-534).

Contract companies which have experienced safety testing of clinical stocks have devised refined methods for detecting putative RCV contaminants in high titre replication-defective vector-stocks. Special attention should be given to potential adventitious agents (including animal e.g., bovine or porcine viruses) inadvertently contaminating reagents used in the manufacture of gene therapy products. This is a serious issue according to the experience of contract QC-companies. This concern applies to both virus and non-virus derived vectors as well as cell-banks and *in vitro* manipulation of patients' cells towards *ex vivo* cell and gene therapy.

4.4   Somatic cell therapy

GMP processing of somatic cells and gene therapy products is required with a view of *ex vivo* gene therapy. Current regulatory framework combines both cell and gene therapy guidelines. A matter of critical importance is the so-called 'spirit of GMP' whereby increased stringency in quality assurance and quality control would be required as the study phases progress. An important question relates to GMP-implementation as a continuum or as distinct steps[67]. In fact, users who have undergone audits systematically ended up being audited on the basis of Full-GMP requirements.

4.5   More controversial subjects

4.5.1

Intra-uterine gene therapy which generates intense debate. Renowned scientists argue in favour of scientific investigation; while some regulators oppose this concept, claiming that permission will never been granted by Medicines Agencies. Others recognise the interest of investigations at the research level while question its putative clinical indications as compared to early treatment in newborns. The latter approach is devoid of maternal risk.

4.5.2

Another area of concern relates to the use of genetically modified xenotransplants designed to counter hyperacute graft-rejection. The main biohazard relates to potential transmission of animal pathogens to humans. This is a serious issue since: (i) the crossing of species barriers has often resulted in extremely severe pathologies in humans; (ii) current donor organ scarcity raises ethical questions as well.

4.5.3

Genetically modified human embryonic stem have been envisioned as a potential alternative research strategy. There nevertheless are obvious significant technical constraints, including building a new organ architecture, and ethical limitations to their use. The latter are the subject of ongoing intense debate at the European Parliament and Council, as well as in the United States.

5. INITIATING A MOVE TOWARD HARMONISATION

There currently exists no legal basis for member states official bodies to initiate a move towards harmonisation of gene therapy regulation. Facilitating interaction between scientists, physicians, companies, the European Commission, regulators and patients' groups has been initiated through the Euregenethy network: (i) A first multidisciplinary forum took place in June 99 in Brussels gathering for the first time all these interested parties; (ii) productive synergy with the European Society of gene therapy, which organises each year a top-scientific meeting. This interaction translates into a satellite session to ESGT-meeting organised by Euregenethy[68]; (iii) the second forum on safety, regulatory and ethical issues in gene therapy took place in June 2000 in Paris[69].

The meeting provided a discussion platform for gene therapy professionals and regulators to promote harmonisation of regulations. Scientific Opinion leaders have reported on regulatory implications of novel technologies regarding biosafety risks and the quality, safety and efficacy characteristics of gene transfer vectors. A significant trend has been initiated in favour of mutual concertation between regulators, companies and scientists towards harmonisation.

A round table assembled one official delegate of regulatory bodies from each country and the EMEA representative, Dr Papaluca-Amati. Under the leadership of Jean-Hugues Trouvin, the regulatory procedure in France has been significantly streamlined in recent years. Applicants for a gene therapy clinical trial now submit to the AFSSAPS a single file containing all the data pertaining to: (i) the description of the GMO; (ii) pre-clinical data on both efficacy and potential toxicity; (iii) the manufacture and control of the product including viral validation and potency; (iv) the design and details of the clinical trial. A number of bodies are involved in the review-process coordinated by the AFSSAPS which holds the statutory authority for approval.

A first question by Pr Gahrton to the panel of regulators was as to whether an application format could be adopted Europe wide and prove as useful to regulators as it would to investigators. This proposal instigated many comments, but raised no formal opposition. There is no legal basis to permit this process to be initiated by regulatory bodies. The initiative will have to come from the user's side. Proposals will be put forward and reviewed in each country. Indeed, the consensus was to start consultations at the national level rather than at the centralised one. Two other questions put to the panel were first, the possibility of designating, like in France, a single regulatory body in each country as the reference interface with applicants and second, the nature and form of current interaction between regulators. Interestingly, there is interaction in the context of European centralised initiatives such as the drafting of the CPMP Notes for Guidance. Altogether, these guidelines result from work in common, bridging national regulatory bodies to centralised authorities. The last issue raised deals with putative setting up of a European Scientific Advisory Committee to provide local committees with an opportunity to refer to Europe wide scientific and technical expertise.

6.   IMPROVING GENE THERAPY OVERSIGHT AND PUBLIC PERCEPTION

Five important areas should be singled out with a view of improving gene therapy oversight and public perception in concert with a broad worldwide context[70]: (i) implementation of clinical trials; (ii) GMOs-regulation; (iii) Public accountability; (iv) serious adverse events reporting and data-base; (v) turning data into knowledge.

6.1   Application format

A unified clinical trial application format consolidated Europe wide will be of significant help as a first step forward; in particular in dealing with rare disorders and transnational studies including those initiated through academia. There currently exists no legal basis for member states official bodies to initiate a move towards harmonisation. The recent approval of a European Directive on GCPs by both the European Council and the European Parliament might provide the appropriate missing link according to the modalities of implementation.

Data relating to both GCPs and GMOs Directives on the one hand and to product quality and safety on the other, need to be consolidated. This should result in the use of a single application file in view of performing a gene therapy clinical trial. It is expected that Guidelines integrating quality, safety arranged along a scientific and pragmatic rationale will be adopted Europe wide and homogeneously implemented. Towards this end sustained interaction is required between the scientific community and regulatory authorities, in particular with regulators in charge with the review of clinical trials in their early phases, at the national level as well as with the European

-Commission DG-Enterprise in charge with pharmaceutical affairs, EMEA and CPMP. The completion of the CPMP Notes for guidance, to which a number of national officers contributed, is bridging the gap *de facto*. Nevertheless, early trial phases are expected to be evaluated on the basis of less stringent criteria than those required for marketing.

6.2   GMOs regulation

In some cases high-level and/or sustained containment of patient might be unnecessary. These aspects pertaining to human health would best be reviewed and authorised by Medicines Agencies as part of a clinical trial application while this specific section could be sent as a notification to official bodies responsible for GMO regulation.

6.3   Transparency and public accountability

There is a requirement to improve transparency and public acceptance, and in doing so build up expertise and liability. Recent accidents in the USA have led to suspicion and enhanced fears relating to both genetically modified organisms in general and their potential therapeutic use in humans, in particular. With regards to dissemination of information; transparency and public accountability, the NIH-RAC plays a major role in the USA in offering a foı um for public debate and in releasing reports. In Europe there is no such centralised institution. The European Group on Ethics has distinct remits; in particular, it does not address scientific issues in details.

Publicly available information includes only one third of currently ongoing clinical trials in Europe as compared to confidential registration by member states regulatory authorities. The collection of data referred to as 'EMEA database' is planned to be confidential according to the European Directive on the implementation of Good Clinical Practice currently being debated at the European Council and Parliament. Alternative source of thorough and fast information need to be envisaged. This is one of the priority focus of Euregenethy.

6.4   Serious adverse events reporting

From a user's standpoint, a pan-European serious adverse events sharing system instantaneously available to investigators using the same technology should be secured. Therefore, generic non-confidential information ought to be extracted and freely released at least to gene therapy professionals. In fact, according to discussions which took place during the last meeting of NIH-RAC (March 8-10th, 2000), a significant move is taking place in the

United States following an extensive consultation at the US-Senate. The setting up of a database of adverse events is currently underway with, for the first time, FDA-willingness to contribute otherwise confidential data to NIH-RAC, OPRR (and/or other bodies to be determined[30,71,72,73,74,75]. In Europe as well, specific communication channels should be placed under the responsibility of official authorities; public release of validated information should be made possible. This requires impartial scrutiny of independent expert assessors which should evaluate key data.

6.5   Euregenethy 2: a referral organisation supported by EC-DG-Research

This scientific network will help provide users and local ethics committees with an opportunity to refer to Europe wide expertise and help with decision making in better identifying risk/benefit ratios. The setting up of such a committee e.g. by the EC-DG-Research will develop efforts with a view to (i) building on data accumulated and turning it into knowledge; (ii) Crossing sources of data to establish potential correlations between a specific technology and the occurrence of serious adverse events attributable to this technology; (iii) establishing generic biodistribution and pharmacokinetic profiles for different vectors. Non-confidential information should be made freely available to gene therapy professionals. An expert database could best be established in order to help better design clinical trials including patients monitoring. For instance, when safety of the vector system is of concern, the use and cross-referencing of preclinical and clinical studies with the same vector and other therapeutic or marker genes (transfected somatic cells), similar doses, schedules and routes of administration is of major importance; in particular where no suitable animal model of the human disease is available.

In fact, gene therapy does not call in essence on a specific regulatory status. There is a requirement to adapt regulation to scientific knowledge as it accumulates, instead. Since gene transfer technologies are broad in spectrum as well as their potential applications, the wide range of underlying scientific issues to be addressed lead to a significant level of complexity. Ultimately, the establishment of a European comprehensive expert-data-base in concert with currently available international databases, might prove useful. This will also contribute to improve public acceptance of genetic technologies at the cutting edge of biomedicine.

ACKNOWLEDGEMENTS

European Commission DG-Research and Euregenethy Partners: Reiner Bolhuis, Klaus Cichutek, Zelig Eshhar, Gösta Gahrton, Miguel Bronchud, Peter Hokland, Dimitri Loakopoulos, Giorgio Parmiani, Cecilia Melani, Michel Perricaudet, Elaine Rankin, Felicia Rosenthal, Heinz Zwierzina, and Bernd Gänsbacher.

REFERENCES

1. Cavazzana-Calvo M, Hacein-Bey S, de Saint-Basile G, Gross F, Yvon E, Nusbaum P, Selz F, Hue C, Certain S, Casanova JL, Bousso P, Le Deist F, Fischer A. Gene therapy of human severe combined immunodeficiency (SCID)-X1 disease. *Science*, 2000; **288**: 669-672.

2. Cohen B. Trials and tribulations. *Nature Genetics* 2000; **24**: 201.

3. Heise C, Hermiston T, Johnson L, Brooks G, Sampson-Johannes A, Williams A, Hawkins L, Kirn D: An adenovirus E1A mutant that demonstrates potent and selective systemic anti-tumoral efficacy. *Nat Med* **6**:1134-9, 2000.

4. Kay MA, Manno. CS., Ragni MV, Larson PJ, Couto LB, McClelland A, Glader B, Chew AJ, J Tai S, Herzog RW, Arruda V, Johnson F, Scallan C, Skarsgard E, Flake AW, High KA: Evidence for gene transfer and expression of factor IX in haemophilia B patients treated with an AAV vector. *Nat Genet.* 2000; **24**: 257-261.

5. Ylä-Herttuala S, Martin JF. Cardio-vascular gene therapy. *Lancet*, 2000; **355**: 213-22.

6. Barinaga M. Asilomar revisited: Lessons for Today? *Science*, 2000; **287**: 1584-85.

7. Commander H. Biotechnology industry responds to gene therapy death. *Nature Med* 2000; **6**: 118.

8. Hollon T. Researchers and regulators reflect on first gene therapy death. *Nature Med* 2000; **6**: 6.

9. Hollon T. Gene therapy investigations proliferate. *Nature Med* 2000; **6**: 235.

10. Marshall E. Gene Therapy on trial. *Science*; 2000: **288**: 951-957.

11. Renault B. Gene Therapy-a loss of innocence. *Nature Med* 2000; **6**: 1.

12. Lehrman S. Virus treatment questioned after gene therapy death. *Nature*, 1999; **401**: 517-518.

13. Temim HM. Safety considerations in somatic gene therapy of human disease with retrovirus vectors. *Hum Gene Ther*, 1990; 1: 11-123.

14. Commission of the European Communities, Council Directive 90/219/EEC of 23 April 1990 on the contained use of genetically modified micro-organisms (O.J. No L 117 of 8. 5. 90).

15. Commission Decision 91/448/EEC of 29 July 1991 concerning the guidelines for classification referred to in Article 4 of Directive 90/219/EEC (O.J. No L 239 of 28.8.91).

16. Cohen-Haguenauer O. Euregenethy: Regulation of Gene Therapy in Europe: a scientific network of users. Biomedical and Health Research Newsletter, European Commission, DGXII, 1998; 9:3-4.

17. European Commission. Council Directive 98/81/EC of 26 October 1998 amending Directive 90/219/EEC on the contained use of genetically modified micro-organisms. Official Journal of the European Commission, 5.12.98, L330/13-31.

18. Commission of the European Communities, Council Directive 90/220/EEC of 23 April 1990 on the deliberate release into the environment of genetically modified organisms (O.J. No L 117 of 8. 5. 90).

19. Commission of the European Communities, Council Directive 90/679/EEC of 26 November 1990 on the protection of workers from risks related to exposure to biological agents at work (seventh individual Directive within the meaning of Article 16(1) of Directive 89/391/EEC) (O.J. No L 374 of 31. 12. 90).

20. Commission of the European Community (III/3477/92), Directorate-General Industry III/E/3, "Production and Quality Control of Medicinal Products Derived by Recombinant DNA Technology."

21. Council Directive 86/609/EEC of 24 November 1986 on the approximation of laws, regulations and administrative provisions of the Member States regarding the protection of animals used for experimental and other scientific purposes (O.J. No L 358 of 18. 12. 86).

22. Council Directive 87/18/EEC of 18 December 1986 on the harmonization of laws, regulations or administrative provisions relating to the application of the principles of good laboratory practice and the verification of their applications for tests on chemical substances (O.J. No L 15 of 17. 1. 87).

23. Council Directive 88/320/EEC of 9 June 1988 on the inspection and verification of Good Laboratory Practice (GLP) (O.J. No L 145 of 11. 6. 88).

24. Council Directive 93/41/EEC of 14 June 1993 repealing Directive 87/22/EEC on the approximation of national measures relating to the placing on the market of high technology medicinal products, particularly those derived from biotechnology (O.J. No L 214 of 24.8.93).

25. European Commission Directive 91/356/EEC of 13 June 1991 laying down the principles and guidelines of good manufacturing practice for medicinal products for human use (O.J.No L 193 of 17.7.91).

26. Cohen-Haguenauer O. Overview of Regulation of Gene Therapy in Europe: A current statement including reference to US Regulation. *Hum Gen Ther* 1995; 6:773-785.

27. Cohen-Haguenauer O. Safety and regulation at the leading edge of biomedical biotechnology. *Current Opinion in Biotechnology* 1996; 7:265-272.

28. CPMP. Note for guidance on Good Clinical Practice. CPMP/ICH/135/95-consolidated FDA: Guidance for Industry. E6 Good Clinical Practice. April 1996.

29. European Commission. Amended proposal for a European Directive on the implementation of good clinical practice in the conduct of clinical trials on medicinal products for human use, Ref: COM (1999) 193 final.

30. Siegel J. Letter to Sponsor of an IND or master file using or producing a gene therapy product. US Department of Health & Human Services. FDA-CBER. Office of Therapeutics Research & review, March 6th, 2000.

31. Smaglik P. NIH tightens up monitoring of gene-therapy mishaps. *Nature* 2000; **404**: 5.

32. US Department of Health and Human Services, National Institutes of Health. Guidelines for Research involving recombinant DNA Molecules (NIH Guidelines). Federal Register, May 11, 1999, (64 FR 25361)-Appendix M: Points to consider in the design and submission of Protocols for the transfer of recombinant DNA molecules into One or More human subjects (Points to consider) pp 92-103

33. Cohen-Haguenauer O. Gene therapy: regulatory issues and international approaches to regulation. *Current Opinion in Biotechnology* 1997; **8**:361-369.

34. Cohen-Haguenauer O. Réglementation de la thérapie génique. *Médecine/Sciences* 1999; 5:682-690.

35. Cichutek K and Krämer I. Gene therapy in Germany and in Europe: Regulatory Issues, *Qual. Assur. J.* **2**, 141-152 (1997).

36. Euregenethy newsletter n°1 - June 1999, Odile Cohen-Haguenauer ed, EDK-Paris.

37. Lindemann A, Rosenthal FM, Hase S, Markmeyer P, Mertelsmann R, for the German Working Group for Gene Therapy. Guidelines for the design and implementation of clinical studies in somatic cell therapy and gene therapy in Germany. *Journal of Molecular Medicine*, 1995; **73**: 207-211.

38. Council Regulation No. (EEC) 2309/93 of 22 July 1993 laying down Community procedures for the authorization and supervision of medicinal products for human and veterinary use and establishing a European Agency for the Evaluation of Medicinal Products (O.J. No. 214 of 24.8.93).

39. Commission of the European Communities, Committee for proprietary medicinal products: Ad hoc Working Party on Biotechnology / Pharmacy: Gene Therapy Products-Quality Aspects in the Production of Vectors and Genetically Modified Somatic Cells. Approved by CPMP December 1994: III/5863/93 Final.

40. CPMP. Note for guidance on the Quality, preclinical and clinical aspects of gene transfer medicinal products. CPMP/BWP/3088/99-draft released for consultation 16 December 1999.

41. Euregenethy network, Cohen-Haguenauer O, Rosenthal F, Bolhuis R, Amate-Blanco J, Carrondo M, Dorsch-Häsler K, Eshhar Z, Gahrton G, Hokland P, Melani C, Rankin E, Sneyers M, Zwierzina H, Cichutek K. Current regulation of gene therapy in European member states: a megachart reporting on heterogeneity while using common headings: a first step in view of harmonisation? *Hum Gen Ther*, 2000, *submitted*.

42. European Parliament & Council Regulation on Orphan drugs 98/0240 (COD) LEX 184. PE-CONS 3637/99 ECO410 SAN 194 CODEC 788.

43. US Department of Health & Human Services. FDA-CBER Guidance for Industry. Guidance for human somatic cell therapy and gene therapy. March 1998.

44. CPMP. Points to consider on human somatic cell therapy CPMP/BWP/41450/98-draft released for consultation 16 December 1999.

45. US Department of Health & Human Services. FDA-CBER Guidance for Industry. Supplemental guidance on testing for replication competent retrovirus in retroviral vector based gene therapy products and during follow-up of patients in clinical trials using retroviral vectors. November 1999.

46. Nolta JA, Dao MA, Kohn DB. Detailed biosafety analyses confirm that there is low risk from MoMuLV retroviral vectors. *J Gene Med* 2000; **2** S4: 16.

47. Kalberer CP, Pawliuk R, Imren S, Bachelot T, Takekoshi KJ, Fabry M, Eaves CJ, London IM, Humphries RK, Leboulch P. Preselection of retrovirally transduced bone marrow avoids subsequent stem cell gene silencing and age-dependent extinction of expression of human beta-globin in engrafted mice. *Proc Natl Acad Sci USA* 2000 May 9; **97**,10:5411-5.

48. Naldini L, Blomer U, Gage FH, Trono D, Verma IM. Efficient transfer, integration, and sustained long-term expression of the transgene in adult rat brains injected with a lentiviral vector. *Proc. Natl. Acad. Sci. USA* 1996; **93**: 11382 – 11388.

49. Naldini L, Blömer U, Gallay P, Ory D, Mulligan RC, Gage FH, Verma IM and Trono D. *In vivo* gene delivery and stable transduction of nondividing cells by a lentiviral vector. *Science* 1996; **272**: 263-267.

50. Kafri T, Blomer U, Peterson DA, Gage FH, Verma IM. Sustained expression of genes delivered directly into liver and muscle by lentiviral vectors. *Nat Genet.* 1997; **17**:314-7.

51. Miyoshi H, Takahashi M, Gage FH, Verma IM: Stable and efficient gene transfer into the retina using an HIV-based lentiviral vector. *Proc Natl Acad Sci USA* **94**:10319-23, 1997.

52. Miyoshi H., Smith K.A., Mosier D.E., Verma I.M., Torbett B.E. Efficient transduction of Human CD34+ cells that mediate long-term engraftment of NOD/SCID mice by HIV vectors. *Science* 1999; **283**: 682-686.

53. Yu, S.F., von Rüden, T., Kantoff, P.W., Garber, C., Seiberg, M., Rüther, U., Anderson, W.F., Wagner, E.F. and Gilboa, E. (1986). Self-inactivating retroviral vectors designed for transfer of whole genes into mammalian cells. *Proc. Natl. Acad. Sci. USA*. **83**, 3194-3198.

54. Naldini L: Lentiviruses as gene transfer agents for delivery to non-dividing cells. *Curr. Opin Biotechnol* **9**:457-63, 1998.

55. Zufferey R, Nagy D, Mandel RJ, Naldini L, Trono D: Multiply attenuated lentiviral vector achieves efficient gene delivery in vivo. *Nat Biotechnol* **15**:871-5, 1997.

56. Kafri T, van Praag H, Gage FH, Verma IM: Lentiviral vectors: regulated gene expression. *Mol Ther* **1**:516-21, 2000.

57. Follenzi A, Ailles LE, Bakovic S, Geuna M, Naldini L: Gene transfer by lentiviral vectors is limited by nuclear translocation and rescued by HIV-1 pol sequences. *Nat Genet* **25**: 217-22, 2000.

58. Naldini L: Lentiviral vectors: design, safety and quality control. *J Gene Med* 2000; **2** S4: 15.

59. Darlix JL, Mangeot P, Negre D, Dubois B, Winter A, Leissner P, Mehtali M, Kaiserlian D, Cosset FL. Development of minimal and safe SIV vectors and their use for the gene transfer of human dendritic cells. *J Gene Med* 2000; **2** S4: 5.

60. Klages N, Zufferey R, Trono D: A stable system for the high-titer production of multiply attenuated lentiviral vectors. *Mol Ther* **2**:170-6, 2000.

61. Escudier B, Griscelli F, Gautier E, Saulnier P, Squiban P, Lamy D, Le Chevalier T, Tursz T. Adenovirus shedding: problems related to patient containment. *J Gene Med* 2000; **2** S4: 6.

62. Grace M. Adenoviral shedding: clinical experience with SCH 58500. *J Gene Med* 2000; **2** S4: 7.

63. Thatcher D and Harris P. RCA issues in the development of CTL102. *J Gene Med* 2000; **2** S4: 19.

64. Hunter WD, Martuza RL, Feigenbaum F, Todo T, Mineta T, Yazaki T, Toda M, Newsome JT, Platenberg RC, Manz HJ, Rabkin SD: Attenuated, replication-competent herpes simplex virus type 1 mutant G207: safety evaluation of intracerebral injection in nonhuman primates. *J Virol* **73**:6319-26, 1999.

65. Heise C, Sampson-Johannes A, Williams A, McCormick F, Von Hoff DD, Kirn DH: ONYX-015, an E1B gene-attenuated adenovirus, causes tumor-specific cytolysis and antitumoral efficacy that can be augmented by standard chemotherapeutic agents. *Nat Med* **3**:639-45, 1997.

66. Markert JM, Medlock MD, Rabkin SD, Gillespie GY, Todo T, Hunter WD, Palmer CA, Feigenbaum F, Tornatore C, Tufaro F, Martuza RL: Conditionally replicating herpes simplex virus mutant, G207 for the treatment of malignant glioma: results of a phase I trial. *Gene Ther* 7:867-74, 2000.

67. Rosenthal FM. GMP processing of somatic cell and gene therapy products. *J Gene Med* 2000; 2 S4: 18.

68. O Cohen-Haguenauer (coordinator), R Bolhuis, M Carrondo, K Cichutek, Z Eshhar, G Gahrton, P Hokland, D Loukopoulos, C Melani, G Parmiani, M Perricaudet, E Rankin, F Rosenthal, H Zwierzina. EUREGENETHY. Regulation of Gene Therapy in Europe: a Scientific Network of Users. *Mol Ther, 2000; 1: S302.*

69. Cohen-Haguenauer O. Second Euregenethy multidisciplinary forum on regulatory issues in gene therapy: Meeting highlights. *J Gene Med* 2000; 2 S4: pp 3-8.

70. Friedmann T. Principles for human gene therapy studies. *Science,* 2000; **287**: 2163-2165.

71. US Department of Health & Human Services. FDA-CBER Letter to Gene Therapy IND Sponsor/Principal Investigator on the report of adverse events. November 6th, 1999.

72. US Department of Health & Human Services. NIH-OBA. Requirements for reporting serious adverse events: request for institutional review. November 22nd, 1999.

73. US Department of Health & Human Services. FDA-CBER and NIH: New initiatives to protect participants in Gene Therapy trials. March 7, 2000.
http://www.fda.gov/bbs/topics/NEWS:NEW00717.html

74. Wadman M. NIH panel to limit secrecy on gene therapy. *Nature,* 1999; **402**: 6.

75. Woo SLC. Policy statement of the American Society of Gene Therapy on reporting of patients adverse events in gene therapy trials. *Mol Ther,* 2000; **1**: 7.

Regulatory Issues for Process Development and Manufacture of Plasmids Under Contract

JOHN M. JENCO
Dow Biopharmaceutical Contract Manufacturing Services, Stony Brook, NY, USA

1.   INTRODUCTION

By now, almost anyone involved in the biological sciences has gone through the process of purifying plasmid DNA from a bacterial host organism to use as a tool in molecular biology. The methodology has become so routine that it is common to purchase one of any number of kits that are designed to make the procedure as simple and as painless as possible. The evolution of plasmid preparation has come to the point that one can achieve success on the first try without having a solid foundation in the process fundamentals.

However, for the purposes of process development, scale-up, and manufacture of plasmid DNA for use in gene therapy, it is an absolute requirement to have a rigorous understanding of the available technology. This is due, in large part, to the regulatory issues that are continually evolving for the use of biologicals in the treatment of various diseases and pathologies. Gene therapy protocols are still being evaluated, and the mechanisms by which they act are not fully understood. A good deal of caution is being prescribed at the research and development level[1], and a number of recommendations have been put forth (Table 1).

Table 1. Recommendations for research and development in gene therapy[1].

- Understand the basic aspects of gene transfer, targeting, and expression
- Understand the mechanisms of disease pathogenesis
- Incorporate high standards of excellence in clinical protocols
- Exercise restraint in public communications
- Execute interdisciplinary training programs
- Subject research to stringent peer review
- Strengthen collaborations between academia and industry
- Coordinate research efforts to reduce duplicity

This same level of caution also applies to process development and manufacture of plasmid DNA intended for use in gene therapy. From this standpoint the goals of this chapter are: to describe the current state of the art in plasmid purification; to define the parameters for successful process development and manufacture of plasmid DNA; and to provide a framework for discussion of the associated regulatory issues.

2.   BIOPHARMACEUTICAL DEVELOPMENT CYCLE

The development of biopharmaceuticals for therapeutic use follows the same path to market as other drugs. The cycle of research to development to production is the same. However, the methodologies that are employed to produce biologicals are vastly different and rely on the complex interplay among biomolecules of all types in living organisms[.]

Biomolecules with potential therapeutic applications are first identified through rigorous scientific research. Cell types for production are identified and then modified using the recombinant techniques of molecular biology. Fermentation parameters are optimised to yield robust expression of the biomolecule in a form that can be readily harvested. Extraction and purification strategies specific for the molecule of interest are designed using the art and science of biochemistry to yield a final product that is pure, stable, and potent. Appropriate analytical techniques are incorporated to verify product integrity. Formulations are developed to deliver the biomolecule to its intended target intact and active so that it will have the desired therapeutic effect[.]

With that in mind, process development aimed at large-scale manufacture of biopharmaceuticals needs to be carefully designed at the bench and pilot scales[.] Processes must be simple, straightforward, reproducible, and scaleable. Toxic chemicals and animal-derived products should be avoided.

Waste streams must be environmentally friendly. In order to assure success and minimise time-to-market, the best strategy is to consider these issues as early in the development process as possible. This is true whether an organisation is manufacturing a product in its own facilities or outsourcing under contract.

A contract manufacturing organisation (CMO) provides development, scale-up, and production services to a variety of clients. A CMO will perform some or all of the manufacture of a product, and is responsible for *maintaining* the facility's overall compliance with applicable product and establishment standards and current good manufacturing practice (cGMP). However, the client bears the final responsibility for *ensuring* compliance since it is the client who will be filing for licensure of a product with the appropriate regulatory agencies . These standards are presented in Table 2.

Table 2. Product and establishment standards and cGMPs[5]

- In-process and release specifications for identity, purity, strength, quality, potency, safety, and efficacy
- Reporting systems for adverse events, errors, and accidents
- Production process development and validation
- Reporting changes to the production process
- Quality assurance oversight and change control for the master and batch production records
- Quality control methodology relative to the production process
- Submission of protocols and samples for lot release
- License application content
- Labeling
- Contracts with establishment(s) performing manufacturing and testing
- Validation, maintenance, and function of facility and all equipment/systems
- Environmental monitoring
- Reporting of all facility changes
- Personnel training

Compliance actions may be taken by the regulatory agencies against both client and contractor for the contractor's failure to comply with cGMP. Therefore, it is in the best interests of both the client and the CMO to design and implement processes that are robust, scaleable, and able to pass regulatory hurdles. This applies to all biopharmaceutical products including plasmid DNA.

In addition to compliance issues, clients and contractors need to have signed written agreements that establish the responsibilities of both parties. The recommended items for such an agreement are outlined in Table 3. Once a contract has been executed, the task of transferring the client's process, analytical methods, and associated documentation can begin.

Table 3. Items to be included in contract manufacturing arrangements[5]

- Identification of the CMO and locations used for manufacture
- Responsibilities of all participants, including contractor's QA, and client supervision and control
- Description of product being shipped to contract facility
- Description of shipping conditions to and from the contract facility
- Description of operations performed at contract facility
- All standard operating procedures relevant to the contract
- Commitment from contractor to inform client of proposed facility changes prior to implementation
- Commitment from contractor to inform client about errors, deviations, and adverse events affecting product
- Description of assessment procedures and schedule performed by client to assure compliance with product and establishment standards and cGMP

3.   TECHNICAL TRANSFER

Once a contract has been established, the first step in the relationship between a client and a CMO is the transfer of the existing process technology and all necessary documentation. The client must provide all necessary details about the current process. The contractor must demonstrate to the satisfaction of the client the ability to understand and assimilate the process, and to perform the process with a high degree of consistency, reproducibility, and uniformity. Ideally, the transfer should be seamless. However, because of differences between the available equipment at the contractor's facility and that of the client, minor changes may need to be made. In addition, parts of the client's process may not be desirable in a multi-product facility or may be unsuitable for large-scale production. In this case, process development will be in order. This is best performed as a collaboration between the client and the contractor in order to make the transfer between development and manufacture as smooth as possible.

The two areas that require the most attention to detail during technical transfer are the process itself, and the analytical methods associated with it. The contractor needs to develop a thorough understanding of the client's process. The client needs to understand that not every aspect of the process as it has been developed is suitable for a manufacturing environment. Some amount of process development may be necessary before full-scale manufacture can begin. Not all methods for plasmid purification will survive regulatory review. The contractor needs to effectively communicate this to the client, and the client needs to be flexible enough to allow for a certain amount of development work. Both parties need to work together to document the process and all analytical methods used. This will ensure the smoothest transition from the client to the contractor, regardless of whether the process is transferred at the bench, pilot, or production scale.

4.   CELL BANKING

Because plasmid DNA is being produced in a cellular host, it is necessary to characterise both the cell line and the vector as completely and specifically as possible[6,7]. Information regarding the cell line should include strain identification and characterisation, origin and history, tests for the presence of adventitious organisms, expiration dating, and handling procedures. Complete information about the sequence of the entire vector should be obtained. This includes selection markers, copy control regions, promoters, and the cDNA insert. Plasmid stability in the host over time needs to be demonstrated, and PCR will likely be the method of choice. Periodic sequence analysis of the cell banks needs to be performed to assure that no changes in critical areas of the plasmid take place.

5.   FERMENTATION

The first step in the process of plasmid production will be fermentation in a suitable host organism. *E. coli* has received the widest use, and there are many strains available. There are limitless combinations of host organisms and plasmids. Therefore, the conditions for fermentation will need to be optimised and tightly controlled for each combination[8-11]. A number of factors need to be addressed including media, temperature, pH, dissolved oxygen, antifoam, and cell density. Each of these can have an impact on cell harvest, as well as primary recovery, yield, and quality of plasmid DNA. For example, the ratio of carbon and nitrogen in the fermentation broth has been shown to directly affect both the quantity and quality of plasmid DNA[12]. Although generic fermentation conditions work in a research environment using shake flasks, they are neither ideal nor scaleable to the pilot and production scale fermentation vessels.

6.   PRIMARY RECOVERY

Following fermentation, the next step will be recovery of the cells from the fermentor. This is typically done by centrifugation using well-established processes and needs no further discussion here. Equipment for large-scale centrifugation directly out of the fermentor is readily available and scaleable. Once the cells have been recovered and washed, the plasmid is then recovered from the cells by lysis, and the lysate is clarified prior to downstream processing.

7.   CELL LYSIS

The choice of one lysis method over another involves consideration of three factors: plasmid size, host strain, and downstream processing [13]. Plasmids greater than 15 kb in size should be released from cells gently, whereas smaller plasmids can withstand more harsh conditions. Lysates of strains that release large amounts of carbohydrate are more difficult to handle downstream. Strains that express endonuclease A (*endA+*) require extra care to avoid degradation of plasmid during subsequent processing steps. Lysis techniques typically fall into one of three categories: chemical, enzymatic, and mechanical.

7.1   Chemical lysis

The basic technique for chemically releasing plasmid from bacterial cells is alkaline lysis [14-16]. Cells are harvested and resuspended in an isoosmotic glucose solution, and a solution of sodium hydroxide (NaOH) and sodium dodecyl sulphate (SDS) is added to lyse the cells. Although the resulting lysate is highly viscous and large in volume, the chemicals used to produce it are relatively innocuous. Because of the high viscosity and the sensitivity of plasmid DNA to shear, mixing techniques require judicious selection of process-scale equipment [9].

There are a number of variations on the basic lysis technique. One involves boiling the bacterial cell pellet in a solution containing Triton X-100 at pH 8.0 instead of SDS in NaOH [17]. This technique is recommended only for smaller plasmids [13], and care must be taken to avoid prolonged exposure to heat in order to prevent irreversible denaturation of plasmid [18]. For very large plasmids (>15 kb), a gentle lysis using SDS and sodium chloride at pH 8.0 may be employed [19].

7.2   Enzymatic lysis

Several enzymes have traditionally been used to prepare plasmid from recombinant bacterial sources. Lysozyme, derived from hen egg white, digests the bacterial cell wall, making lysis by chemical agents more efficient [13]. Since this is an animal-derived enzyme, it raises the issue of potential transmission of infectious diseases. Although the risk of transmission is minimal, it can be avoided altogether by not using it at all.

7.3   Mechanical lysis

A number of mechanical techniques are available for the disruption of bacterial cells to release plasmid[20]. These include homogenisation, sonication (hard to scale up), nitrogen cavitation, and extrusion (e.g. French press, Microfluidizer). However, because of its large size and extended conformations, plasmid DNA is sensitive to destruction by shear forces. It may be possible to use mechanical methods when the size of the plasmid is small or if condensation reagents are used to keep the plasmid molecules compact[21]. These methods would have to be evaluated on a case by case basis. However, as a rule, the gentler chemical lysis techniques tend to give higher yields although process volumes tend to be larger.

8.   PURIFICATION

There are a number of techniques available to purify plasmid DNA[22]. Once the crude lysate has been processed and clarified, the choices come down to the conventional techniques of fractionation, chromatography and ultrafiltration. Each must be thoroughly assessed in terms of the impact on the process.

8.1   Fractionation

Techniques for lysate fractionation include precipitation, centrifugation, and filtration. Each of these methods can be used alone or in combination to reduce the levels of contaminants and to reduce the volume of lysate prior to the next steps. Contaminant removal can be performed using precipitation followed by either centrifugation or filtration and harvesting the liquid phase. Volume reduction can be performed using precipitation followed by centrifugation and harvesting the pellet. Volume reduction can also be accomplished by filtration alone using filters with appropriate molecular weight cut-offs.

Once the plasmid has been released from the cells, it is necessary to purify it further from the crude lysate. Chemical fractionation techniques have been utilised to produce a bulk preparation that has been enriched in plasmid and reduced in other host cell contaminants. Potassium acetate is typically added to both neutralise the NaOH and precipitate SDS-protein complexes[15,16]. Centrifugation or filtration is used to remove precipitated material. Alcohol and polyethylene glycol[23] have been used to selectively precipitate plasmid DNA[13,24,25] from the clarified lysate. Extraction with phenol and chloroform has been used to remove contaminants. Ultracentrifugation on cesium chloride gradients[16,26-28], with or without ethidium bromide, has been used as a polishing step[21]. Compaction agents and chaotrophic agents have also been used[29]. It has been noted, however, that the benefits of fractionation techniques are often accompanied by a loss in the yield of plasmid[11,29].

Processes that rely on chemicals that are mutagenic, carcinogenic, or otherwise toxic should not be included in a manufacturing process. This includes ethidium bromide, caesium chloride, phenol, and chloroform. Although these chemicals may have utility in the analysis of plasmid DNA, they should be avoided in the production of plasmid destined for gene therapy.

Enzymes such as bovine pancreatic RNase have been used to selectively digest host cell RNA[13]. However, the use of animal-derived enzymes and the risk of disease transmission again becomes an issue. The use of such enzymes should be avoided, and may even be altogether unnecessary[29].

8.2   Chromatography

Typically, chemical fractionation is not sufficient to completely purify plasmid DNA to homogeneity. Chromatographic techniques have been developed that take advantage of separations based on plasmid size, charge, and affinity. These include combinations of anion exchange[21,22,29-33], gel filtration[34-38], hydroxyapatite[39-46] and reverse phase[47] chromatography. Displacement chromatography[48,49] and silica[50,51] have also been used, but not as widely. Each method has advantages and disadvantages in terms of process development, manufacturing, and regulatory issues.

8.3   Ultrafiltration

Ultrafiltration is used for one of two purposes – volume reduction and/or buffer exchange. Plasmid DNA does not appear to pose any unusual problems in either case, either from a processing or regulatory standpoint. However, the size and conformation of plasmid DNA does present certain hydrodynamic considerations as it seems to behave more as a nanoparticulate than a solute molecule[52].

9. ANALYTICAL METHODS

Purified plasmid DNA used in any therapeutic application must be tested for identity, purity, and potency using appropriate analytical methods. Table 4 lists a battery of tests that may be employed. A combination of methods will be necessary for final release testing[6,7,9,53,54]. It should be noted here that these specifications are based on predicted dosing regimens. They will no doubt vary among different gene therapy protocols, especially for parameters such as endotoxin and other host cell contaminants. Additionally, these specifications will evolve with further advances in our understanding of gene therapy.

Table 4. Final release specifications for plasmid used in gene therapy

Parameter	Available tests	Typical specifications
Identity	Restriction map, PCR sequencing, molecular weight by agarose gel or HPLC	In agreement with predicted size and sequence
Homogeneity	Agarose gel electrophoresis	>90-95% supercoiled
Purity	Spectrophotometry	$A_{260/280}$=1.75-1.85
Endotoxin	LAL assay	<100 EU/mg plasmid DNA
Host cell RNA	Northern blot	Undetectable
Host cell genomic DNA	Southern blot, dot blot, PCR	<10 µg/mg of plasmid DNA
Protein	Western blot, BCA assay, ELISA, HPLC	<10 µg/mg of plasmid DNA
Sterility	USP	No growth
Potency	*In vitro* transfection assay	Positive expression compared to working reference standard

9.1 Identity

Plasmid identity testing is the most challenging aspect for in-process analysis. Three parameters have been identified as particularly crucial. The first is plasmid sequence. Polymerase chain reaction (PCR) sequencing of the entire plasmid is the only way to positively identify a plasmid and distinguish it from anything else. It will also be the most sensitive means of identifying contamination by any other plasmid, especially if the only difference between one plasmid and the other is the cDNA insert. Restriction mapping may yield qualitative information, but it is conceivable that two different plasmids digested with the same combination of restriction enzymes may produce indistinguishable banding patterns on an agarose gel.

The second part of identity testing is confirmation of the molecular weight. Again, it is possible that two different plasmids can have identical molecular weights. However, this is a simple test to run by either agarose gel electrophoresis or HPLC. It will function to corroborate the more sensitive PCR test, but it can also be used as a preliminary test before committing to the more sensitive and more time consuming method.

Lastly, there is the issue of plasmid conformation. There is considerable debate regarding the presence of nicked, relaxed, or open circular plasmid. From the perspective of both quality and efficacy, there are no firm conclusions regarding the distinction from supercoiled or closed circular plasmid[53]. This area requires further study. However, since methodologies are available that can separate supercoiled plasmid from the nicked form, the perceived problem can be addressed within the process. Analysis is also simple since nicked and supercoiled forms can be separated and identified on agarose gels.

9.2   Purity

Purity analysis will consist of testing for host cell contaminants and process residuals. Sensitive and specific tests are available[55]. The issue of nicked vs. supercoiled forms (discussed above) has also been regarded as one of purity rather than identity.

9.2.1   Host cell contaminants

Host cell contaminants – genomic DNA, RNA, protein, and endotoxin – are the biggest problem affecting plasmid purity. The goal of process development and manufacturing is to effectively achieve removal of these contaminants on a consistent and reproducible basis. Although there are a number of options available to produce pure plasmid DNA, not every option is effective at removing every host cell contaminant. Typically, a combination of methodologies will need to be employed in order to achieve acceptable levels of these contaminants. Although it is generally thought that most plasmids can be purified using the same process, each plasmid should be evaluated on its own merit in terms of how it performs using "generic" purification protocols. Despite the chemical uniformity of DNA, plasmid size, sequence, and conformation may contribute to subtle yet significant differences in purity and the strategies used to achieve it.

It is important to remember that two classes of host cell contaminants are chemically similar to plasmid, i.e. RNA and genomic DNA. Processes must be carefully designed to achieve selective removal of these contaminants. Analytical methods must be specific as well to avoid ambiguity. Endotoxin is difficult to deal with as it carries the same charge as DNA and often copurifies with it[56]. This is extremely important since endotoxin can affect transfection of target cells[50,57]. Endotoxin levels are also important from a drug safety perspective. The final class of host cell contaminants – proteins – must be removed in order to prevent the development of an inflammatory or immune response in the host. Testing for host cell proteins should be sensitive and specific[58].

9.2.2   Process residuals

If a process is designed that uses any quantity of a hazardous, toxic, or otherwise noxious chemical, then its complete removal becomes an issue. Not only will a removal strategy need to be incorporated, but appropriate testing will be necessary as well to demonstrate the effectiveness of the process. As described above, traditional plasmid purification methods have utilised a number of such reagents. This includes phenol, chloroform, caesium chloride, ethidium bromide, and isopropanol. The best approach would be to avoid these chemicals altogether and thereby avoid the complications that they will cause.

9.3   Potency

It is one thing to purify plasmid DNA to homogeneity. It is another to purify plasmid DNA that is also active. Adequate testing for plasmid activity must be incorporated into a manufacturing scheme. The goal of gene therapy is to incorporate DNA into a specific cell type and express a specific protein. Therefore, an appropriate in-process test would involve *in vitro* transfection of a relevant cell type, analysis of transfection efficiency, and a demonstration of specific protein expression.

10.   FORMULATION

The choice of formulation for a plasmid depends largely on the intended use and the means of delivery. Delivery of plasmid to its intended target(s) is an area that still requires a great deal of research. Studies have been conducted that simply use "naked" DNA, incorporate DNA into cationic liposomes, or utilise proprietary delivery systems. From a contract manufacturing standpoint, formulation is a client-specific issue.

11.   CHANGEOVER

Changeover is the process of removing all traces of one process in order to prepare for the next.   Cleaning, sanitisation, maintenance, and environmental clearance are all part of changeover.   Because a contract manufacturer is involved with the production of a number of biopharmaceuticals for a number of clients, facility and equipment changeover becomes an important issue.   It is not permissible to contaminate one process with the residuals from the previous process.   It is more desirable to invest the time in changeover procedures from the very beginning to ensure no cross-contamination between different products. This is especially important for plasmid DNA.   Since the whole notion of gene therapy is to impart functionality to a cell, it is extremely important to show that one purification run does not contain plasmid from the previous purification run.   This is the only way to guarantee that a subpopulation of cells has not been given another, possibly contrary, function.

11.1   Same client, same cell line, AND same vector

In the situation where consecutive runs are being performed for the same client, using the same cell line and expressing the same plasmid, carryover from one run to the next is restricted to host cell contaminants and process residuals.   Equipment cleaning protocols must be designed and validated to adequately remove these classes of contaminants after each run.   Routine procedures, reagents, and schedules for regeneration, re-equilibration, cleaning in place, and sanitisation of column resins must be in place[4]. Achievable limits should be established using the same analytical methods for the detection of contaminants and residuals during the purification process.

11.2   Different client, different cell line, OR different vector

Changeover between non-equivalent runs becomes an important issue in contract manufacturing of plasmid DNA[4,6].   Whenever two consecutive runs are different in any way, the amount of effort that must be exerted becomes greater in order to reduce, minimise, or eliminate the potential for cross-contamination.   Cleaning procedures need to be more vigorous, and residue testing becomes more critical.   It is still important to demonstrate the consistent removal of host cell contaminants and process residuals from the previous run.   However, it is even more important to show that product from the prior run has been eliminated as well.   Testing of product residuals on equipment surfaces and in wash and rinse samples will need to be carried out. The most sensitive test for this purpose will be PCR. Column resins must be dedicated to a single product or disposed of altogether between runs to eliminate the risk of cross-contamination[59,60].

For equipment, a full changeout of all wetted surfaces would be the most effective means of eliminating cross-contamination between equivalent runs. However, for both fermentation and purification, that would involve every component except the controllers. The most reasonable strategy would be an effective clean in place program for non-disposable elements of the systems (glass, stainless steel, probes), and a changeout of all disposable elements, (seals, O-rings, in-line filters)[4]. Plant sanitisation should also be performed.

12.   VALIDATION

In contract manufacturing, validation is not only absolutely necessary, but it is absolutely good business. A contract manufacturer typically deals with a number of clients at any given time. Validation is the only resource that will give a client the assurance that a contractor can consistently reproduce a process as it was designed, and that the process is unaffected by any other process performed for any other client. Validation is the only means of confirming and documenting that a process meets the necessary in-process specifications that document identity, purity, and potency of a plasmid DNA preparation or any other biopharmaceutical[2-4,7,53].

12.1   Raw materials

Two issues are important regarding raw materials in a contract manufacturing facility – identity and segregation. First, routine identity testing of common raw materials according to the USP will ensure that the label on the container correctly identifies the contents of that container. It will also provide assurance that the contents have not been contaminated in any way. Second, adequate segregation of raw materials that are specific to a given client's process is necessary to avoid mix-ups and confusion between two separate and distinct processes.

12.2   Facility – HVAC and water

Since plasmid DNA, like other biopharmaceuticals, is destined for parenteral use, tight controls must be maintained over the facility. Air handling equipment must be designed and monitored to demonstrate efficacy of removal of airborne particulates. Air and surface sampling protocols should be in place. Water purification equipment must also be designed and monitored to ensure removal of inorganic and organic contaminants, viable organisms, and endotoxin. Conductivity, total organic carbon, bioburden, and LAL assays need to be conducted. Long-term trend analysis of both of these systems will be necessary to demonstrate to the client that the facility is tightly controlled, and that prior processes have not had an adverse effect on the facility. In addition, testing for the absence of recombinant organisms from previous runs should also be conducted in order to show that the preparation of one plasmid is not likely to be contaminated with another.

12.3   Equipment – maintenance, cleaning, and operation

By far, the most critical area for preparation of plasmid for use in gene therapy will be the actual equipment that is exposed to the upstream and downstream processes. This will include the fermentors, buffer and media tanks, centrifuges, process tanks, chromatography columns and equipment, and ultrafiltration devices. Since plasmid DNA presents the possibility of becoming incorporated into a target cell or host organism, it is necessary to show that all plasmid from a previous run has been removed from equipment surfaces and will not contaminate the next run. Cleaning protocols will have to be designed to show that they are effective in removing any residual plasmid. Testing protocols should be specific for the last plasmid known to have come in contact with processing equipment.

For fermentation equipment, it will be necessary to show that organisms from previous fermentations do not survive cleaning and sterilisation programs. Buffer and media tanks should not have residues from prior contents. Centrifuges, process tanks, ultrafiltration devices, and chromatography equipment should show no evidence of both viable organisms and residual plasmid. Chromatography resins should either be discarded after a single run or dedicated to the production of a single plasmid for a single client. In the latter case, the cleaning, regeneration, and sanitisation cycles should show no evidence of microbial contamination.

12.4   Documentation

One of the least appreciated and most necessary operations for any manufacturer is documentation. This is especially true for contract manufacturers since more than one operation may be taking place in a facility at any given time, depending on the level of segregation between rooms and equipment. From buffer recipes to batch records to final release specifications, documentation is the only means by which information about past operations can be retrieved and reconciled. Maintenance of a paper trail is required by the regulatory agencies for process investigations and recall purposes.

12.5   Training

Each client's process should be regarded as a separate entity, and training records should reflect that. There may be fundamental similarities between two different processes (i.e. chromatography on an anion exchange resin), but there will also be subtle differences (i.e. pH 6.8 vs. pH 8.6). A good training program will enable personnel to become familiar with these differences and provide a level of confidence to the client that the process will run as intended.

12.6   Quality and change control

Because of the ramifications associated with permanent incorporation of plasmid DNA sequence into a host cell genome, it is critical to maintain control over the process, the facility, the equipment, and the personnel. Any change, no matter how insignificant it may appear, should not be taken lightly. Changes to a process may result in an undesirable effect on the final product. The most important part of the process will be the sequence of the plasmid itself. Although recombinant organisms have been designed so as to minimise the risk of mutation in the plasmid DNA, it is not advisable to assume that the plasmid will remain constant over time. Periodic analysis of the stability of the plasmid sequence should be performed on the cell banks, the fermentation, and the purification. Final release should depend on the confirmation of the plasmid sequence at each of these steps. Since a specific technique for such an analysis is available – namely, PCR – this should not pose any unusual or burdensome challenges.

13.   SUMMARY

The process of purifying plasmid DNA for gene therapy, whether performed in-house or contracted out, requires a great deal of care and diligence. On the one hand, the process must be robust, scaleable, and consistent from run to run. On the other hand, it is vital to take the time needed to accurately determine the identity of the product, its purity, and its potency. In short, quality control and quality assurance become paramount in order to provide the necessary level of safety.

When a client and a contractor enter an agreement to produce plasmid DNA intended for gene therapy, both parties must understand the impact of the process on the quality of the product. It is their mutual obligation to ensure that the process does not compromise the product, and that quality control, and quality assurance receive the greatest attention. There will always be more than one way to purify plasmid DNA. However, to produce plasmid DNA that meets the criteria of identity, purity, potency, and safety requires a commitment from both client and contractor to strictly and scrupulously adhere to the principles of cGMP.

ACKNOWLEDGEMENTS

Many thanks to James Hayward, Vishva Rai, David Watkins, Susan Dexter, and Roger Kelley at Dow BCMS and The Collaborative Group for their encouragement, support, and comments in preparing this manuscript.

REFERENCES

1. Orkin SH, Motulsky AG. Report and recommendations of the panel to assess the NIH investment in research on gene therapy. Bethesda, MD: National Institutes of Health; 1995.

2. Wu-Pong S, Rojanasakul Y, editors. *Biopharmaceutical Drug Design and Development.* Totowa, NJ: Humana Press; 1999. 435 p.

3. Avis KE, Wu VL, editors. Biotechnology and Biopharmaceutical Manufacturing, Processing, and Preservation. Volume 2. Buffalo Grove, IL: Interpharm Press, Inc.; 1996. 386 p.

4. Sofer G, Hagel L. *Handbook of Process Chromatography: A Guide to Optimization, Scale-up, and Validation.* San Diego, CA: Academic Press; 1997. 387 p.

5. Cooperative manufacturing arrangements for licensed biologics. Rockville, MD: Center for Biologics Evaluation and Research; 1999, draft.

6. Points to consider on plasmid DNA vaccines for preventive infectious disease indications. Rockville, MD: Center for Biologics Evaluation and Research; 1996. 96N-0400.

7. Guidance for human somatic cell therapy and gene therapy. Rockville, MD: Center for Biologics Evaluation and Research; 1998.

8. Clewell DB, Helinski DR. Effect of growth conditions on the formation of the relaxation complex of supercoiled ColE1 deoxyribonucleic acid and protein in *Escherichia coli.* *Journal of Bacteriology* 1972; **110**: 1135-46.

9. Marquet M, Horn NA, Meek JA. Process development for the manufacture of plasmid DNA vectors for use in gene therapy. *BioPharm* 1995; **8**(9): 26-37.

10. Lahijani R, Hulley G, Soriano G, Horn NA, Marquet M. High-yield production of pBR322-derived plasmids intended for human gene therapy by employing a temperature-controllable point mutation. *Human Gene Therapy* 1996; 7(16): 1971-80.

11. Prazeres DM, Ferreira GN, Monteiro GA, Cooney CL, Cabral JM. Large-scale production of pharmaceutical-grade plasmid DNA for gene therapy: problems and bottlenecks. *Trends In Biotechnology* 1999; 17(4): 169-74.

12. O'Kennedy RD, Baldwin C, Keshavarz-Moore E. Effects of growth medium selection on plasmid DNA production and initial processing steps. *Journal of Biotechnology* 2000; 76(2-3): 175-183.

13. Sambrook J, Fritsch EF, Maniatis T. Molecular Cloning. A Laboratory Manual. Cold Spring Harbor, NY: Cold Spring Harbor Press; 1989.

14. Birnboim HC, Doly J. A rapid alkaline extraction procedure for screening recombinant plasmid DNA. *Nucleic Acids Research* 1979; 7: 1513-23.

15. Birnboim HC. A rapid alkaline extraction method for the isolation of plasmid DNA. *Methods in Enzymology* 1983; 100: 243-55.

16. Ish-Horowicz D, Burke JF. Rapid and efficient cosmid cloning. *Nucleic Acids Research* 1981; 9: 2989-98.

17. Holmes DS, Quigley M. A rapid boiling method for the preparation of bacterial plasmids. *Analytical Biochemistry* 1981; 114: 193-7.

18. Vinograd J, Lebowitz J. Physical and topological properties of circular DNA. *Journal of General Physiology* 1966; 49: 103-25.

19. Godson GN, Vapnek D. A simple method of preparing large amounts of $\phi\chi 174$ RF I supercoiled DNA. *Biochimica et Biophysica Acta* 1973; 299: 516-523.

20. Carlson A, Signs M, Lierman L, Boor R, Jim Jem K. Mechanical disruption of Eschericia coli for plasmid recovery. *Biotechnology and Bioengineering* 1995; 48: 303-315.

21. Murphy JC, Wibbenmeyer JA, Fox GE, Willson RC. Purification of plasmid DNA using selective precipitation by compaction agents. *Nature Biotechnology* 1999; 17: 822-3.

22. Davis HL, Schleef M, Moritz P, Mancini M, Schorr J, Whalen RG. Comparison of plasmid DNA preparation methods for direct gene transfer and genetic immunization. *Biotechniques* 1996; 21: 92-4, 96-9.

23. Lis JT. Fractionation of DNA fragments by polyethylene glycol induced precipitation. *Methods in Enzymology* 1980; 65: 347-53.

24. Zasloff M, Ginder GD, Felsenfeld G. A new method for the purification and identification of covalently closed circular DNA molecules. *Nucleic Acids Research* 1978; 5: 1139-52.

25. Klein RD, Selsing E, Wells RD. A rapid microscale technique for isolation of recombinant plasmid DNA suitable for restriction enzyme analysis. *Plasmid* 1980; **3**: 88-91.

26. Bauer W, Vinograd J. The interaction of closed circular DNA with intercalative dyes. I. The superhelix density of SV40 DNA in the presence and absence of dye. Journal of *Molecular Biology* 1968; **33**: 141-71.

27. Clewell DB, Helinski DR. Supercoiled circular DNA-protein complex in Escherichia coli: purification and induced conversion to an opern circular DNA form. *Proceedings of the National Academy of Sciences USA* 1969; **62**: 1159-66.

28. Radloff R, Bauer W, Vinograd J. A dye-buoyant-density method for the detection and isolation of closed circular duplex DNA: the closed circular DNA in HeLa cells. *Proceedings of the National Academy of Sciences USA* 1967; **57**: 1514-21.

29. Ferreira GN, Cabral JM, Prazeres DM. Development of process flow sheets for the purification of supercoiled plasmids for gene therapy applications. *Biotechnology Progress* 1999; **15**: 725-31.

30. Hines RN, KC OC, Vella G, Warren W. Large-scale purification of plasmid DNA by anion-exchange high-performance liquid chromatography. *Biotechniques* 1992; **12**: 430-4.

31. Chandra G, Patel P, Kost TA, Gray JG. Large-scale purification of plasmid DNA by fast protein liquid chromatography using a Hi-Load Q Sepharose column. *Analytical Biochemistry* 1992; **203**: 169-72.

32. McClung JK, Gonzales RA. Purification of plasmid DNA by fast protein liquid chromatography on superose 6 preparative grade. *Analytical Biochemistry* 1989; **177**: 378-82.

33. Schorr J, Moritz P, Seddon T, Schleef M. Plasmid DNA for human gene therapy and DNA vaccines. Production and quality assurance. *Annals of the New York Academy of Sciences* 1995; **772**: 271-3.

34. Vo-Quang T, Malpiece Y, Buffard D, Kaminski PA, Vidal D, Strosberg AD. Rapid large-scale purification of plasmid DNA by medium or low pressure gel filtration. Application: construction of thermoamplifiable expression vectors. *Bioscience Reports* 1985; **5**: 101-11.

35. Raymond GJ, Bryant PKd, Nelson A, Johnson JD. Large-scale isolation of covalently closed circular DNA using gel filtration chromatography. *Analytical Biochemistry* 1988; **173**: 125-33.

36. Horn NA, Meek JA, Budahazi G, Marquet M. Cancer gene therapy using plasmid DNA: purification of DNA for human clinical trials [see comments]. *Human Gene Therapy* 1995; **6**(5): 565-73.

37. Gómez-Márquez J, Freire M, Segade F. A simple procedure for large-scale purification of plasmid DNA. *Gene* 1987; **54**: 255-9.

38. Bywater M, Bywater R, Hellman L. A novel chromatographic procedure for purification of bacterial plasmids. *Analytical Biochemistry* 1983; **132**: 219-24.

39. Markov GG, Ivanov IG. Hydroxyapatite column chromatography in procedures for isolation of purified DNA. *Analytical Biochemistry* 1974; **59**: 555-63.

40. Meinke W, Goldstein DA, Hall MR. Rapid isolation of mouse DNA from cells in tissue culture. *Analytical Biochemistry* 1974; **58**: 82-8.

41. Ivanov IG, Venkov PV, Markov GG. Isolation of DNA from yeast by chromatography on hydroxyapatite. *Preparative Biochemistry* 1975; **5**: 219-28.

42. Colman A, Byers MJ, Primrose SB, Lyons A. Rapid purification of plasmid DNAs by hydroxyapatite chromatography. *European Journal of Biochemistry* 1978; **91**: 303-10.

43. Popa LM, Repanovici R, Iliescu R, Angelescu S. Isolation of plasmid pBR313 and pBR322 DNAs by different methods. *Revue Roumaine de Biochimie* 1980; **17**(4): 279-283.

44. Bachvarov DR, Ivanov IG. Large scale purification of plasmid DNA. *Preparative Biochemistry* 1983; **13**: 161-6.

45. Yamasaki Y, Yokoyama A, Ohnaka A, Kato Y, Murotsu T, Matsubara K. High-performance hydroxyapatite chromatography of nucleic acids. *Journal of Chromatography* 1989; **467**: 299-303.

46. Gagnon P, Frost R, Tunon P, Ogawa T. Ceramic hydroxyapatite - a new dimension in chromatography of biological molecules. Hercules, CA: Bio-Rad Laboratories; 1997. US/EG Bulltein 2156.

47. Green AP, Prior GM, Helveston NM, Taittinger BE, Liu X, Thompson JA. Preparative purification of supercoiled plasmid DNA for therapeutic applications. *BioPharm* 1997; **10**(5): 52-62.

48. Freitag R. Displacement chromatography for biopolymer separation. *Nature Biotechnology* 1999; **17**(3): 300-2.

49. Freitag R, Vogt S. Separation of plasmid DNA from protein and bacterial lipopolysaccharides using displacement chromatography. *Cytotechnology* 1999; **30**: 159-167.

50. Weber M, Moller K, Welzeck M, Schorr J. Short technical reports. Effects of lipopolysaccharide on transfection efficiency in eukaryotic cells. *Biotechniques* 1995; **19**(6): 930-40.

51. Neudecker F, Grimm S. High-throughput method for isolating plasmid DNA with reduced lipopolysaccharide content. *BioTechniques* 2000; **28**(1): 106-108.

52. Lyddiatt A, DA OS. Biochemical recovery and purification of gene therapy vectors. *Current Opinion in Biotechnology* 1998; **9**(2): 177-85.

53. Marquet M, Horn NA, Meek JA. Characterization of plasmid DNA vectors for use in human gene therapy, part 1. *BioPharm* 1997; **10**(5): 42-50.

54. Marquet M, Horn NA, Meek JA. Characterization of plasmid DNA vectors for use in human gene therapy, part 2. *BioPharm* 1997; **10**(6): 40-45.

55. Lahijani R, Duhon M, Lusby E, Betita H, Marquet M. Quantitation of host cell DNA contaminate in pharmaceutical-grade plasmid DNA using competitive polymerase chain reaction and enzyme-linked immunosorbent assay. *Human Gene Therapy* 1998; **9**(8): 1173-80.

56. Wicks IP, Howell ML, Hancock T, Kohsaka H, Olee T, Carson DA. Bacterial lipopolysaccharide copurifies with plasmid DNA: implications for animal models and human gene therapy. *Human Gene Therapy* 1995; **6**(3): 317-23.

57. Cotten M, Baker A, Saltik M, Wagner E, Buschle M. Lipopolysaccharide is a frequent contaminant of plasmid DNA preparations and can be toxic to primary human cells in the presence of adenovirus. *Gene Therapy* 1994; **1**(4): 239-46.

58. Anicetti VR, Fehskens EF, Reed BR, Chen AB, Moore P, Geier MD, Jones AJ. Immunoassay for the detection of E. coli proteins in recombinant DNA derived human growth hormone. *Journal of Immunological Methods* 1986; **91**: 213-24.

59. Kim AI, Hebert SP, Denny CT. Cross-contamination limits the use of recycled anion exchange resins for preparing plasmid DNA. *Biotechniques* 2000; **28**(2): 298.

60. Fogel BL, McNally MT. Trace contamination following reuse of anion-exchange DNA purification resins. *Biotechniques* 2000; **28**(2): 299-302.

Risk Assessment in Gene Therapy

AKSHAY ANAND AND SUNIL K. ARORA
Dept. of Immunopathology, Post Graduate Institute of Medical Education and Research, Chandigarh, India

1. INTRODUCTION

Gene therapy is an attractive and promising technology which holds bright prospect of curing the genetic as well as other diseases by introducing the synthetic (correct) gene directed and intended to replace the defective gene responsible for that particular disease. A number of methods to introduce the gene are being developed for which various viral and non-viral vectors are being improvised to act as safe vehicles or carriers. With more and more research being done on ushering the human civilisation to the era of gene based therapies, it has become an ethical issue if alteration of genome is right apart from questioning the risks associated with such genetic intervention. The preparation of the final draft of human genome project has further boosted the morale of prospective gene therapists. There is a call for weighing the pros and cons of gene therapy before it is adopted in practice. While progress in medicine has solved many problems facing humanity, it has led to many new ones. There is no reason to believe that this ambivalence will cease to exist with respect to future application of gene technology in medicine. A sober and differential assessment of the chances and risks of this new technology is therefore necessary.

2.   BIOLOGICAL

Among the biological risks associated with gene therapy, the impending danger of reversion of pathogenicity of viral vectors looms large. This may lead to disease and death.   Viruses, used as vectors, are genetically engineered to remove genes responsible for pathogenicity, however, these can revert back to pathogenic state endangering the patient.   Like the acquisition of resistance by bacteria and some tumours due to widely prescribed antibiotics and drugs it is conceivable to anticipate hazards associated with development of resistance (at genetic level) for world wide practice of gene therapy.   An interference into the very gene pool of an individual may be an abortive attempt to hijack the regulation of the metabolism of the body that has evolved over millions of years driven by both chemical and physical forces. This may manifest in the form of various biological disorders.   An attempt to tinker with the vital genome may endanger the individual to new diseases and tumours/cancers hitherto unknown. The insertion of the gene transcript in a tumour suppressor loci or similar sequence elsewhere may lead to many unanticipated disorders. Till date, knowledge about the causation of a certain defect or a mutation in a gene is almost negligible, and any attempt to replace it by introducing the correct gene, may, at theoretical level, appear very appealing, however, the risk of playing with the carefully preserved genome may loom large over the entire human population.   Moreover, the likelihood of replaced/defective transcript recombining at another complementary sequence cannot be ruled out. With the success of human genome project the entire sequence of much sought after genome governing our very existence is being deciphered. Almost all diseases, ailments and metabolic regulatory machinery involved in the homeostasis of a cell owe their description to this revered sequence. With the public access to this vital knowledge through various public domains this vital data will have the potential to determine susceptibility to a certain disease apart from yielding information about I.Q. temperament and other aspects of personality. The tendency to manipulate the DNA of an embryo by germ line therapy will be overpowering. It is already assuming an important place in public eye given the controversial and unprecedented issue it raises being ethically unacceptable. Visualisation of a world, in post genomic era, is likely to be totally bereft of present definition, understanding and conceptualisation of health, spirituality, and law.   The long term implications of gene therapy may be devastating. The eradication of disease, certainty in health may transform the willingness to struggle and feel euphoric which may result in disillusionment. Among other known systems of therapy, gene therapy has yielded very few successful results despite claims to the contrary.   Nevertheless growing coteries of scientists are generating an imagined world of genetically controlled state. It may be worth taking a second look at the system of therapy that genetic intervention promises to offer. The complex milieu of genetic networking borne out of physical forces manifest at the level of atoms and electrons, need to be exhaustively unravelled and the integration of their interaction with other cells and tissues need to be studied in detail before any huge budgeting proposals of gene therapy are entertained. To give a small example, the gene delivery mechanisms like the basic molecular processes involved in the uptake of foreign DNA are largely unknown and yet the gene delivery systems are not only being used but being religiously claimed to be regularly improvised.

2.1   Mutagenicity testing

Research scientists are yet to experiment on the risks of gene therapy by exhaustive genetic testing particularly in the context of causation of new mutations after.   Without this important data any adventure genetic intervention at genetic level may prove counterproductive.   A battery of available tests like SOS, Ames test, Microsomal Assay, etc. (mutation detection assays) are available and better ones can be devised, which can be performed on individuals and animals undergoing experimental gene therapy. The biological incorporation of the corrected gene can be tested by PCR to assess the risks of genetic intervention. The onset of gene therapy may lead to the shift in focus of pharmaceutical industries more on the common diseases like Cancer and AIDS rather than the rarer diseases whereby the latter would remain unattended. The current discussion was catalysed by tragic death of Tessa Gelsinger, an 18 yrs old patient with OTC (Ornithine transcarboxylase) deficiency, who died apparently as a result of experimental gene therapy in University of Philadelphia. This leads us to reflect on non conclusive nature of pre-clinical studies which even could not predict the discovery that diet medication fen-phen is associated with potentially life threatening cardio vascular damage.

The Primary health care workers shall need adequate training for management and treatment of such diseases before the technology is smoothly transferred from lab to clinic.

3.   ECOLOGY

The nature has successfully maintained the sensitive balance between its various species by the diversity in their unique genomes. Any mediation by manipulation of the genome may disturb this intricate balance. Already a wide range of pathogenic bacteria have mushroomed possessing antibiotic resistance genes.

Similarly, the altered genome may become host for unknown pathogens hitherto reported. Scientists can direct their work on the immunological studies associated with gene therapy. Such studies are warranted in order to augment the present therapeutic approaches with safe and successful genetic mediation.

4.   PHILOSOPHICAL AND ETHICAL ASPECTS

The role of gene therapy in interfering the process of evolution provokes adequate reconsideration and its application should thus be restricted to only serious diseased states. There may be social and philosophical problems with widespread use of gene therapy. Perhaps of even greater concern are repeated predictions of a golden age for gene therapy in near future. The very definition of disease, health shall undergo a major paradigm shift in the future generations. There is still disagreement about the scope or application of the concepts of health, disease or normality to gambling, sexual promiscuity, pre-menstrual syndrome, hyperactivity or homosexuality. The debate over the classification of these behaviours and traits has been heated and fierce. Uncertainty about whether or not to treat short stature in children, low blood sugar and hypertension have also produced heated controversies as to their disease and health classification and consequently, the appropriateness or inappropriateness of therapeutic intervention. The intricate problem of how to fit new knowledge about heredity into present categories of disease, normality and health can, perhaps, be forestalled by arguing that the sole therapeutic goal of the human genome project is somatic gene therapy for human disease.

4.1   What is disease?

Screening for eligibility for government benefits to certifying persons as fit to play sports or serve in the military, the fight against disease occupies centre stage in what people expect health care providers to do. So if it is possible to become clearer about what disease is then it may be possible to have a better understanding of the boundaries of what is and is not licit with respect to the application of new knowledge arising from the human genome project and gene therapy trials.

E.A. Murphy cogently observed in his book, 'The Logic of Medicine':

".... The clinician has tended to regard the disease as that state in which the limits of the normal have been transgressed."

For example many physicians believe that blood pressure readings which vary from normal for specific age groups within the population are, in themselves, indicative of disease. For American physicians, variations skewed towards higher numbers are disturbing. For German physicians, both high and low numbers are equally likely to be diagnosed as disease. Critics of 'disease as abnormality' approach point out that there is nothing inherent about difference that makes a particular biological, chemical or mental state a disease. Moreover, since variation is an omnipresent feature of human beings, it is especially odd to argue that extremes of variation are somehow indicative of disease. Indeed, critics of the view that equate difference with disease note that this equation has, throughout the history of medicine, led to the classification of differences with respect to race, gender and ethnicity as diseases which in turn has been the basis for unfair and even harmful interventions against persons suffering from nothing more than a darker skin colour or the presence of ovaries. After all, those who are unusually smart, strong, fast or prolific are not classified as diseased. Subjectivity and a lack of consensus could bode ill especially for the uses to which new knowledge of human heredity might be put since applications might be controlled by the powerful or the economically privileged to advance their own values. The problem with linking values and abnormalities is that not all states commonly recognised, as diseases are necessarily indicative of difference or abnormality. Nor are all dysfunctions or impairments always disvalued. For example, every human being suffers from the common cold, acne, and anxiety to dental caries. Those who do not wish to have children may rejoice to discover that they have ovaries that are incapable of ovulation, lack a uterus or possess testes that cannot create sperm. Someone born with one kidney may remain entirely indifferent to and even unaware of this dysfunction.

The stance that these in clinical genetics adopt towards assessing the significance of difference and abnormality at the level of genome, the role of values in defining genetic disease and the need to link genetic disease to dysfunction will play pivotal roles in what is done with the knowledge generated by ongoing work to map and sequence the genome.

It is instructive to look and see how the disease is currently defined in order to try and forecast how new knowledge about human heredity will be absorbed into clinical and gene therapeutic practices. Not so long ago a woman at a large medical centre was informed by a genetics counsellor that the foetus she was carrying possessed an abnormal chromosome – the child had xyy syndrome – an extra y chromosome. The couple was told, in the counselling sessions, some researchers had proven criminality to this chromosome abnormality besides having some correlation between this and tall physical stature. The couple decided to abort the pregnancy. Is xyy syndrome a disease? Does it merit abortion? OCA (Oculo Cutaneous albinism) is associated with extreme sensitivity to light (photophobia), nystagmus, severe impairment in visual acuity and a greatly increased risk of squamous cell skin cancer. But unless albinism is a disease, why should any one try to detect it much less provide information about it to parents?

For the present it seems clear that the first wave of new information about the human genome is likely to bear a fair amount of news about human differences that many have uncertain unknown phenotypic consequences. Clinical genetics is still in its infancy. As such, it ought proceed with great caution in labelling states or variations as abnormal much less diseases. And it should assert clearly that the central goal of human clinical genetics is the prevention or amelioration of disease not the improvement of the genome. It is important to note that abjuring eugenics as a proper goal of clinical genetics is not the same thing as foregoing any effort to meddle or intervene with the genetics of reproductive cells.

4.2   Eugenics

The major reason for not undertaking germ line interventions is that hereditary information that is of value, not for the individual but for the species, may be lost. If lethal or disabling genes are removed from a certain individual's gametes it may be that benefits conferred on the population when these genes recombine with other, non-lethal genes will be lost. The argument against germ line therapy is that no one would really want to use it for the purposes for eugenics. But this is patently false. Even putting aside Germany's three decades long embrace of race hygiene and eugenics there are examples of private and government organisations avidly and unabashedly pursuing eugenics goals. The government of Singapore instituted numerous eugenics policies during the 1980s including a policy of providing financial incentives to 'smart' people to have more babies. The California-based Repository for Germinal Choice, known more colloquially as the Nobel prize sperm bank, has assigned itself the mission of seeking out and storing gametes from men selected for their scientific, athletic or entrepreneurial acumen. Their sperm is made available for use by women of high intelligence for the expressed purpose of creating genetically superior children. Should scientists or clinicians really promise never to try to eliminate or modify the genetic messages contained in a sperm or an egg. If that message contains instructions which may cause sickle cell disease, Lesch-Nyhan Syndrome or retinoblastoma.

5. CONCLUSIONS

The greatest challenge to securing continuing funding for genome project and gene therapy trials does not originate from concern about privacy, confidentiality, or coercive genetic testing. It is eugenics manipulating the human genome in order to improve or enhance the human species, that is the real source of worry. Promising not to do anything that remotely hints of germ line engineering where eugenics is the goal, is relatively easy for those connected with the genome project since none of them believe that anyone is even remotely close to knowing how to alter the germ lines of a human being, much less whether germ line engineering will actually work.

The grim history of eugenically inspired social policy tells why it is important to protest and even prohibit the activity of the noble sperm bank or to vehemently criticise the birth incentive policies of Singapore. It does not provide an argument against allowing voluntary, therapeutic efforts using germ line manipulations to prevent certain and grievous harm from befalling future persons.

It is at best cruel to argue that some people must bear the burden of genetic disease in order to allow benefits to accrue to the group or species. At best, genetic diversity is an argument for creating a gamete bank to preserve diversity. It is hard to see why an unborn child has any obligation to preserve the genetic diversity of the species at the price of a grave harm or certain death.

ACKNOWLEDGEMENTS

We are grateful to all our colleagues for their help in the critical analysis of the manuscript and also Dr. Usha Datta for encouragement.

REFERENCES

1. Barinaga M, Asilomar revisited: lessons for today? *Science* 2000 Mar 3; **287** (5458): 1584-5.

2. Bayertz K: Ethical aspects of gene therapy and molecular genetic diagnostics. *Cytokines Mol Ther* 1996 Sep; **2**(3): 207-11.

3. Butler D: call: for risk/benefit study of gene therapy *Nature*, 1994 Dec 22-29; **372** (6508): 716.

4. Collins FS: Shattuck lecture – medical and societal consequences of the Human Genome Project. *N Engl J Med* 1999 Jul 1; **341**(1); 28-37.

5. FriedMann T, Medical ethics: Principles for human gene therapy studies, *Science* 2000 Mar 24; **287** (5461): 2163.

6. Gordon JW: Germline alteration by gene therapy: assessing and reducing the risks. *Mol Med Today* 1998 Nov; **4**(11): 468-70.

7. Ledley FD: Assessing risk. *Human Gene Ther* 1995 May; **6**(5): 551-2.

8. Rosenberg LE, Schechter AN: Gene therapist, heal thyself, *Science* 2000, Mar 10; **287** (5459): 1751.

Index

9 780306 466809

Printed in the United States
135638LV00002B/83/A